European Consortium for
Mathematics in Industry 3

Neunzert (Ed.)

Proceedings of the Second European Symposium
on Mathematics in Industry

European Consortium for Mathematics in Industry

Edited by
Michiel Hazewinkel, Amsterdam
Helmut Neunzert, Kaiserslautern
Alan Tayler, Oxford
Hansjörg Wacker, Linz

ECMI Vol. 3

Within Europe a number of academic groups have accepted their responsibility towards European industry and have proposed to found a European Consortium for Mathematics in Industry (ECMI) as an expression of this responsibility.

One of the activities of ECMI is the publication of books, which reflect its general philosophy; the texts of the series will help in promoting the use of mathematics in industry and in educating mathematicians for industry. They will consider different fields of applications, present casestudies, introduce new mathematical concepts in their relation to practical applications. They shall also represent the variety of the European mathematical traditions, for example practical asymptotics and differential equations in Britain, sophisticated numerical analysis from France, powerful computation in Germany, novel discrete mathematics in Holland, elegant real analysis from Italy. They will demonstrate that all these branches of mathematics are applicable to real problems, and industry and universities in any country can clearly benefit from the skills of the complete range of European applied mathematics.

Proceedings of the

Second European Symposium on Mathematics in Industry

ESMI II
March 1–7, 1987 Oberwolfach

Edited by
Helmut Neunzert, Kaiserslautern

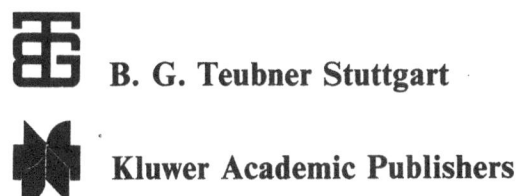

B. G. Teubner Stuttgart

Kluwer Academic Publishers

Editor:

Prof. Dr. Helmut Neunzert
Department of Mathematics
University of Kaiserslautern, W.-Germany

Distributors:

Continental Europe (excluding U. K.)
B. G. Teubner GmbH, P. O. Box 80 10 69, D-7000 Stuttgart-80

United States and Canada
Kluwer Academic Publishers, 101 Philip Drive, Norwell, MA 02061, USA

All other countries (including U. K.)
Kluwer Academic Publishers, P. O. Box 322, NL-3300 AH Dordrecht

CIP-Titelaufnahme der Deutschen Bibliothek

European Symposium on Mathematics in Industry : (March 1-7, 1987 Oberwolfach)
Proceedings of the Second European Symposium on Mathematics in Industry / ed. by
H. Neunzert
Stuttgart : Teubner
Dordrecht ; Boston ; Lancaster : Kluwer Acad. Publ. 1988
  (European Consortium for Mathematics in Industry ; Vol. 3)
  ISBN-13: 978-94-010-7838-2   e-ISBN-13: 978-94-009-2979-1
  DOI: 10.1007/ 978-94-009-2979-1
NE: Neunzert, Helmut (Hrsg.) ; European Consortium for Mathematics in Industry:
European Consortium for ...
Auf d. Haupttitels. auch: ESMI ...

Library of Congress Cataloguing in Publication Data

CIP data appear on separate card.

Preface

"Mathematics in Industry" – since the volume containing the proceedings of the 1985 Oberwolfach conference was published*), this subject has become more fashionable in Europe, America and also in the third world. The Europeans have come closer to each other: They formed a European Consortium for Mathematics in Industry, abbreviated ECMI. This ECMI supported mainly by mathematicians from Amsterdam, Bari, Eindhoven, Firenze, Kaiserslautern, Limerick, Linz, Paris, Oxford and Trondheim has become a legal entity with a rapidly growing number of members. It has organized a common, really European postgraduate programme, establishes contact between industry and universities and organizes other conferences everywhere in the world.

Industrial mathematics is a special method to get interesting problems; a special attitude of curiosity for technical or economical questions; a general rather broad knowledge in all branches of mathematics; but it always remains real mathematics. Our first proceedings contained many articles about "why and how to start". Now we are more selfconfident about our ideas: These proceedings include only examples of "how to do". It is a pleasure to see how many different kinds of good mathematics are applied to so many different problems from industry. Part of the selection criteria for this volume was that some of the applications of what is usually considered ivory tower mathematics be represented.

It was hard for the editor to find an order among these 20 articles. Should I use the normal classification into pde, ode, signalprocessing, optimization, systemtheory, CAD in order to give a hint about the main mathematical content, or should I put more emphasis on the different fields of applications: The second idea would end up with the trivial classification, one member in each class, which shows the enormous variety of applications of mathematics. How can we combine coal slurries (which are modelled by a nonlinear diffusion equation) with oscillations in gear boxes (where we find dynamical systems with strange attractors); the problem of stabilizing large structure like space stations (with a boundary control for hyperbolic differential equations) with the simulation of chemical plants (where tricky nonlinear equation solvers are used); modelling of mosfets in semiconductor industry with calculating the shape of a pipe laid by a barge in off shore technology: If anyone still believes that industrial mathematics means very specialized mathematics that has lost the generality and beauty of "normal" mathematics, he may become convinced that he is wrong just by looking through the book.

Mathematicians who try to prove that even very pure mathematics can be extraordinarily useful in modern technology have to know something from both sides. They must be aware of what is going on in university mathematics and they must also have at least some ideas about the mathematical needs in industry. Reading the articles in this volume should give an idea of how stimulating practical problems

---

*) Neunzert, H. (Ed.): Proceedings of the Conference Mathematics in Industry, October 24–28, 1983 Oberwolfach, B. G. Teubner Stuttgart 1984

can be for mathematical research and how useful some fields of mathematics, which are considered to be not-applied, are. Finally I hope that many of our industrial-mathematics-friends find it rewarding to study the book in order to get better solutions for their own problems.

Kaiserslautern, autumn 1987   H. Neunzert

Table of Contents

VIII

MAP DYNAMICS IN GEARBOX MODELS

R. VILELA MENDES

Instituto de Física e Matemática
Av. Gama Pinto, 2 - 1699 Lisboa Codex, Portugal

*M. de Faria** and *L. Streit*

Research Center Bielefeld-Bochum-Stochastics
Univ. Bielefeld, Postfach 8640, 4800 Bielefeld 1, FRG

ABSTRACT

Maps associated to the impact dynamics of simplified one- and two-step gearbox models are derived.
The dynamical features of the models are studied by analytical and numerical methods, some conclusions being drawn concerning the nature of the attractors, noise generation and overload.

*Permanent address: Area de Matemática, Universidade do Minho
4700 Braga, Portugal.

This paper is in final form and no version of it will be submitted for publication.

1. <u>INTRODUCTION</u>

The study of impulsive motion in dynamical systems has a long tradition in plasma and accelerator physics. The effects of external impulsive forces and of the impacts resulting from the free play between the several components of a system is now also becoming a subject of interest in mechanical enginee-ering.

Following the pioneering work of Pfeiffer and Kücükay[1-3], Hongler and Streit[4] have identified in a simplified gearbox model a dynamical map of the family of the Fermi acceleration maps[5], thus giving an interpretation of the seemingly chaotic behaviour observed in numerical simulations. In this paper we analyse in more detail the maps associated to the dynamics of simple gearbox models.

In sect 2a. we analyse a one-step model and find that both in the no-load and in the heavy load cases one can obtain expli-cit two-dimensional maps describing the dynamics of the model These maps are obtained under what seem to be physically rea-sonable simplifying assumptions, their explicit form being adequate for an analytical study of the dynamics.

In sect. 2b a similar analysis is carried out for a two-step model. Numerical simulation using the derived maps in-dicates that in this case there is a wide range of regimes in the symbolic dynamics of impact sequences. This will make the analytical study of the dynamics much harder than in the one-step model and an heavy reliance on numerical simulation seems to be unavoidable in this case.

In sect. 3 we analyse the dynamics of the maps derived in
sect. 2a and attempt to draw some conclusions which, although
obtained from oversimplified models, might be of some interest
in the study of more complex models and as an inspiration for
further study leading to designing criteria for gearboxes.
Finally in sect. 4 we report on preliminary results concerning
a numerical study of the two-step model.

2. THE MODEL MAPS

In this section we derive maps which describe the dynamics
in two simple one-step and two-step gearbox models. The emphasis
will be in obtaining maps which, although capturing the essential
dynamical complexity of the physical system, admit nevertheless
a sufficiently compact mathematical description. This implies
that some simplifying assumptions and limiting situations
will have to be considered. The benefit one extracts from not
taking explicitely into account all the physical parameters is
that, as we will see in sect. 3, one is able to go beyond numerical simulation of the models and derive analytical pre-
dictions and estimates on the dynamical behaviour of the models.
If our simplified models have the same dynamical complexity
as the actual physical systems (a likely possibility which
might be checked experimentally), then the analytical results
may serve as a guide to obtain designing criteria for gearboxes.

4

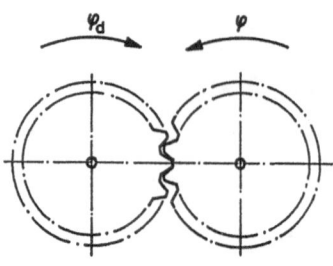

Fig.1

2.a One-step model

Here one considers a system with one degree of freedom ϕ, the angle of the driving wheel ϕ_d being forced from outside by an excitation function

$$\phi_d(t) = e(t) = \omega t + E\, f(\Omega t) \qquad (2.1a)$$

where f is a periodic function. Two concrete examples will be considered, namely

$$f_1(x) = \cos(x) \quad \text{and} \qquad (2.1b)$$

$$f_2(x) = \begin{cases} 1 - 4x(\text{mod } 2\pi)/2\pi & \text{for } x(\text{mod } 2\pi) < \pi \\ -3 + 4x(\text{mod } 2\pi)/2\pi & \text{for } x(\text{mod } 2\pi) \geq \pi \end{cases} \qquad (2.1c)$$

f_1 being used for the analytical studies and f_2 for some of the numerical simulations.

Between impacts of the driving on the driven wheel this one follows the dynamical equation.

$$\Theta\ddot{\phi} = T - d\dot{\phi} \qquad (2.2)$$

where Θ is the moment of inertia coefficient, T the load and oil drag torque and d the damping parameter. The variable that controls the impacts between the two wheels is

$$s(t) = e(t) - k\phi(t) \qquad (2.3)$$

k being the transmission ratio.

s(t) lies in the tolerance interval

$$- \Delta/2 \leq s(t) \leq \Delta/2 \qquad (2.4)$$

At the time of the nth impact the state of motion is changed by

$$\dot{s}(t_n^+) = \dot{e}(t_n^+) - k\dot{\phi}(t_n^+) = - \varepsilon (\dot{e}(t_n^-) - k\dot{\phi}(t_n^-)) = -\varepsilon\dot{s}(t_n^-) \qquad (2.5)$$

where ε is an impact damping parameter and t^-, t^+ denote the times before and after the impact. Assuming that the driving system has a large inertia $\dot{e}(t_n^+) = \dot{e}(t_n^-) = \dot{e}(t_n)$

We now consider two extreme situations:

The first is the case of driving without a load. Between impacts resistance to the motion is only due to the oil drag and, as long as one is only interested in the speed at the next impact $\dot{\phi}(t_{n+1}^-)$, the effect of the (small) T and d coefficients may be approximated by appropriately increasing the impact damping coefficient ε. Then between impacts $\ddot{\phi} \approx 0$ and one obtains from (2.5)

$$\dot{\phi}(t_{n+1}^-) = -\varepsilon\dot{\phi}(t_n^-) + \frac{1+\varepsilon}{k} \dot{e}(t_n) \qquad (2.6)$$

we now compute the time between two impacts. At the nth impact $s(t_n) = \alpha_n \Delta/2$, where $\alpha_n = \pm 1$. Then

$$(\alpha_{n+1} - \alpha_n)\Delta/2 = e(t_{n+1}) - e(t_n) - k\dot{\phi}(t_{n+1}^-)(t_{n+1}-t_n)$$

Using a linear approximation for $e(t_{n+1}) - e(t_n)$ one obtains

$$t_{n+1} \simeq t_n + \frac{\alpha_{n+1} - \alpha_n}{e(t_n) - k\phi(t_{n+1}^-)} \frac{\Delta}{2}$$

Hence in this approximation $\alpha_{n+1} = -\alpha_n$. Taking into account the time nonlinearity of $e(t)$ and the effects of the oil drag, sucessive impacts of the some type are of course possible, however for the model without load it is the sequence of alter nating $\pm \Delta/2$ impacts that indeed controls the dominant dyna mical features.

Defining $\qquad\qquad \dot{\phi}(t_n^-) = \frac{\omega}{k} + \frac{J_n}{k}$ $\qquad\qquad$ (2.7a)

$$\Omega t_n = \theta_n \qquad\qquad (2.7b)$$

one obtains the map

$$J_{n+1} = -\varepsilon J_n + (1 + \varepsilon) E \Omega \; f'(\theta_n)$$

$$\theta_{n+1} = \theta_n + \left| \frac{\Delta\Omega}{J_{n+1} - E\Omega \; f'(\theta_n)} \right|$$

$\qquad\qquad$ (2.8)

that describes the impact mechanics of the one step model without load. This is the map we will actually use for our study of the dynamics of the one-step model in sect.3. Alter- natively one might also use a map written in the impact varia bles, namely

$$t_{n+1} = t_n + \left| \frac{\Delta}{\varepsilon \dot{s}(t_n^-)} \right|$$

$$\dot{s}(t_{n+1}^-) = -\varepsilon\, \dot{s}(t_n^-) + \dot{e}(t_{n+1}) - \dot{e}(t_n)$$

where linearization of $e(t_{n+1}) - e(t_n)$ is also used to compute t_{n+1} explicitly.

\# The second limiting situation which we analyse in the one step model is the case of a heavy constant load. Then it is reasonable to assume that all the relevant impacts take place at $s = +\Delta/2$. From $s(t_{n+1}) = s(t_n) = +\Delta/2$, considering $\Theta\,\ddot{\phi} = T$ ($T < 0$) between the impacts and using (2.5), one now obtains

$$\dot{\phi}(t_{n+1}^-) = -\varepsilon\, \dot{\phi}(t_n^-) + \frac{1+\varepsilon}{k}\, \dot{e}(t_n) + (t_{n+1} - t_n)\, \frac{T}{\Theta}$$

$$0 = e(t_{n+1}) - e(t_n) + \{k\varepsilon\dot{\phi}(t_n^-) - (1+\varepsilon)\dot{e}(t_n)\}\,(t_{n+1} - t_n) - \frac{kT}{2\Theta}\,(t_{n+1}-t_n)^2$$

Using a linear approximation for $e(t_{n+1}) - e(t_n)$ and the same change of variables as in (2.7) one obtains the map

$$
\boxed{
\begin{aligned}
J_{n+1} &= \varepsilon J_n + (1 - \varepsilon)E\Omega\, f'(\theta_n) \\[2mm]
\theta_{n+1} &= \theta_n + \frac{2\,\Theta\Omega}{kT}\,(J_{n+1} - E\Omega f'(\theta_n))
\end{aligned}
}
$$
(2.9a)

$$T < 0$$

In this case we may however use a second order approximation for $e(t_{n+1}) - e(t_n)$ and still obtain an explicit esti_ mate for $t_{n+1} - t_n$

$$t_{n+1} \simeq t_n + \varepsilon\{\dot{e}(t_n) - k\dot{\phi}(t_n^-)\}\{\tfrac{1}{2}\ddot{e}(t_n) - \frac{kT}{2\Theta}\}^{-1}$$

With the same change of variables and the excitation function (2.1) one obtains the map,

$$\theta_{n+1} = \theta_n - \varepsilon\Omega \{ -E\Omega f'(\theta_n) + J_n \}\{\frac{k(-T)}{2\theta} - \frac{E\Omega^2}{2} f(\theta_n)\}^{-1}$$

$$J_{n+1} = - \varepsilon J_n + (1+\varepsilon)E\Omega f'(\theta_n) + \frac{Tk}{\Omega\theta} (\theta_{n+1} - \theta_n)$$

$$T < 0$$

$$(2.9b)$$

Under the hypothesis of a heavy load there is no singularity in (2.9b). The map (2.9b) gives a more accurate description of the dynamics because it takes into account the second time derivative of the excitation. However, for a heavy load (large negative T) it actually reduces to (2.9a), and we will later use this simpler form.

The approximations used in the explicit computation of the time difference $t_{n+1} - t_n$ introduce an unphysical feature in the maps (2.9), which should be corrected. In fact, after $\dot{\phi}(t_n^-)$ and t_n are computed from $\dot{\phi}(t_{n-1}^-)$ and t_{n-1} using Eqs. (2.9), it may happen that $\dot{e}(t_n) - k\dot{\phi}(t_n^-)$ is a negative quantity. This would lead to $t_{n+1} - t_n < 0$ which is unphysical. What it means is that the impact does not really occur at t_n (computed from (2.9)) but some time later which we denote by t_n'. Assume that at the time t_n the impact variable $s(t)$ is already close to $+\Delta/2$, but the hit does not occur simply because the velocity difference $\dot{e}(t_n) - k\dot{\phi}(t_n)$ is negative. Then the actual hitting time t_n' may be estimated from

$$\dot{e}(t_n)(t_n' - t_n) \simeq k\dot{\phi}(t_n^-)(t_n' - t_n) + \frac{Tk}{2\theta} (t_n' - t_n)^2$$

which leads to

$$t_n' - t_n \simeq \frac{2\Theta}{kT} (\dot{e}(t_n) - k\dot{\phi}(t_n^-)) \qquad (2.10)$$

Denote by M_1 the map (2.9) and by M_2 the map obtained from (2.10), namely

$$M_2 \begin{cases} J_{n+1} = - J_n + 2E\Omega \, f'(\theta_n) \\ \\ \theta_{n+1} = \theta_n - \frac{2\Theta\Omega}{kT} [J_n - E\Omega \, f'(\theta_n)] \end{cases} \qquad (2.11)$$

Then the dynamics is defined globally as follows:

$$\begin{pmatrix} J_n \\ \theta_n \end{pmatrix} \xrightarrow{M_1} \begin{pmatrix} J_{n+1} \\ \theta_{n+1} \end{pmatrix} \quad \text{if} \quad (M_1 J_n) - E\Omega \, f'(M_1\theta_n) \leq 0$$

$$\begin{pmatrix} J_n \\ \theta_n \end{pmatrix} \xrightarrow{M_2 \circ M_1} \begin{pmatrix} J_{n+1} \\ \phi_{n+1} \end{pmatrix} \quad \text{if} \quad (M_1 J_n) - E\Omega \, f'(M_1\theta_n) > 0$$

For the map (2.9a) the composite map $M_2 \circ M_1$ is

$$J_{n+1} = -\epsilon J_n - (1 - \epsilon)E\Omega \, f'(\theta_n) + 2E\Omega \, f'(M_1\theta_n)$$

$$\theta_{n+1} = \theta_n + \frac{2\Theta\epsilon\Omega}{kT}\{J_n - E\Omega \, f'(\theta_n)\} -$$

$$- \frac{2\Theta\Omega}{kT}\{\epsilon J_n - E\Omega \, f'(M_1\theta_n) + (1 - \epsilon)E\Omega \, f'(\theta_n)\}$$

The condition $J_k - E\Omega \, f'(\theta_k) \leq 0$ is required at each step to validate the impact at time t_k. If after the application of

M_2 the condition is still not satisfied then the map M_2 should be iterated until the first p for which $(M_2^p \circ M_1 J_n) - E\Omega f'(M_2^p \circ M_1 \Theta_n) \leq 0$

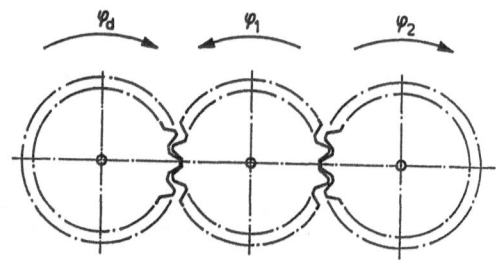

φ_d φ_1 φ_2

2.b Two-step model

Fig.2

There are now two impact variables

$$s_1(t) = e(t) - k_1\phi_1(t) \qquad (2.12a)$$

$$s_2(t) = \phi_1(t) - k_2\phi_2(t) \qquad (2.12b)$$

lying in the tolerance intervals

$$- \Delta_1/2 \leq s_1(t) \leq \Delta_1/2$$

$$- \Delta_2/2 \leq s_2(t) \leq \Delta_2/2$$

and two types of impacts, one between the driving and the wheel 1 and another between the wheels 1 and 2. The boundary condition for impacts of the first type is

$$\dot{s}_1(t_n^+) = - \epsilon \dot{s}_1(t_n^-) \qquad (2.13)$$

and for the second type

$$\dot{\phi}_1(t_n^+) = \frac{1}{\Theta_1 + \Theta_2} \{(\Theta_1 - \epsilon \Theta_2)\dot{\phi}_1(t_n^-) + (1+\epsilon) \Theta_2 k_2\dot{\phi}_2(t_n^-)\} \qquad (2.14a)$$

$$k_2 \dot{\phi}_2(t_n^+) = \frac{1}{\Theta_1 + \Theta_2} \{(1+\varepsilon) \Theta_1 \dot{\phi}_1(t_n^-) + (\Theta_2 - \varepsilon \Theta_1) k_2 \dot{\phi}_2(t_n^-)\} \qquad (2.14b)$$

Θ_1 and Θ_2 being the moments of inertia of the wheels 1 and 2. As in the one-step model we consider the no-load and the heavy load cases for which we assume $\ddot{\phi}_1 \approx \ddot{\phi}_2 \approx 0$ and $\ddot{\phi}_1 \approx 0, \ddot{\phi}_2 = T/\Theta_2$.

\# Let in the first case all the variables $\phi_i(t_n)$, $\dot{\phi}_i(t_n^+)$ be known after an impact at t_n. Consider the four quantities $\Delta t_{\pm}^i$

$$e(t_n + \Delta t_{\pm}^1) - k_1 \dot{\phi}_1(t_n^+) \Delta t_{\pm}^1 = \pm \Delta_1/2 + k_1 \phi_1(t_n) \qquad (2.15a)$$

$$\Delta t_{\pm}^1 \approx \{\pm \Delta_1/2 + k_1 \phi_1(t_n) - e(t_n)\}/\{\dot{e}(t_n) - k_1 \dot{\phi}_1(t_n^+)\} \qquad (2.15b)$$

$$\Delta t_{\pm}^2 = \{\pm \Delta_2/2 + k_2 \phi_2(t_n) - \phi_1(t_n)\}/\{\dot{\phi}_1(t_n^+) - k_2 \dot{\phi}_2(t_n^+)\} \qquad (2.15c)$$

where linearization of $e(t) - e(t_n)$ was used to approximate $\Delta t_{\pm}^1$ in (2.15b). Then the time of the next impact is obtained from

$$t_{n+1} = t_n + \min \{\Delta t_{\pm}^i > 0\} \qquad (2.16)$$

Once t_{n+1} is found the new values of the variables after the impact are obtained from

$$\phi_i(t_{n+1}) = \phi_i(t_n) + (t_{n+1} - t_n) \dot{\phi}_i(t_n^+) \qquad (2.17)$$

and

$= \text{If} \quad \min \{ \Delta t_{\pm}^{i} > 0 \} = \Delta t_{\pm}^{1}$

$$\dot{\phi}_1(t_{n+1}^{+}) = -\varepsilon \dot{\phi}_1(t_n^{+}) + \frac{1+\varepsilon}{k_1} \dot{e}(t_{n+1}) \qquad (2.18a)$$

$$\dot{\phi}_2(t_{n+1}^{+}) = \dot{\phi}_2(t_n^{+})$$

$= \text{If} \quad \min \{ \Delta t_{\pm}^{i} > 0 \} = \Delta t_{\pm}^{2}$

$$\dot{\phi}_1(t_{n+1}^{+}) = \frac{1}{\Theta_1 + \Theta_2} \{ (\Theta_1 - \varepsilon \Theta_2) \dot{\phi}_1(t_n^{+}) + (1+\varepsilon) \Theta_2 k_2 \dot{\phi}_2(t_n^{+}) \}$$

$$(2.18b)$$

$$k_2 \dot{\phi}_2(t_{n+1}^{+}) = \frac{1}{\Theta_1 + \Theta_2} \{ (1+\varepsilon) \Theta_1 \dot{\phi}_1(t_n^{+}) + (\Theta_2 - \varepsilon \Theta_1) k_2 \dot{\phi}_2(t_n^{+}) \}$$

Eqs. (2.15-18) completely define the no-load impact dynamics of the two-step model. They are well suited for a numerical iteration scheme. However for analytical study they are not very appropriate.

Instead of studying the dynamics in its full generality one might consider to restrict oneself to the embedded subdy- namics associated to particular impact sequences. If one iden tified the typical or more frequent such sequences, the corres ponding subdynamics would, at least, give a lower estimate of the dynamical complexity of the actual system. Complementary to the study of the embedded subdynamics is of course the study of the symbolic dynamics of the impact sequences.

To obtain some insight on the usefulness of the concept of embedded subdynamics we have used a numerical simulation of the maps (2.17-18) to study the change of the impact sequence structure with the excitation amplitude. The results some of which are reported in sect. 4 display a wide range of impact sequence patterns. The implication is that the subdynamics associated to each particular impact sequence has a narrow range of usefulness and therefore an heavy reliance on numerical simulation seems to be unavoidable in the study of the two-step model.

For the case with load one assumes

$$\ddot{\phi}_1 \approx 0 \quad \text{and} \quad \ddot{\phi}_2 = T/\Theta_2 . (T < 0) .$$

Then

$$\Delta t_{\pm}^1 \approx \{\pm \Delta_1/2 + k_1\phi_1(t_n) - e(t_n)\}/\{\dot{e}(t_n) - k_1\dot{\phi}_1(t_n^+)\}$$

$$\frac{\Theta_2}{k_2 T} \Delta t_{\pm}^2 = \dot{\phi}_1(t_n^+) - k_2\dot{\phi}_2(t_n^+) \mp \{[\dot{\phi}_1(t_n^+)-k_2\dot{\phi}_2(t_n^+)]^2 + \frac{2k_2 T}{\Theta_2}[\phi_1(t_n)-$$

$$- k_2\phi_2(t_n) \mp \frac{\Delta_2}{2}]\}^{1/2}$$

and the time of the next impact t_{n+1} is found from

$$t_{n+1} = t_n + \min \{\Delta t_{\pm}^i > 0\}$$

The variables after the impact are

$$\phi_1(t_{n+1}) = \phi_1(t_n) + (t_{n+1} - t_n)\dot{\phi}_1(t_n^+)$$

$$\phi_2(t_{n+1}) = \phi_2(t_n) + (t_{n+1} - t_n)\dot{\phi}_2(t_n^+) + \frac{T}{2\Theta_2}(t_{n+1}-t_n)^2$$

and

If $\min\{\Delta t_{\pm}^i > 0\} = \Delta t_{\pm}^1$

$$\dot{\phi}_1(t_{n+1}^+) = -\varepsilon\,\dot{\phi}_1(t_n^+) + \frac{1+\varepsilon}{k}\,\dot{e}(t_{n+1})$$

$$\dot{\phi}_2(t_{n+1}^+) = \dot{\phi}_2(t_n^+) + \frac{I}{\Theta_2}(t_{n+1} - t_n)$$

If $\min\{\Delta t_{\pm}^i > 0\} = \Delta t_{\pm}^2$

$$\dot{\phi}_1(t_{n+1}^+) = \frac{1}{\Theta_1 + \Theta_2}\{(\Theta_1 - \varepsilon\Theta_2)\dot{\phi}_1(t_n^+) + (1+\varepsilon)\Theta_2(\dot{\phi}_2(t_n^+) + \frac{I}{\Theta_2}\Delta t_{\pm}^2)\}$$

$$\dot{\phi}_2(t_{n+1}^+) = \frac{1}{\Theta_1 + \Theta_2}\{(1+\varepsilon)\Theta_1\dot{\phi}_1(t_n^+) + (\Theta_2 - \varepsilon\Theta_1)(\dot{\phi}_2(t_n^+) + \frac{I}{\Theta_2}\Delta t_{\pm}^2)\}$$

3. DYNAMICS OF THE ONE-STEP MODEL

3.a No-load dynamics

We consider first the no-load map (2.8). With the change of variables

$$J_n = (-1)^n(1+\varepsilon)E\Omega\,x_n \tag{3.1}$$

$$\theta_n + n\pi = 2\pi y_n$$

and $f(\Omega t) = \cos(\Omega t)$ one obtains

$$x_{n+1} = \varepsilon\,x_n + \sin(2\pi y_n)$$

$$y_{n+1} = y_n + \frac{1}{2} + \left|\frac{\mu/\varepsilon}{x_{n+1} + x_n}\right| \tag{3.2}$$

where $\mu = \frac{\Delta}{2\pi E}$

The Jacobian of the map (3.2) is not equal to one in the

non-dissipative case ($\varepsilon = 1$). This is because x and y are not canonical variables and is a familiar feature in Fermi acceleration when one considers collisions with the moving wall[5]. However it is in these variables that the map has a simple explicit form.

The map (3.2) is invariant for the transformation $x \to -x, y \to y + \frac{1}{2}$. Therefore it suffices to study it for $x > 0$. In Figs. 3a-i we show the structure of the attractor set (for $x > 0$) numerically obtained for $\varepsilon = .99$ and μ in the range 1.-100. One sees some attractive invariant circles, some chaotic attractors and some circles breaking into chaotic attractors. To interpret this structure one obtains analytic approximations to the invariant circles.

If there is a continuous function Φ such that for $y \varepsilon S^1$

$$x_n = \Phi(y_n) \implies x_{n+1} = \Phi(y_{n+1})$$

then the map (3.2) has an invariant circle. Using the map this condition is equivalent to the existence of a continuous solution to the functional equation

$$\varepsilon \Phi(y) + \sin(2\pi y) = \Phi\{y + \frac{1}{2} + \left| \frac{\mu/\varepsilon}{(1+\varepsilon)\Phi(y) + \sin 2\pi y} \right| \} \qquad (3.3)$$

For $(1 - \varepsilon)$ small an approximate solution is

$$\Phi_n(y) = \pm \frac{\mu}{n\varepsilon(1+\varepsilon)} - \frac{1}{(1+\varepsilon)} \sin(2\pi y) \qquad (3.4)$$

which becomes exact in the limit $\varepsilon \to 1$

Eq. (3.4) for Φ_n approximates well the invariant circle, which in the figure we denote by I_n, and also gives a good estimate of the location of the corresponding chaotic attractor. For each fixed n when μ decreases to approximately $4.3\,n^2$ (if $\varepsilon \lesssim 1$) the circle I_n breaks down giving rise to a chaotic attractor. When μ is increased the invariant circle moves outward but lags behind the approximate position (3.4) for $\varepsilon < 1$. To first order this deviation from (3.4) is given by

$$\Psi_n(y) = \Phi_n(y) + f(y) \qquad (3.5)$$

with

$$f(y) = \frac{-(1-\varepsilon)\mu}{n} \frac{\mu^2 - \pi n^2 \cos 2\pi y}{(2\pi n^2 \cos 2\pi y)^2} \qquad (3.6)$$

valid for $y \approx 0, 1/2$.

Because of this lagging behind, the fixed point P_n near $(x = \frac{\mu}{2n+1}$, $y = \frac{1}{2})$ catches up with I_n as μ increases, and when this happens I_n vanishes without giving rise to a chaotic attractor.

The attracting set which we denote by II in the figure is of a different nature of those of class I. It corresponds to the approximate solution to Eq. (3.3)

$$\Phi_\delta(y) \simeq \frac{\mu}{4\delta} - \frac{1}{2\cos 2\pi\delta} \sin 2\pi (y-\delta) \qquad (3.7)$$

which only becomes exact in the limit of large $|\Phi|$ and $\varepsilon \to 1$.

Eq. (3.7) approximates fairly well the attracting set of class II shown in Figs. 3a-d for δ in the range .1 - .2. This set disappears at $\mu \simeq 7.5$, being hit by the hyperbolic fixed point P_0.

Recalling that x is proportional to the speed of the wheel at the time of the impacts, it is of importance for the applications to realize that in this model one has several stable speed regimes, with the highest one being typically a factor of two above the lowest one. Which regime the system will choose will depend on which bassin of attraction one starts from in the initial conditions.

Let us analyse the frequency and intensity of the rattling noise associated to one of the regimes which we denote I_n (corresponding to the approximation Φ_n of Eq. (3.4)). From (3.4) and (3.2) one obtains

$$y_{j+1} = y_j + \frac{1}{2} + n$$

Using now (3.1) and (2.7b) one concludes that the time separation between two sucessive impacts is

$$t_{j+1} - t_j = \frac{2\pi}{\Omega} (n + 1/2)$$

Hence the impact frequency f_{ϕ_n} is related to the excitation frequency f ($\Omega = 2\pi f$) by

$$f_{\phi_n} = \frac{f}{n+1/2} \qquad (3.8)$$

The largest rattling frequency that is possible is controlled by the smallest n for which ϕ_n is stable. This implies that when $\mu = \Delta/2\pi E$ grows the largest frequency decreases. If one insures that the lowest speed regime is chosen avoiding high speed initial conditions, then the generated impact frequency will be the smallest possible. On the other hand as in general the low speed attractor is a chaotic attractor, the generated noise will have its spectrum broadened, loosing the characteristics of a sharp pitch. One should however never forget that (at least in this model) there are other stable regimes higher up in speed space.

We now compute the parameter dependence of the noise intensity. The noise intensity being proportional to the energy lost in the collisions between the teeth of the wheels it should be proportional to $(\dot{s}(t))^2$ at the time of the impact. Using

$$\dot{s}(t) = \dot{e}(t) - k\dot{\phi}(t) = - E\Omega \, \sin(\Omega t) - J(t)$$

one obtains at the time of the ℓth impact in the regime approximated by ϕ_n

$$\dot{s}(t_\ell) = -(-1)^\ell E\Omega \frac{\mu}{n\epsilon} = -(-1)^\ell \frac{\Omega \Delta}{2\pi n\epsilon}$$

Therefore the noise intensity is proportional to $\frac{\Omega^2 \Delta^2}{n^2}$.

An interesting feature of this regime is that (for fixed n) the noise intensity does not depend directly on the intensity E of the excitation. One should however not forget that E controls the μ parameter and that decreasing E, μ increases and when μ increases larger n's appear as stable regimes. From our estimate of the noise intensity one also concludes on the convenience of locking in a low speed regime where n is the largest.

For the attractor of class II, anproximated by ϕ_δ of Eq.(3.7), the calculation follows similar steps, namely

$$t_{j+1} - t_j \approx \frac{2\pi}{\Omega} \left(\frac{1}{2} + \frac{4\delta}{\epsilon(1+\epsilon)} \right)$$

leding to

$$f_{\phi_\delta} \approx \frac{f}{2\delta + 1/2}$$

for small 1-ϵ. For $\delta \approx .1-.2$ one has $f_{\phi_\delta} \approx (1.4-1.1)f$. In this regime the impact frequency is higher or of the order of the excitation frequency, whereas in the ϕ_n regime it is smaller than f. For the noise intensity one computes the average

$$\overline{(\dot{s}(t_\ell))^2} \approx E^2\Omega^2 \left\{ \frac{\Delta^2}{(2\pi)^2 E^2 4\delta^2} + \frac{3}{2} + \frac{1}{2\cos^2(2\pi\delta)} \right\}$$

for small ε. There are now terms that depend directly on E^2.

3.b Heavy load dynamics

With the change of variables $E\Omega x = J$ $\quad 2\pi y = \theta$ and $f(\Omega t) = \cos(\Omega t)$ the maps M_1 of Eq. (2.9a) and M_2 of Eq. (2.11) become

$$M_1 \begin{cases} x_{n+1} = \varepsilon x_n - (1-\varepsilon) \sin(2\pi y_n) \\[2mm] y_{n+1} = y_n + \nu\{x_n + \sin(2\pi y_n)\} \end{cases} \tag{3.9}$$

$$M_2 \begin{cases} x_{n+1} = -x_n - 2\sin(2\pi y_n) \\[2mm] y_{n+1} = y_n - \dfrac{\nu}{\varepsilon}\{x_n + \sin(2\pi y_n)\} \end{cases} \tag{3.10}$$

where
$$\nu = \frac{\varepsilon E \Theta \Omega^2}{\pi k T} \qquad (\nu < 0)$$

the dynamics being defined by $M = M_2^p \circ M_1$, where $p = 0,1,...$ is the smallest p for which $(Mx_n) + \sin(2\pi M y_n) \leq 0$.

The dynamics defined by the map $M = M_2^p \circ M_1$ constrains the physical phase space points to the region

$$L = \{(x,y) : x + \sin 2\pi y \leq 0\}$$

Define also the regions

$$L_I = \{(x,y): x \leq -1\} \quad \text{and}$$

$$L_\varepsilon = \{(x,y) : x \geq 1 - 2/\varepsilon \, , \, x + \sin 2\pi y \leq 0\}$$

If $x_n < -1$ then from (3.9) $|x_{n+1}| \leq |x_n|$ and on iteration one approaches the line $x = -1$. As long as $x_n < 1 - \frac{2}{\varepsilon}$; x_{n+1} falls in the region L_I. For $x_n \in L_\varepsilon$, $|M_1 x_n| \leq 3 - 2\varepsilon$. When $M_1 x_n$ then falls in the "forbidden region" $(x + \sin(2\pi y) > 0)$ one still has the bound $M_2^p \circ M_1 x_n > -3$.

Depending on the value of ν there are typically two types of asymptotic behaviour. The first, for small $|\nu|$, is dominated by the fixed points of M_1 (Fig. 4a-b), whereas in the second, for large $|\nu|$, the asymptotic points are concentrated in a neighbourhood of the line $x = -1$ (Fig. 4c-d).

The map M_1 has a line $x = -\sin 2\pi y$ of fixed points. The linearized map TM_1 has eigenvalues $\lambda_1 = 1$ and $\lambda_2 = \varepsilon + 2\pi\nu \cos(2\pi y)$ to which correspond the eigenvectors

$$V_1 = \begin{vmatrix} -2\pi\nu \cos(2\pi y) \\ 1 \end{vmatrix} \qquad V_2 = \begin{vmatrix} (\varepsilon - 1)/\nu \\ 1 \end{vmatrix}$$

Depending on the values of ν and y the direction V_2 is contracting or expanding. The fixed points become unstable when

$$\cos(2\pi y) < \frac{-1 + \varepsilon}{2\pi |\nu|} \qquad (3.11a)$$

or
$$\cos(2\pi y) > \frac{1 + \varepsilon}{2\pi |\nu|} \qquad (3.11b)$$

The first inequality explains why the fixed points in the central y region are always absent from the asymptotic beha viour and the second implies that as $|\nu|$ increases more of remaining   fixed points become unstable as well.

Of   practical interest for the load dynamics of the one-
-step model is the maximum value that the impact   variable
$\dot{s} = \dot{e} - k\dot{\phi}$ can reach, because this is related to the overload
on the wheels.   Taking into account that $M_2^p \circ M_1 x > -3$ when
$x \in L_\varepsilon$,   one concludes that $|\dot{s}| < 4E\Omega$ in the steady-state regime.   Transient behaviour may of course lead to stronger
impacts.

4. TWO-STEP MODEL

In this section we report on preliminary results obtained
by numerical simulation of the two-step   model   without load
(Eqs. 2.15-18).

We use the change of variables

$$\phi_1(t_n) = \frac{\omega}{k_1} t_n + \frac{E}{k_1} x_n^{(1)} \; ; \; \phi_2(t_n) = \frac{\omega}{k_1 k_2} t_n + \frac{E}{k_1 k_2} x_n^{(2)} \qquad (4.1a)$$

$$\dot{\phi}_1(t_n^+) = \frac{\omega}{k_1} + \frac{E\Omega}{k_1} \dot{x}_n^{(1)} \; ; \; \dot{\phi}_2(t_n^+) = \frac{\omega}{k_1 k_2} + \frac{E\Omega}{k_1 k_2} \dot{x}_n^{(2)} \qquad (4.1b)$$

$$\tau_n = \frac{1}{2\pi} \Omega t_n \quad ; \quad A_1 = \frac{\Delta_1}{2E} \; ; \; A_2 = \frac{\Delta_2 k_1}{2E} \qquad (4.1c)$$

and from Eqs. (2.15-18) we obtain

$$\cos\left[2\pi(\tau_n + \Delta\tau_\pm^1)\right] - x_n^{(1)} - \dot{x}_n^{(1)}(2\pi\Delta\tau_\pm^1) = \pm A_1 \qquad (4.2a)$$

$$x_n^{(1)} - x_n^{(2)} + 2\pi\Delta\tau_\pm^2(\dot{x}_n^{(1)} - \dot{x}_n^{(2)}) = \pm A_2 \qquad (4.2b)$$

$$\tau_{n+1} = \tau_n + \min\ \{\Delta\tau_\pm^i > 0\} \qquad (4.3)$$

$$x_{n+1}^{(i)} = x_n^{(i)} + 2\pi(\tau_{n+1} - \tau_n)\ \dot{x}_n^{(i)} \qquad (4.4)$$

If $\min\ \{\Delta\tau_\pm^i > 0\} = \Delta\tau_\pm^1$

$$\dot{x}_{n+1}^{(1)} = -\varepsilon\ \dot{x}_n^{(1)} + (1+\varepsilon)\ f'(2\pi\tau_{n+1}) \qquad (4.5a)$$

$$\dot{x}_{n+1}^{(2)} = \dot{x}_n^{(2)} \qquad (4.5b)$$

If $\min\ \{\Delta\tau_\pm^i > 0\} = \Delta\tau_\pm^2$

$$\dot{x}_{n+1}^{(1)} = \frac{1}{\Theta_1 + \Theta_2}\ \{(\Theta_1 - \varepsilon\ \Theta_2)\ \dot{x}_n^{(1)} + (1+\varepsilon)\Theta_2\ \dot{x}_n^{(2)}\} \qquad (4.5c)$$

$$\dot{x}_{n+1}^{(2)} = \frac{1}{\Theta_1 + \Theta_2}\ \{(1+\varepsilon)\ \Theta_1\dot{x}_n^{(1)} + (\Theta_2 - \varepsilon\ \Theta_1)\ \dot{x}_n^{(2)}\} \qquad (4.5d)$$

The system of equations 4.2-5 was numerically studied for an excitation function $f = f_2$ (of Eq. 2.1c) and $\Theta_1 = \Theta_2 = 1$. The initial condition is $x^{(1)} = x^{(2)} = 0$ with $\dot{x}^{(1)}$ and $\dot{x}^{(2)}$ chosen at random. After the transients die out the motion is followed for 500 iterations and then a new set of initial conditions is chosen.

The results, contained in Figs. 5a-f, show the nature of the asymptotic motion in this model. In the figures $x \equiv \dot{x}^{(1)}$, $y \equiv \dot{x}^{(2)}$ and $\theta \equiv \tau$.

There are two types of asymptotic motion, one that corres
ponds to an approximately quasi-periodic motion and the other
to an irregular motion centered on the region of low velocities. The size of the irregular region grows with the value of
the parameters A_i. We have also related the type of motion
to the nature of the impact sequences. Denoting by A,B,C,D
the impacts of type $\Delta\tau_+^1$, $\Delta\tau_-^1$, $\Delta\tau_+^2$, $\Delta\tau_-^2$ one finds that the
quasi-periodic motion corresponds to long stretches of CD
impacts (Fig. 6a) whereas an irregular impact pattern is found
to be associated to the irregular motion in velocity space
(Fig. 6b).

REFERENCES AND FOOTNOTES

[1] F.Pfeiffer ; "Mechanische Systeme mit unstetigen Ünbergängen", Ingenieur-Archiv, 54 (1984) 232.

[2] F.Pfeiffer and F.Küçükay; "Eine erweiterte mechanische Stosstheorie und ihre Anwendung in der Getriebedynamik", VDI-Zeitschr. Bd 127 (1985) 341.

[3] F.Küçükay and F.Pfeiffer ; "Über Rasselschwingungen in KFZ--Schaltgetrieben", Ingenieur-Archiv 56 (1986) 25.

[4] M.-O.Hongler and L.Streit; "On the origin of chaos in gearbox models", BiBoS preprint, 1986.

[5] A.J.Lichtenberg and M.A. Lieberman; "Regular and Stochastic Motion", Appl. Math. Sc. 38, Springer-Verlag, 1983, chapter 3.

FIGURE CAPTIONS

1. One step model

2. Two step model

3. Attractor set for the one step model without load

4. Asymptotic set for the one step model with an heavy load

5. Asymptotic velocity distribution in the two step model

6. Regular and irregular impact sequences.

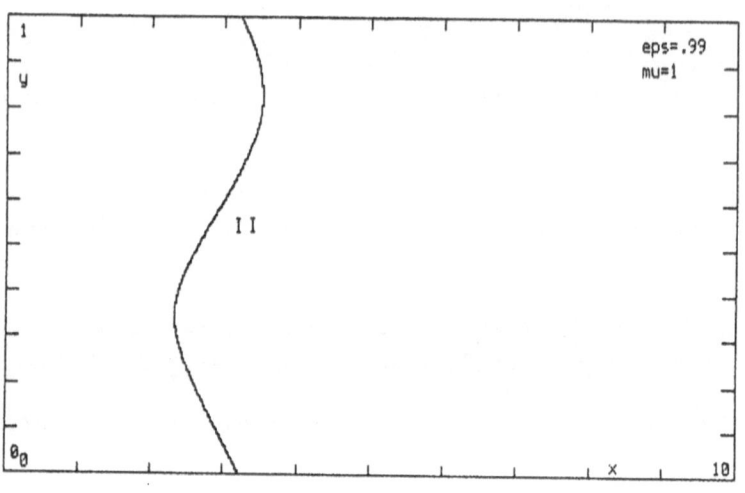

FIG. 3A

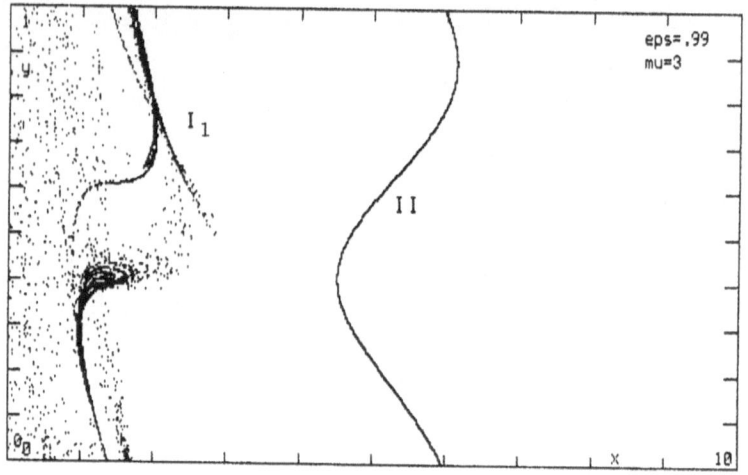

FIG. 3B

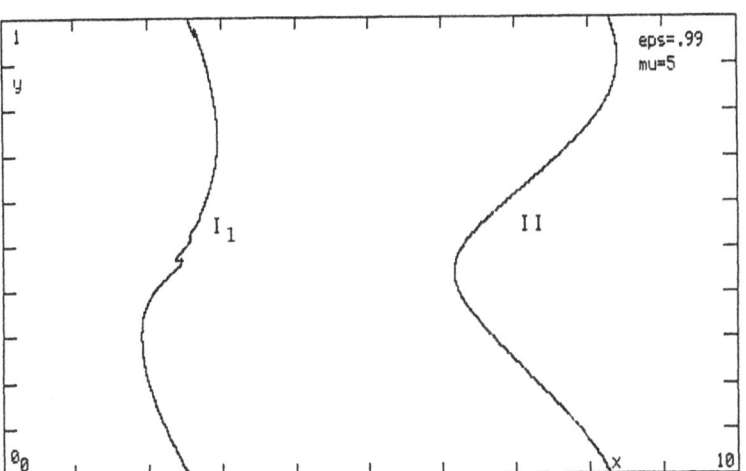

FIG. 3c

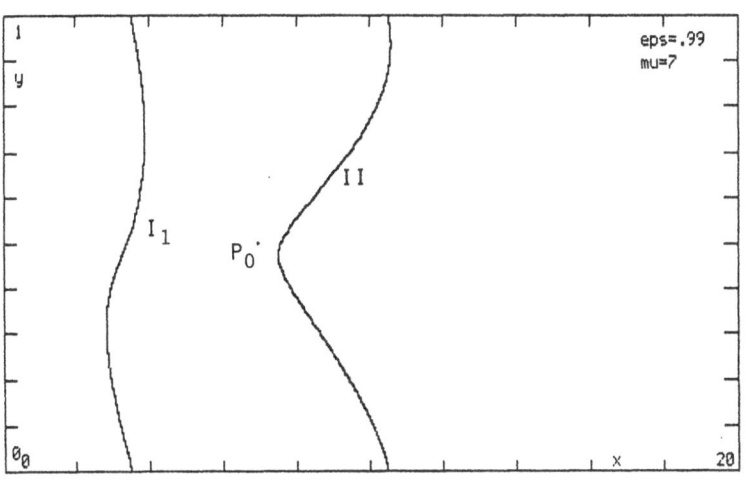

FIG. 3d

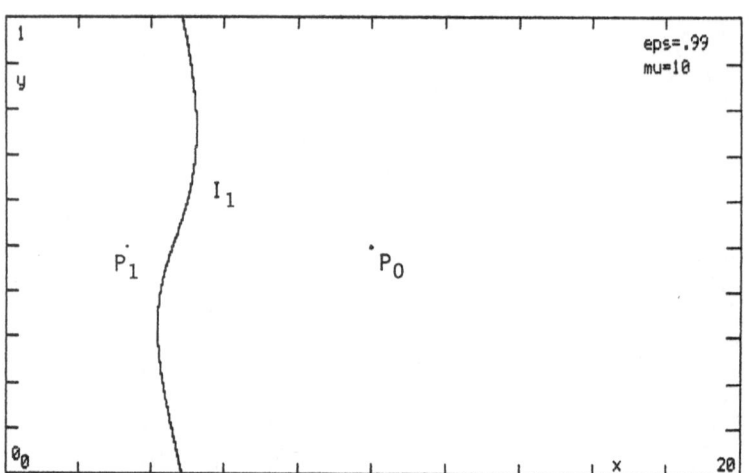

FIG. 3E

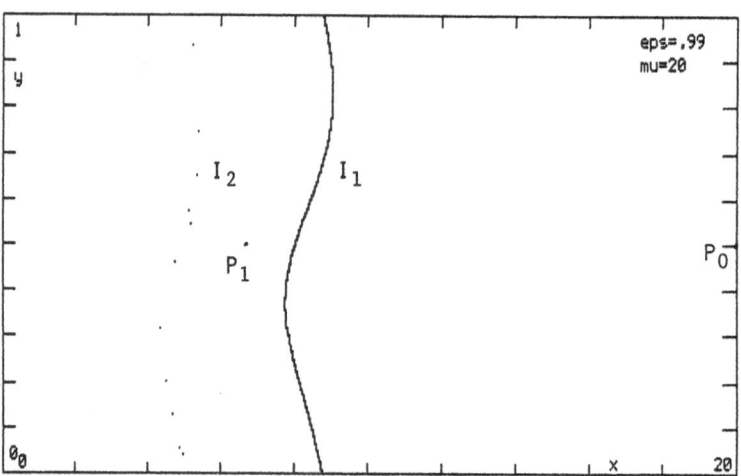

FIG. 3F

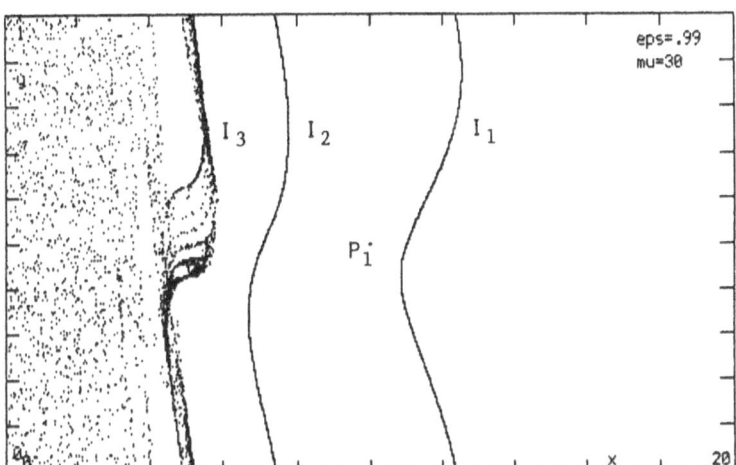

FIG. 3G

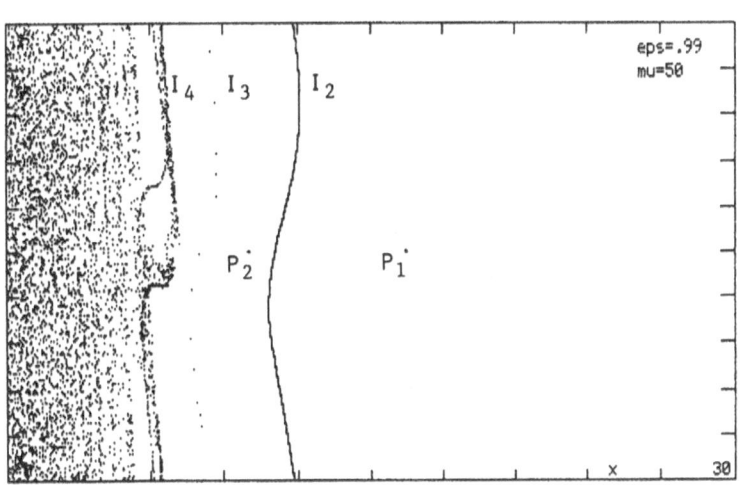

FIG. 3H

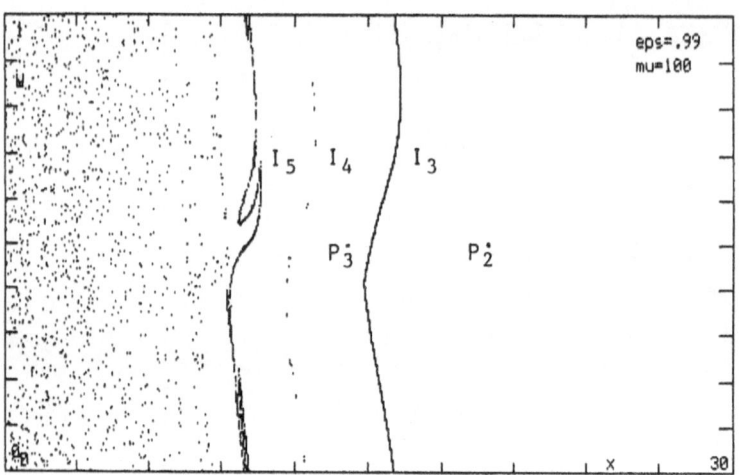

FIG. 31

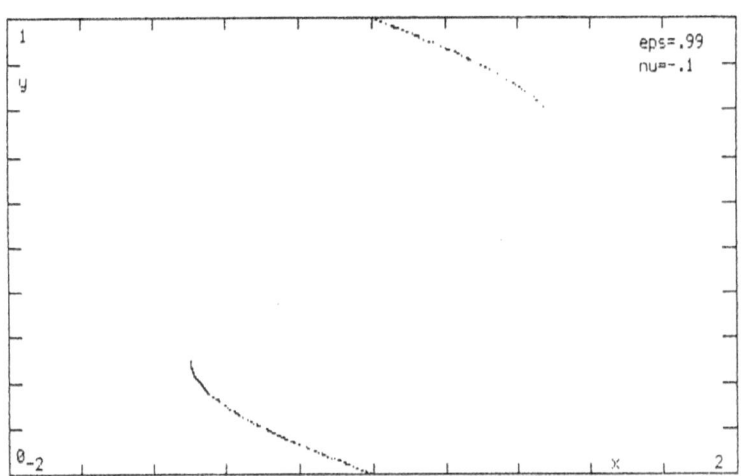

FIG. 4A

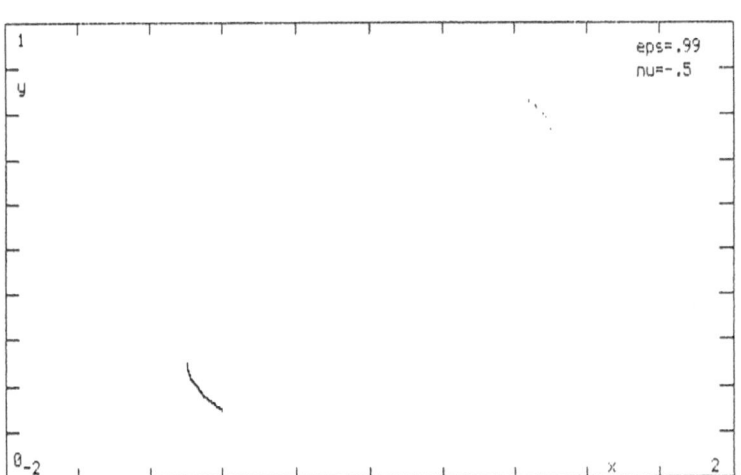

FIG. 4B

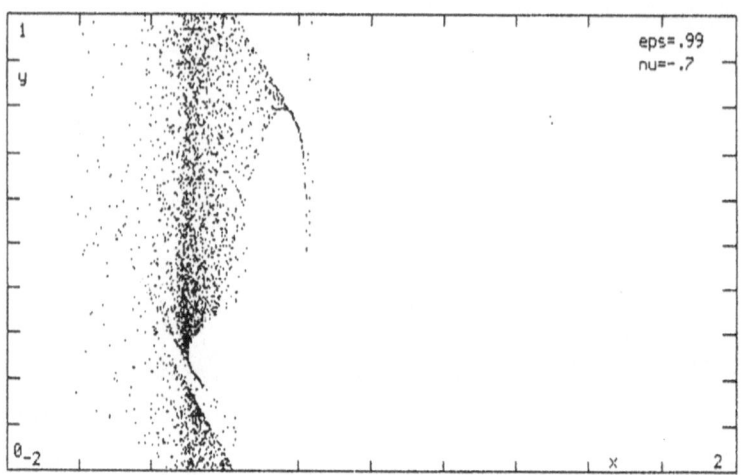

FIG. 4c

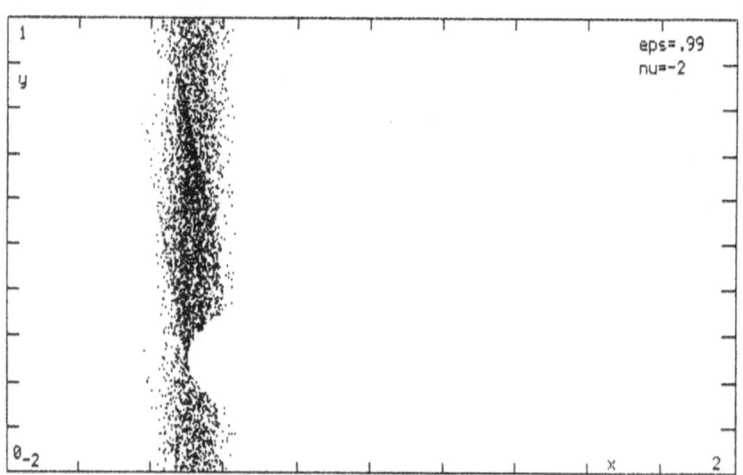

FIG. 4d

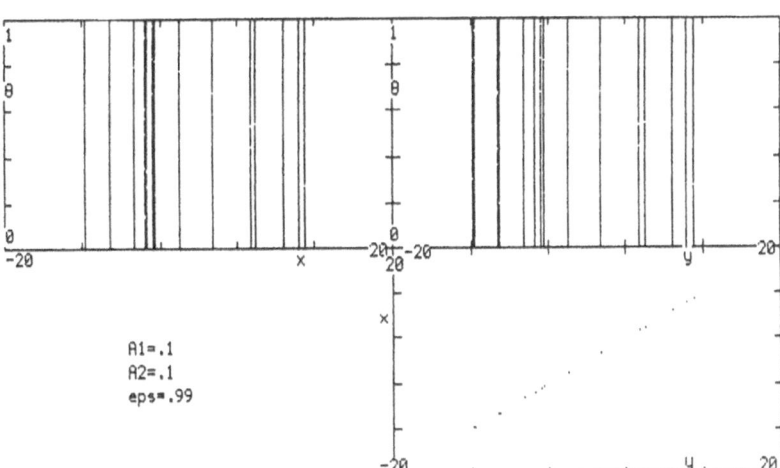

A1=.1
A2=.1
eps=.99

FIG. 5A

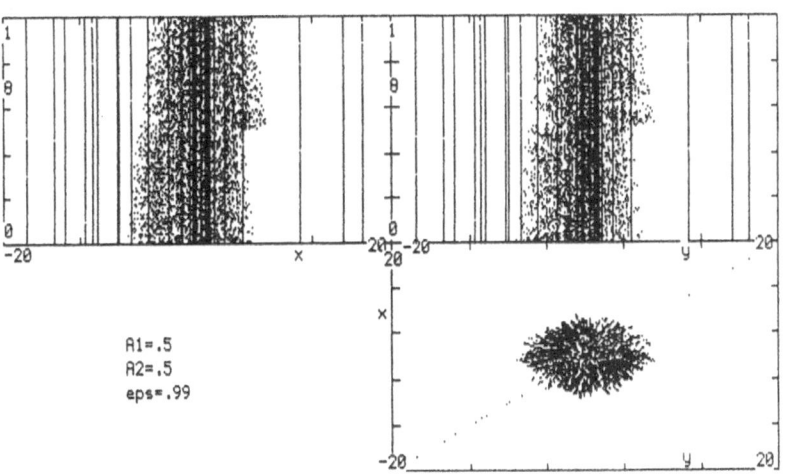

A1=.5
A2=.5
eps=.99

FIG. 5B

34

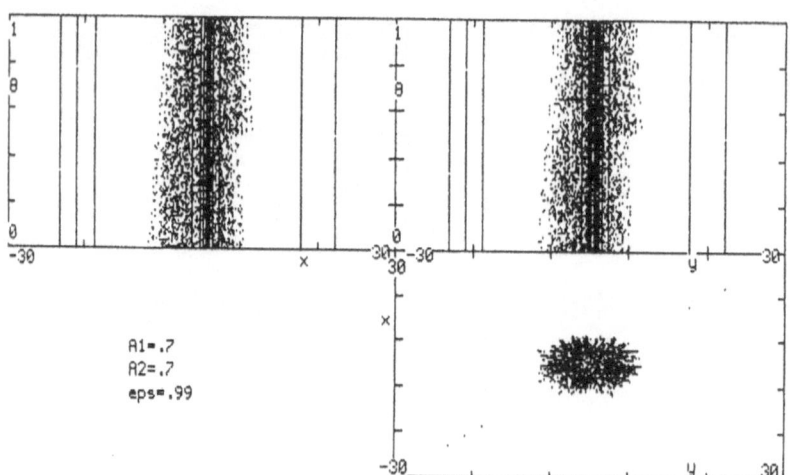

A1=.7
A2=.7
eps=.99

FIG. 5c

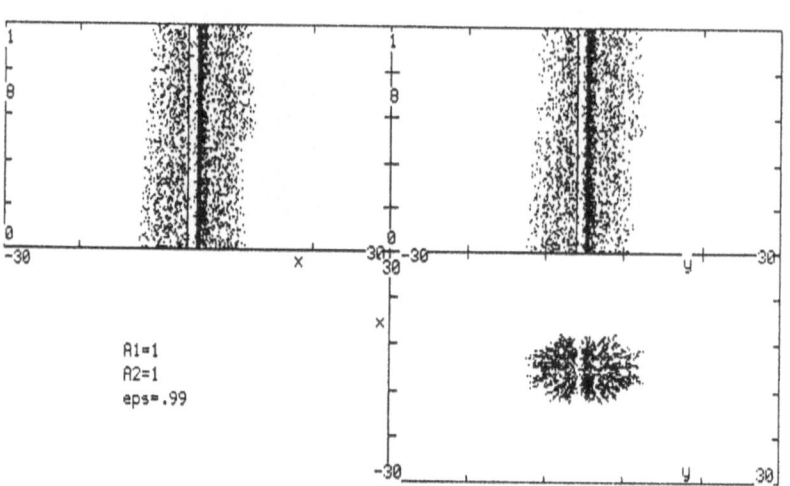

A1=1
A2=1
eps=.99

FIG. 5D

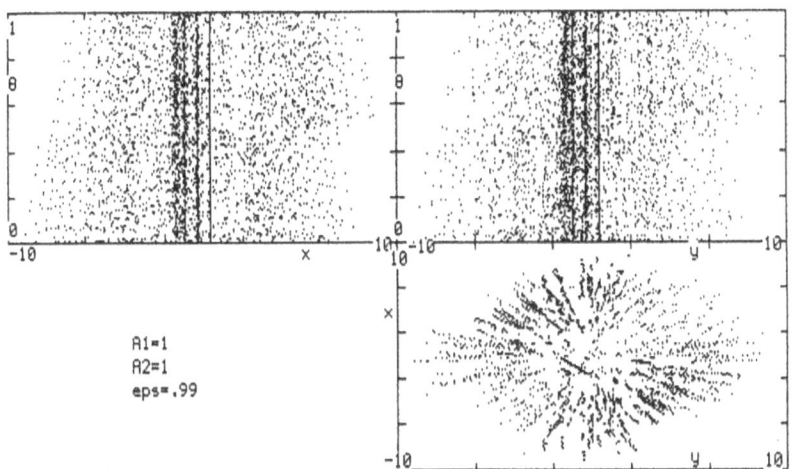

A1=1
A2=1
eps=.99

FIG. 5E

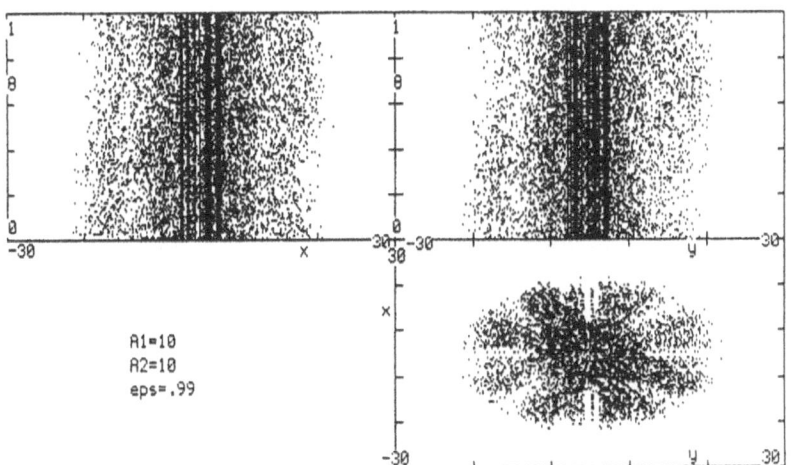

A1=10
A2=10
eps=.99

FIG. 5F

A1=.1   A2=.1   eps=.99

```
DCDCDCDCDCDCDCDCDCDCDCDCDCDCDCDCDCDCDCDCDCDCDCDCDCDCDCDCDCDCDCDCDCDCDCDCDCDCDCDCDCD
CDCDCDCDCDCDCDCDCDCDCDCDCDCDCDCDCDCDCDCDCDCDCDCDCDCDCDCDCDCDCDCDCDCDCDCDCDCDCDCDCDC
DCDCDCDCDCDCDCDCDCDCDCDCDCDCDCDCDCDCDCDCDCDCDCDCDCDCDCDCDCDCDCDCDCDCDCDCDCDCDCDCDCD
CDCDCDCDCDCDCDCDCDCDCDCDCDCDCDCDCDCDCDCDCDCDCDCDCDCDCDCDCDCDCDCDCDCDCDCDCDCDCDCDCDC
DCDCDCDCDCDCDCDCDCDCDCDCDCDCDCDCDCDCDCDCDCDCDCDCDCDCDCDCDCDCDCDCDCDCDCDCDCDCDCDCDCD
CDCDCDCDCDCDCDCDCDCDCDCDCDCDCDCDCDCDCDCDCDCDCDCDCDCDCDCDCDCDCDCDCDCDCDCDCDCDCDCDCDC
DCDCDCDCDCDCDCDCDCDCDCDCDCDCDCDCDCDCDCDCDCDCDCDCDCDCDCDCDCDCDCDCDCDCDCDCDCDCDCDCDCD
CDCDCDCDCDCDCDCDCDCDCDCDCDCDCDCDCDCDCDCDCDCDCDCDCDCDCDCDCDCDCDCDCDCDCDCDCDCDCDCDCDC
DCDCDCDCDCDCDCDCDCDCDCDCDCDCDCDCDCDCDCDCDCDCDCDCDCDCDCDCDCDCDCDCDCDCDCDCDCDCDCDCDCD
CDCDCDCDCDCDCDCDCDCDCDCDCDCDCDCDCDCDCDCDCDCDCDCDCDCDCDCDCDCDCDCDCDCDCDCDCDCDCDCDCDC
DCDCDCDCDCDCDCDCDCDCDCDCDCDCDCDCDCDCDCDCDCDCDCDCDCDCDCDCDCDCDCDCDCDCDCDCDCDCDCDCDCD
CDCDCDCDCDCDCDCDCDCDCDCDCDCDCDCDCDCDCDCDCDCDCDCDCDCDCDCDCDCDCDCDCDCDCDCDCDCDCDCDCDC
DCDCDCDCDCDCDCDCDCDCDCDCDCDCDCDCDCDCDCDCDCDCDCDCDCDCDCDCDCDCDCDCDCDCDCDCDCDCDCDCDCD
CDCDCDCDCDCDCDCDCDCDCDCDCDCDCDCDCDCDCDCDCDCDCDCDCDCDCDCDCDCDCDCDCDCDCDCDCDCDCDCDCDC
DCDCDCDCDCDCDCDCDCDCDCDCDCDCDCDCDCDCDCDCDCDCDCDCDCDCDCDCDCDCDCDCDCDCDCDCDCDCDCDCDCD
CDCDCDCDCDCDCDCDCDCDCDCDCDCDCDCDCDCDCDCDCDCDCDCDCDCDCDCDCDCDCDCDCDCDCDCDCDCDCDCDCDC
DCDCDCDCDCDCDCDCDCDCDCDCDCDCDCDCDCDCDCDCDCDCDCDCDCDCDCDCDCDCDCDCDCDCDCDCDCDCDCDCDCD
CDCDCDCDCDCDCDCDCDCDCDCDCDCDCDCDCDCDCDCDCDCDCDCDCDCDCDCDCDCDCDCDCDCDCDCDCDCDCDCDCDC
DCDCDCDCDCDCDCDCDCDCDCDCDCDCDCDCDCDCDCDCDCDCDCDCDCDCDCDCDCDCDCDCDCDCDCDCDCDCDCDCDCD
CDCDCDCDCDCDCDCDCDCDCDCDCDCDCDCDCDCDCDCDCDCDCDCDCDCDCDCDCDCDCDCDCDCDCDCDCDCDCDCDCDC
DCDCDCDCDCDCDCDCDCDCDCDCDCDCDCDCDCDCDCDCDCDCDCDCDCDCDCDCDCDCDCDCDCDCDCDCDCDCDCDCDCD
CDCDCDCDCDCDCDCDCDCDCDCDCDCDCDCDCDCDCDCDCDCDCDCDCDCDCDCDCDCDCDCDCDCDCDCDCDCDCDCDCDC
DCDCDCDCDCDCDCDCDCDCDCDCDCDCDCDCDCDCDCDCDCDCDCDCDCDCDCDCDCDCDCDCDCDCDCDCDCDCDCDCDCD
CDCDCDCDCDCDCDCDCDCDCDCDCDCDCDCDCDCDCDCDCDCDCDCDCDCDCDCDCDCDCDCDCDCDCDCDCDCDCDCDCDC
DCDCDCDCDCDCDCDCDCDCDCDCDCDCDCDCDCDCDCDCDCDCDCDCDCDCDCDCDCDCDCDCDCDCDCDCDCDCDCDCDCD
CDCDCDCDCDCDCDCDCDCDCDCDCDCDCDCDCDCDCDCDCDCDCDCDCDCDCDCDCDCDCDCDCDCDCDCDCDCDCDCDCDC
DCDCDCDCDCDCDCDCDCDCDCDCDCDCDCDCDCDCDCDCDCDCDCDCDCDCDCDCDCDCDCDCDCDCDCDCDCDCDCDCDCD
CDCDCDCDCDCDCDCDCDCDCDCDCDCDCDCDCDCDCDCDCDCDCDCDCDCDCDCDCDCDCDCDCDCDCDCDCDCDCDCDCDC
DCDCDCDCDCDCDCDCDCDCDCDCDCDCDCDCDCDCDCDCDCDCDCDCDCDCDCDCDCDCDCDCDCDCDCDCDCDCDCDCDCD
CDCDCDCDCDCDCDCDCDCDCDCDCDCDCDCDCDCDCDCDCDCDCDCDCDCDCDCDCDCDCDCDCDCDCDCDCDCDCDCDCDC
DCDCDCDCDCDCDCDCDCDCDCDCDCDCDCDCDCDCDCDCDCDCDCDCDCDCDCDCDCDCDCDCDCDCDCDCDCDCDCDCDCD
```

FIG. 6A

A1=10   A2=10   eps=.99

```
BDCDCDCDCDCDBCACACBDBCACADBDBCACADBDCACDBDCDCACDBDCABDBDCACDBDCACDBDC
ACDBDCACDBDCACDBDCACDBDCACDBDCACDBDCACDBDCACDBDCACDBDCACDBDCACDBDBCAC
DBACDBDACABDCDCDCABDBDACABDBDCACDCDBDACACDBDBCACADBDBACABDBACDCDCDCDC
DCDBACADBDCDCABDCABDCABDCACBDBDCACDCDBDACACDBDCDCACBDBDCACDBACDBDCACA
DBDCABDCACBDBDCACDCDBDCACADBDCDCACBDBDCACADBDCABDCABDCABDCABDCACBDBDC
ACDCDBDCACADBDCABDCABDCACDBDBCACDBDCACDBDCACADBDBACDCDBACADBDCACDBDBC
ACDCDBDACACBDBDCACADBDCACDCDBDACACBDBACDCDBACDBDCACADBDBCACABDBACDCDC
DBDACACDBDBCACADBDCDCABDCABDCACBDBDCACADBDCDCACBDBDCACADBDBCACABDBDAC
ACDBDBCACADBDBACACBDBACDCDCDCDBACADBDBCACADBDBACACBDBDACACBDBDACACBDB
DCACADBDCDCACBDBDCACADBDBCACADBDBCACADBDBACACBDBDCACADBDBCACADBDBCACA
DEDBACABDCDCDCDCDCABDBACABDBDACABDBACABDBACABDBACABDBACACBDBACDCDCD
BACABDBDACACBDBDACDCDCDBACDCDBACDCDCDBACACBDBCACABDBCACADBDCABDCABDCAB
DCDCACDBDBCACABDBDCACBDBCACACBDBCACABDBCACABDBACDCDBDBCACDCDBDA
CACADBDCDCACDBDBCACADBDCACADBDBACABDBACABDCDCDCDCABDBCACABDBACABDBAC
ACABDBACABDBACADBDBACABDBACACBDBACABDBDACABDBCACABDBACDCDCDCDBACDCDBA
CDCDCDBDACACBDBDACABDCDCABDCDCDCABDCDCABDCDCABDCDCABDCDCABDCDCACBDBDA
CACDBDBCACADBDBCACADBDCDCABDCABDCABDCABDCDCDCABDCDCACBDBDCACDCDBDCACA
DBDCACDBDCACDBDCACDBDCABDCACDBDCACDBDCACDBDCACDBDBCACDBACDBDCACDBACDB
DCACDBDCACADBDCACDBDCACDBACDBDCACDBDCACDBDCACDBDCACADBDCACDBDB
CACDCDBDCACADBDCACDBDCACBDBDCACDBDACACDBDCACDBDBCACDBACDBDCACDBAC
DCDBDACACDBDBCACADBDBCACADBDBACACBDBDACACBDBDCACADBDBCACDBACDBDCACADB
DCABDCACBDBDACABDCDCABDBACACBDBACADBDBACABDCDCDCABDBCACABDBDACACBDBDA
CACDBDBCACADBDBCACADBDBACACBDBACDCDCDCDBACADBDBACACBDBACDCDCDBACADBDBACACBD
BACDCDBACDCDBDACACBDBDACACBDBDACACBDBDCACADBDCDCACBDBDCACADBDBCACADBD
BACACBDBDACACDBDBCACDCDBACDBACDCDBACDCDBACDBACDCDBACDCDBDACACDBDBCACD
BDCACDBDCACDCDBDCACDBDCACDBDCACDBDCACDBDCDCDBDCACDBDCACDBDCACDBDCACDBDCAC
DBDCACDBDCACADBDCACDBDCACDBDCACBDBDBCACDBDCACDBDCACDBDCACDBDCACADBDCACBDBDBCACDCDB
```

FIG. 6B

ON AN OFF-SHORE PIPE LAYING PROBLEM

R.M.M. Mattheij   S.W. Rienstra

The final version of this paper will be submitted for publication elsewhere

Abstract

In this paper we consider a model for determining the shape of a pipe laid by a barge. We investigate how the solution of a resulting second order non-linear differential equation depends critically on the (unknown) vertical bottom reaction force; by this we can explain some difficulties met when approximating the solution numerically. An important part of the analysis is based on studying a pendulum analogy in a time dependent gravity field, by which we obtain existence and uniqueness results.

1. Introduction

The transportation of gas or oil from offshore wells requires the installation of pipelines to connect the well to some mainland station.Typically such pipes are 16-36 inch in diameter and may be up to several hundreds of kilometers long. Usually these pipes are laid in relatively shallow ($\sim$ 50 m) and deep ($\sim$ 300 m) water, although recently technology has been developed to handle very deep water (300-1000 m and more). Part of the technology is in the material of which the pipes are being made (high quality steel), but very much depends on the laying technique too. The main problem to be dealt with is buckling due to too high bending stresses during laying; pipe collapse due to a too high water pressure becomes also important at greater depths, but we shall not consider that problem here.

At present the pipes are being laid using a so called laybarge where the pipe sections are welded together, and then gradually released off the barge to the sea bed. If the welding ramp is horizontally positioned, the pipe is released via a (circular) stinger. According to the shape of the suspended pipeline this technique is called S-lay. If the welding ramp is in vertical position the technique is called J-lay. Here we shall concentrate (although not in principle) on the S-lay technique, which is the usual one for not very deep water. The analysis we give below is equally applicable for J-lay configurations though.

Usually, the pipes are relatively thin-walled and, without additional weight, would float on the water. This weight is provided by application of a concrete coating, giving the pipes a net specific gravity of about 1.05 to 1.30 times that of water. (The value depends on the product to be transported, and the presence of sea currents.) This non-negligible own weight, especially in deep water, tends to give the pipe during laying a shape with such a high curvature, that without precautions the pipe would buckle. Therefore, the laybarge is equipped with tension machines to apply a certain horizontal tension, just sufficient to stretch the S curve and reduce the bending stresses to a safe level.

The practical question now is: given a certain geometry, how much tension is needed and, further-more, which is the optimal position of the stringer. On the one hand, repairing a buckled pipe is very expensive, but on the other hand, so are the tension machines and other equipment, so rather accurate calculations are required. Given the large number of parameters involved this problem has to be solved again for each new case, and a mathematical model including a computer program is obviously required (Ref. 1,2,3,4,5,6,7,8,9). This will be the subject of the present paper.

The problem as such arose from requests from offshore companies, who were interested in a routine efficient and fast enough for a small computer, to be used on board in order to be able to respond immediately to unforeseen variations in (e.g.) steel quality, concrete quality, or other problem parameters. In trying to develop a program we hit severe numerical problems. In particular, since the problem is nonlinear, we were often not able to make the Newton method involved converge. In search for the phenomena behind this problem we encountered some interesting mathematical questions. An explanation of the aforementioned problem was the presence of a 'close' family of other solutions. We shall treat this problem to some extent as we believe that it nicely demonstrates why numerical black boxes should be treated with great care, also in industrial problems; another reason is that the fairly messy nonlinear structure is elucidated substantially by invoking a mechanical analogy (§3). The existence of the desired solution as a limiting case of the family of close solutions is treated in §4. Finally, in §5 we consider some more mathematical questions, arsising as mere off shoots, in particular concerning the bending energy.

2. The model

In this section we shall derive a model for the problem outlined above and indicate some questions related to this problem. In figure 1 we have sketched the configuration of a laybarge. The pipe is guided into the water by a stinger having a certain uniform (adjustable) radius of curvature R. Typically three forces are exerted on the pipe: a) gravity, due to the not negligible net weight, b) a horizontal tension H, applied by the tension machines mounted on the vessel which is supposedly anchored, and c) the (unknown) vertical bottom reaction force V.

By denoting:

(2.1)

$$s = \text{arclength}$$

$$Q = \text{effective pipe weight per unit length}$$

$$M = \text{bending moment}$$

$$EI = \text{flexural rigidity}$$

we have (Fig.2; cf. ref. 10,11)

(2.2a)
$$M = EI \frac{d\psi}{ds} \quad (\text{Bernoulli} - \text{Euler law})$$

(2.2b)
$$\frac{dM}{ds} = H \sin \psi - V_s \cos \psi \quad (\text{equilibrium of moments})$$

(2.2c)
$$\frac{dH}{ds} = 0 \quad (\text{horizontal tension remains constant})$$

(2.2d)
$$\frac{dV_s}{ds} = Q.$$

From (2.2) we derive the basic equation

(2.3)
$$EI \frac{d^2\psi}{ds^2} = H \sin \psi - (Qs - V) \cos \psi$$

where we note that

$$x(s) = \int_0^s \cos \psi \, ds', \quad y(s) = \int_0^s \sin \psi \, ds'.$$

Besides ψ and V, also the free length L (from bottom to stinger departure point) is unknown. This requires <u>four</u> boundary conditions in total to be specified. The ones we will adopt here are not exactly corresponding to the configuration as drawn in fig.1 (which would include a free boundary at the stinger, cf. ref.8), but somewhat simplified, however, without altering the characteristics of the problems discussed.

We will consider in principle:

$$(2.4a) \qquad \psi(0) = 0, \quad \frac{d\psi(0)}{ds} = 0 \quad \text{(free end supported by the bottom)}$$

$$(2.4b) \qquad \frac{d\psi(L)}{ds} = -\frac{1}{R} \quad \text{(stinger curvature)}$$

$$(2.4c) \qquad \int_0^L \sin\psi(s)\, ds = D \quad \text{(the total depth)}$$

We now like to find the shape of the pipe, given a horizontal force H. This force is needed to give the pipe a low enough curvature to prevent it from buckling. This buckling usually happens when the combined stress due to bending, tension, and water pressure exceeds the yield stress, i.e., the maximal stress that the steel can afford without plastic deformation.

We first reduce the problem parameters to a basic set. Introduce dimensionless quantities

$$(2.5a) \qquad t = s/D_0, \quad l = L/D_0, \quad d = D/D_0$$

$$(2.5b) \qquad h = HD_0^2/EI, \quad q = QD_0^3/EI, \quad v = VD_0^2/EI$$

For convenience we introduced an unspecified length D_0, in order to prevent limiting cases as $Q = 0$, $H = 0$, $L \to \infty$, $D \to \infty$ to produce artificially singular, and therefore less interpretable problems. So in general the number of dimensionless parameters may be reduced by one, and we may assume, for example, $h = 1$ or $q = 1$. So we obtain

$$(2.6a) \qquad \ddot{\psi} = h\sin\psi - (qt - v)\cos\psi$$

with boundary conditions

$$(2.7a) \qquad \psi(0) = 0, \quad \dot{\psi}(0) = 0$$

(2.7b) $\dot{\psi}(l) < 0$ given

(2.7c) $\displaystyle\int_0^1 \sin\psi \, dt = d$

By augmenting (2.6a) with

(2.6b) $\dot{v} = 0$

(2.6c) $\dot{l} = 0$

we have a fourth order ODE (2.6) with four BC (2.7). When this system is 'fed' into a BVP routine on a computer, one needs to specify initial estimates for $\psi, \dot{\psi}, v$, and l, as one has to solve by iteration, say using Newton's method. For problems with values of $l \approx d$ this turns out to be increasingly and, eventually, prohibitively difficult. In order to produce a reliable program we therefore investigate what phenomenon is possibly causing these troubles. First, we note that for longer pipes there will be a considerable part with negligible bending stress where the pipe behaves like a catenary ($\ddot{\psi} \approx 0$; cf. ref. 8). Only at the ends, in boundary layers, the $\ddot{\psi}$-term is of importance. If l is large enough, the pipe will be, in the catenary part, nearly vertical (i.e. $\psi \approx \pi/2$). Obviously, the entire solution can be built up by singular perturbation techniques (e.g. ref. 8). Our interest here, however, concentrates on what may happen in the first part where ψ grows from 0 to $\pi/2$; in particular for which values of v this vertical shape (for l large) is obtained. Therefore we derive the following subproblem:

Find in the family of solutions of (2.6a) with (2.7a) the

(2.8) 'critical' $\psi(t, v)$ (with corresponding v) which satisfies

the condition $\psi \to \pi/2$ for $t \to \infty$.

(For short, we will here and below always imply $\psi := \psi(\mathrm{mod}2\pi)$).

For convenience later we have denoted the dependence on v explicitly. Since we have no dependence on l any more, we have a third order problem now.

For the subsequent analysis, it is advantageous to introduce the first integral of (2.6a), which amounts to the elastic free energy density of the bent pipe (ref. 11)

(2.9) $u(t) = \dot{\psi}^2/2 = h - h\cos\psi - (qt - v)\sin\psi + q\displaystyle\int_0^t \sin\psi \, dt'$

The total energy is then

$$(2.10) \qquad\qquad E = \int_0^\infty u(t)dt$$

Intuitively we may expect the problem (2.8) not to have a unique solution (and, likewise, also (2.6), (2.7)), as the pipe could possibly make several bends and loops. In fig. 3 we have given some indication of the intrinsical behaviour of ψ (in cartesian coordinates) for various values of v. This observation is the clue to the numerical difficulties mentioned earlier: we shall show that the most natural solution of (2.8) (with a minimal energy E) is dense in a family of solutions which are, at least numerically, close to the required one of (2.6) with (2.7) if d is large. This then explains why it is hard, sometimes impossible, to single out the desired one. To do this we shall employ a useful analogy which is introduced in the next section.

3. Pendulum analogy

The variety and character of the solutions of (2.6a) would be much easier predicted and classified if we can find an analogous problem, described by the same equation, but easier interpretable.

If $q = 0$ such an analogy exists indeed, and is actually a classic result due to Kirchhoff (cf. ref.10). It consists of a mathematical pendulum in a gravity field $(-h, v)$, where ψ now denotes the angle of the pendulum with the positive y-axis, and t is time; (fig. 4a). For given v, ψ will oscillate between $\psi = 0$ and its mirror in the line $(-h, v)$: $\psi = 2\pi - 2\,\text{arctg}(v/h)$. If $v = 0$ the pendulum will remain stationary in its (unstable) initial position $\psi = 0$, as the oscillation period tends to infinity for $v \to 0$.

For $q \neq 0$ it is possible to generalise the pendulum analogy if we use a more general (admittedly more artificial) time dependent gravity field $(-h, v - qt)$; (fig. 4b). Then, as long as $v - qt > 0$, the force field is directed to the right half (i.e. $x > 0$), and ψ will tend to return to the interval $(0, \pi)$; however, once $v - qt$ has changed sign and increases with t without bound in negative direction, the force field points to the left, and, as time increases, forces ψ to remain in a decreasing interval around $\psi = 3\pi/2$; (fig. 5a). If v is large, this process takes some time, and ψ starts to oscillate around $\psi = \pi/2$; (fig. 5b).

In a similar fashion, ψ starts to oscillate between 0 and 2π if h is large; however, if q is large, ψ is almost immediately forced to its final phase of oscillations around $\psi = 3\pi/2$.

This behaviour is reflected in fig. 6 where several (numerical) solutions of eq. (2.6a) are plotted (in cartesian coordinates (x, y)). From these pictures it is immediately clear that the generalised pendulum analogy, supported by some continuity arguments, provides very simply the evidence of the existence of multiple solutions, particularly if v is large enough.

Another interesting result that can be derived from the analogy is the nature of the critical solutions, with $\psi \to \pi/2$ for $t \to \infty$. If $h = 0$, $\psi = \pi/2$ is clearly a solution, but an increasingly unstable one if $v - qt < 0$. If $h \neq 0$, $\psi = \pi/2$ is not a solution any more, although if $t \to \infty$ there may exist a (delicate) path for ψ to approach $\pi/2$ in such a way that the tangential component of the large force $(-h, v - qt)$ is never large enough to remove ψ from $\pi/2$. (The existence of such a solution is not trivial and we will return to this subject in the next section.) From the unstable nature of the solution $\psi \to \pi/2$, with $v = v_c$ say, it is clear that near v_c there are infinitely many v with solutions that, after some time, will move away from the neighbourhood of $\pi/2$, and will eventually approach $3\pi/2$ (though in an oscillatory way). Especially in a numerical contex this is important because it implies that for only very small variations in v we may have several different solutions.

4. Existence of the 'critical' solutions of (2.8)

We now investigate the existence of the solution $\psi(t, v_c)$, where v_c is such that we have the intuitively most likely solution of (2.8), as sketched in fig. 7. In the next section we shall show that this solution has minimal bending energy.

Two particular properties of the sought solution are

solution

(4.1a) $0 < \psi(t, v_c) < \pi/2$ $0 < t < \infty$

(4.1b) $0 < \dot{\psi}(t, v_c)$ $0 < t < \infty$

In the original problem setting, we considered v as a parameter to be determined along with the solution from an ODE formulation (via $\dot{v} = 0$). Here we shall look for such values of v that the ODE (2.6a) with (2.7a) has a solution with the properties (4.1) by fine tuning v. Inspired by (4.1), let us define two sets of v-values giving rise to two families of solutions ψ of (2.6a), (2.7a)

(4.2a)
$$V_1 = \{v > 0 | (\exists t_1 < \infty)\, (\dot{\psi}(t_1, v) = 0) \wedge (\psi(t, v) < \pi/2,\ 0 < t < t_1)\}$$

(4.2b)
$$V_2 = \{v > 0 | (\exists t_2 < \infty)\, (\psi(t_2, v) = \pi/2) \wedge (\dot{\psi}(t, v) > 0,\ 0 < t < t_2)\}$$

The sets V_1 and V_2 are constructed such that either (4.1a) or (4.1b) does not hold.

Property 4.3. (i) $V_1 \cap V_2 = \emptyset$, (ii) $v_c \in \mathbb{R}_+ \setminus (V_1 \cup V_2)$

Proof. (i) Suppose $v \in V_1 \cap V_2$, then we obviously have $t_2 = t_1$. Hence $\psi(t_2, v) = \pi/2$, $\dot{\psi}(t_2, v) = 0$. From (2.6a) this implies that $\ddot{\psi}(t_2, v) = h$. If $h = 0$ then all derivatives of ψ vanish, so ψ is constant, which is a contradiction. If $h > 0$, ψ must have a (local) minimum at $t = t_2$, implying $\dot{\psi}(t, v) < 0$ for $t < t_2$. This, however, would only occur if $v \notin V_2$, so the result follows by contradiction. ◇

(ii) Let $v > 0$, $v \notin V_1 \cup V_2$, then

$$\left[\forall t\{(\dot{\psi} \neq 0) \vee (\exists \tau < t \text{ with } \psi(\tau) = \pi/2)\}\right] \wedge \left[\forall t\{(\psi \neq \pi/2) \vee (\exists \tau < t \text{ with } \dot{\psi}(\tau) = 0)\}\right]$$

However, both for $t_2 < t_1$ and $t_1 < t_2$ the requirements $\dot{\psi}(t_1) = 0$ and $\psi(t_2) = \pi/2$ are mutually exclusive. So (4.1) implies that ψ converges as $t \to \infty$ to, say, a constant $c < \pi/2$. Then $\ddot{\psi} \to -(qt - v)\cos c \neq 0$, which give a contradiction. Hence $\psi \to \pi/2$ for $t \to \infty$ if v is element of the complement of $V_1 \cup V_2$ (which, however, may be empty). ◇

We now finally arive at

Theorem 4.4. V_1, V_2, and the complement of $V_1 \cup V_2$ are not empty; moreover $\sup(V_1) = v_c = \inf(V_2)$.

Proof. (a) Let v be small (i.e. close to zero) and consider ψ for t small. Then we may linearize (2.6a) to obtain

$$\ddot{\psi} = h\psi - qt + v.$$

For this elementary equation it easily follows that for some small value of t, say $t = t_1$, we have $\dot{\psi}(t, v) = 0$. By choosing v small enough, ψ can be kept as small as we like along $[0, t_1]$; in any case small enough for the linearization, and smaller than $\pi/2$, which thus yields a nonempty V_1.

(b) To prove that V_2 is nonempty, we look for large enough v. To start, we observe that for $t \to 0$ such that $\psi \to 0$ we have $\psi \simeq \frac{1}{2}vt^2$. So for $v \to \infty$ we have $\psi = 2/v$ at $t = 2/v$. Furthermore, using (2.9), we have along $0 < t < v/q$ (as long as $\psi < \pi$) $\frac{1}{2}\dot{\psi}^2 > 0$, so $\dot{\psi} > 0$, and ψ increases. Again using (2.9), we thus obtain for $2/v < t < 3v/4q$ (and, of course, ψ not close to π) $\frac{1}{2}\dot{\psi}^2 > (v - qt)\sin\psi > (v - qt)2/v > \frac{1}{2}$. So $\dot{\psi} > 1$, $\psi = \int_0^1 \dot{\psi}dt > t$, and therefore ψ can be made arbitrarily large (at least larger than $\pi/2$) before $\dot{\psi}$ vanishes.

(c) From the foregoing and property 4.3 we conclude that $\sup(V_1) = v_c = \inf(V_2)$ as follows. First, we observe that if a v_c exists, it is a single point: consider a solution $\psi = \psi_c + \phi$, $v = v_c + \epsilon$, in the neighbourhood of the critical solution (ϕ and ϵ small). Then for large enough t, when $\psi_c \to \pi/2$, we have

$$\ddot{\phi} = (qt - v_c)\phi$$

with exponentially growing solutions. So near v_c there can be no other critical v's.

Finally, only the existence of v_c remains to be shown. For this, we consider a sequence of angles, increasing to $\pi/2$. With each angle we select a corresponding solution ψ_n of which $\dot{\psi}_n$ vanishes for the first time at this angle (at time $t = t_{1,n}$, say). So by definition, the corresponding sequence $\{v_n\}$ is a subset of V_1. Since $\ddot{\psi}(t_{1,n}) < 0$ we have with (2.6a)

$$qt_1 > v + h\,\mathrm{tg}\psi(t_{1,n})$$

so $t_{1,n} \to \infty$, and we have thus constructed a sequence in V_1 converging to v_c.   ◇

5. Further investigations of the equation (2.8)

It is interesting to note that we should expect a sequence of critical v values, of which v_c, considered in the previous section is just the first member. Intuitively this follows immediately from the pendulum analogy as mentioned in §3. Here we like to look for a more precise (though qualitative) characterization.

Before we do that we remark that the existence of another such 'critical v' will complicate the numerical problem, as sketced in §2, even more; for if we might zoom in on that wrong value, during some iterative procedure (say due to a completely off-value initial estimate for v), we not only encounter the troubles associated with so many close solutions, but also the fact that our solution is completely wrong, if it converges at all.

Let us recall the energy density relation (2.9). We can characterise a solution $\psi(t, v)$ by its energy

$$(5.1) \qquad E_l(v) = \int_0^l \frac{1}{2}\dot{\psi}^2(\tau, v)\, d\tau = \frac{1}{2}\int_0^{\psi(l,v)} \dot{\psi}(\tau, v)\, d\psi$$

In our problemsetting we are interested more particularly in the case $l \to \infty$ and the question therefore arises whether $E_\infty(v)$ exists. This question is addressed in the following two properties.

Property 5.2. $E_\infty(v_c)$ is finite.

Proof. If t is sufficiently large, we have

$$\psi(t, v_c) = \pi/2 - \text{arctg}\frac{h}{qt - v_c} \simeq \pi/2 - h/qt$$

So $\dot{\psi}(t, v_c) = h/qt^2 + O(t^{-3})$, i.e. $[\dot{\psi}(t, v_c)]^2 = O(t^{-4})$. Hence $\frac{1}{2}\int_0^l \dot{\psi}^2(\tau, v_c)d\tau$ converges as $l \to \infty$. ◇

Property 5.3. If $\lim_{t \to \infty} \psi(t, v) \neq \pi/2$ then $\lim_{t \to \infty} \psi(t, v) = 3\pi/2$.

Moreover $\lim_{l \to \infty} E_l(v)$ does <u>not</u> exist.

Proof. As we have seen in §3, $3\pi/2$ is the only alternative attractor. So let $\psi(t) = [3\pi/2 + \phi(t)] \mod (2\pi)$, then

$$\ddot{\psi} = -qt\sin\phi = -qt\phi.$$

The latter equation is therefore approximately equal to the Airy equation (cf. ref.12) with appropriate scaling of variable. So we can write

$$\phi(t) = \alpha Ai(-t/q^{\frac{1}{3}}) + \beta Bi(-t/q^{\frac{1}{3}}),$$

for some $\alpha, \beta \in \mathbb{R}$. Hence $\dot{\phi}(t) = \alpha\dot{Ai} + \beta\dot{Bi} \sim t^{\frac{1}{4}}\sin(t^{\frac{3}{2}})$, so that $[\dot{\phi}(t)]^2 = t^{\frac{1}{2}}\sin^2(t^{\frac{3}{2}})$. Since (with $t = \tau^{\frac{2}{3}}$)

$$\int_\tau^T t^{\frac{1}{2}}\sin^2(t^{\frac{3}{2}})\, dt = \int_{\tau^{\frac{3}{2}}}^{T^{\frac{3}{2}}} \sin^2(\tau)\, d\tau,$$

we see that for $T \to \infty$ this integral and thus $E_T(v)$ (see 5.1)) diverges. ◇

Theorem 5.4. Considered as a function of v, $E_\infty(v)$ has a local minimum at $v = v_c$.

Proof. (a) Denote by $\tilde{t}_1(v)$ the first time that $\ddot{\psi} = 0$ (to become negative). Since $\ddot{\psi}_c$ starts positive and finally becomes negative, it follows that $T = \lim_{v \to v_c} \tilde{t}_1(v) < \infty$, and $\lim_{v \to v_c} \psi(\tilde{t}_1, v) < \pi/2$. Hence there is a subset $\hat{V}_2 \subset V_2$ with

$$\ddot{\psi}(\tilde{t}_1) = 0 \ \wedge \ \psi(\tilde{t}_1) < \pi/2 \ \text{ for } v \in \hat{V}_2.$$

Let $v \in \hat{V}_2$. Then by definition there is a time t_2 where $\psi = \pi/2$. Here, $\ddot{\psi} = h > 0$, so there is another time $\hat{t}_2$ with $\tilde{t}_1 < \hat{t}_2 < t_2$, where $\ddot{\psi} = 0$ to become positive. Now denote by $\dot{\phi}(\psi, w)$ the moment (or 'speed') $\dot{\psi}$ as a function of the angle, and consider along $[0, \pi/2]$ four sectors:

I: $[0, \psi(T, v_c)]$ $(\ddot{\psi}(T, v_c) = 0$, force becomes negative after $t = T$ for $v = v_c)$

II: $[\psi(T, v_c), \psi(\tilde{t}_1, v)]$

III: $[\psi(\tilde{t}_1, v), \ \psi(\hat{t}_2, v)]$ $(\ddot{\psi}$ becomes positive after $t = \hat{t}_2)$

IV: $[\psi(\hat{t}_2, v), \pi/2]$

In I the 'force' $\ddot{\psi}$ is larger for the choice v than for v_c. Hence with v the point $\psi(T, v_c)$ is reached earlier and with a higher speed. In II this effect is reinforced as for the case v_c the motion is slowed down in contrast to the case v. In III we have

$$qt(v) < qt(v_c),$$

so for any angle $\alpha \in$III, $h \sin \alpha - (qt(v) - v) \cos \alpha > h \sin \alpha - (qt(v_c) - v_c) \cos \alpha$. Hence, despite the 'force' being negative, it is less negative for v than for v_c.

Finally, in IV the motion for v is accelerated again, in contrast to v_c. Summarising we thus have found: for any $\psi \in (0, \pi/2)$ is $\dot{\phi}(\psi, v) > \dot{\phi}(\psi, v_c)$. From (5.1) we deduce

$$(*) \qquad E_\infty(v_c) < \frac{1}{2} \int_0^{\pi/2} \dot{\phi}(\psi, v) \, d\psi < \lim_{l \to \infty} E_l(v).$$

(b) Let $v \in V_1$ (see (4.1a)) be very close to v_c. At the backward swing the pendulum passes the point $\psi = 0$ again, at time $t = t_3$, say. We have there $\ddot{\psi}(t_3, v) = -(qt_3 - v)$ (which is of course < 0). Now let the maximal 'force' $\ddot{\psi}$ at $\ddot{\psi}(t, v_c)$ on $[0, \psi(T, v_c)]$ be m. Then it is possible by choosing t_3 large enough to make the 'force' $\ddot{\psi}$ at $\psi(\phi, v)$ on $[0, -\psi(T, v_c)]$ larger in absolute value than m (by choosing v sufficiently close to v_c); on $[-\psi(T, v_c), -\pi/2]$ the force for ψ is definite negative. We conclude from this that for such a choice of v the 'speed' $\dot{\phi}$ on $[0, -\pi/2]$ is larger in modulus than $\dot{\phi}(\psi, v_c)$ on $[0, \pi/2]$, whence

$$(\ast\ast) \qquad E_\infty(v_c) < \frac{1}{2} \int_0^{-\pi/2} \dot{\phi}(\psi, v)\, d\psi < \lim_{l \to \infty} E_l(v).$$

From (*) and (**) we deduce that $E_\infty(v_c)$ is a local minimum. $\diamond$

Remark. The proof for the case $v < v_c$ (part (b)) of the preceding theorem can also be based on employing the fact that for v sufficiently close to v_c $\lim_{t \to \infty} \psi(t, v) \to -\pi/2$ and using this fact in combination with Properties 5.2 and 5.3.

The question remains which of the solutions converging to $\pi/2$ is the best. Apart from the solution $\psi(t, v_c)$ characterised by (4.1) we may expect e.g. another solution $\psi(t, v_{c2})$ say with $v_{c2} > v_c =: v_{c1}$, having a terminal value $\lim_{t \to \infty} \psi(t, v_{c2}) = 5\pi/2$; i.e. this solution is still monotonic, passes $\pi/2$ to come at rest only after completing another cycle. Similarly there are v_{ci} to be expected such that $\lim_{t \to \infty} \psi(t, v_{ci}) = \pi/2 + (i-1)2\pi$, $\dot{\psi}(t) > 0$. Existence will not be considered here, but basically we anticipate it to be similar as for the case of v_{c1}. Also, for $v < v_{c1}$ we may conjecture a sequence of 'v-values', such that after an initial interval $\dot{\psi} > 0$, there is a point $t(v)$ where $\dot{\psi}(t(v)) = 0$, $0 < \psi(t(v), v) < \pi/2$, and $\dot{\psi} < 0$ for $t > t(v)$, $\lim_{t \to \infty} \psi(t, v) \to -\pi/2 - 2k\pi$. It is interesting to realise, however, that whatever other critical value of v may exist, $v_c = v_{c1}$ is optimal in the sense of minimal energy, as is shown in

Theorem 5.5. $E_\infty(v_c)$ is the global minimum.

Proof. v_{c2} should apparently be larger than v_{c1}, as follows from the construction of the set V_2. If $v_{c2} \in \hat{V}_2$ then it follows from part (a) of the proof of Theorem 5.4 that $\frac{1}{2} \int_0^{\pi/2} \dot{\phi}(\psi, v_{c2})\, d\psi < E_\infty(v_c)$.

If $v_{c2} \in V_2 \setminus \hat{V}_2$ then it is simple to see that $\dot{\phi}(\psi, v_{c2}) > \dot{\phi}(\psi, v_{c1})$ like in the aforementioned proof and thus that again $\frac{1}{2} \int_0^{\pi/2} \dot{\phi}(\psi, v_{c2})\, d\psi < E_\infty(v_{c1})$. $\diamond$

50

Conclusion

In this paper we have considered how the theoretical shape of a pipe depended on the value of the bottom reaction force V. The 'natural shape' for some critical value of $v = v_c$ for a long pipe was shown to have minimal bending energy. However, there are values of v arbitrarily close to v_c, where this shape is completely different. This then explains (at least partially) why the nonlinear boundary value problem resulting from the pipe laying model is hard to solve without special techniques. A possible way out for this last problem is to use appropriate continuation techniques (e.g. in the length). Further investigations are currently under way.

References

1. D.A. Dixon, D.R. Rutledge, Stiffened Catenary Calculations in Pipeline Laying Problem. Journal of Engineering for Industry 90 (1), 1969, pp. 153-160.

2. J.T. Powers, L.D. Finn, Stress Analysis of Offshore Pipelines During Installation, Offshore Technology Conference 1071, 1969.

3. J.R. Wilkins, Offshore Pipeline Stress Analysis, Offshore Technology Conference 1227, 1970.

4. G.C. Daley, Physical Interpretations of the Instabilities Encountered in the Deflection Equations of the Unconstrained Pipeline, Offshore Technology Conference 1933, 1974.

5. P. Terndrup Pedersen, Equilibrium of Offshore Cables and Pipelines During Laying, International Shipbuilding Progress, Vol.22, Dec. 1975, pp.399-408.

6. G. Maier, Optimization of Stinger Geometry for Deepsea Pipelaying, Journal Energy Resources Tech. 104, pp.294-301.

7. I. Konuk, Higher Order Approximations in Stress Analysis of Submarine Pipelines, ASME paper 80-PET-72, 1980.

8. S.W. Rienstra, Analytical Approximations For Offshore Pipelaying Problems, Proceedings of the First International Conference on Industrial and Applied Mathematics ICIAM 87, Paris 1987, Contributions from the Netherlands. (WI Tract 36, ed. by A.H.P. van der Burgh and R.M.M. Mattheij, 1987).

9. M.J. Brown, L. Elliott, Pipelaying From a Barge, Mathematical Engineering For Industry 1 (1), 1987, pp.33-46.

10. R. Frisch-Fay, Flexible Bars, Butterworths, London 1962.

11. L.D. Landau, E.M. Lifshitz, Theory of Elasticity, Pergamon, Oxford, 1970.

12. M. Abramowitz, I.A. Stegun (eds.), Handbook of Mathematical Functions, National Bureau of Standards 1964.

R.M.M. Mattheij, Faculteit Wiskunde en Informatica, Technische Universiteit Eindhoven, Postbus 513, 5600 MB Eindhoven, The Netherlands.

S.W. Rienstra, Wiskundige Dienstverlening, Faculteit Wiskunde en Natuurwetenschappen, Katholieke Universiteit, Toernooiveld, 6525 ED Nijmegen, The Netherlands.

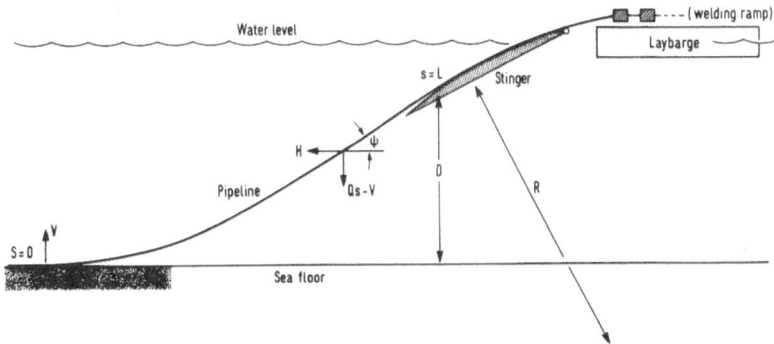

Figure 1.   Sketch of the pipelay problem

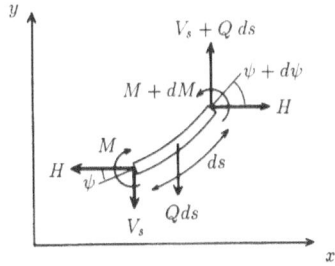

Figure 2   Equilibrium of pipe element

52

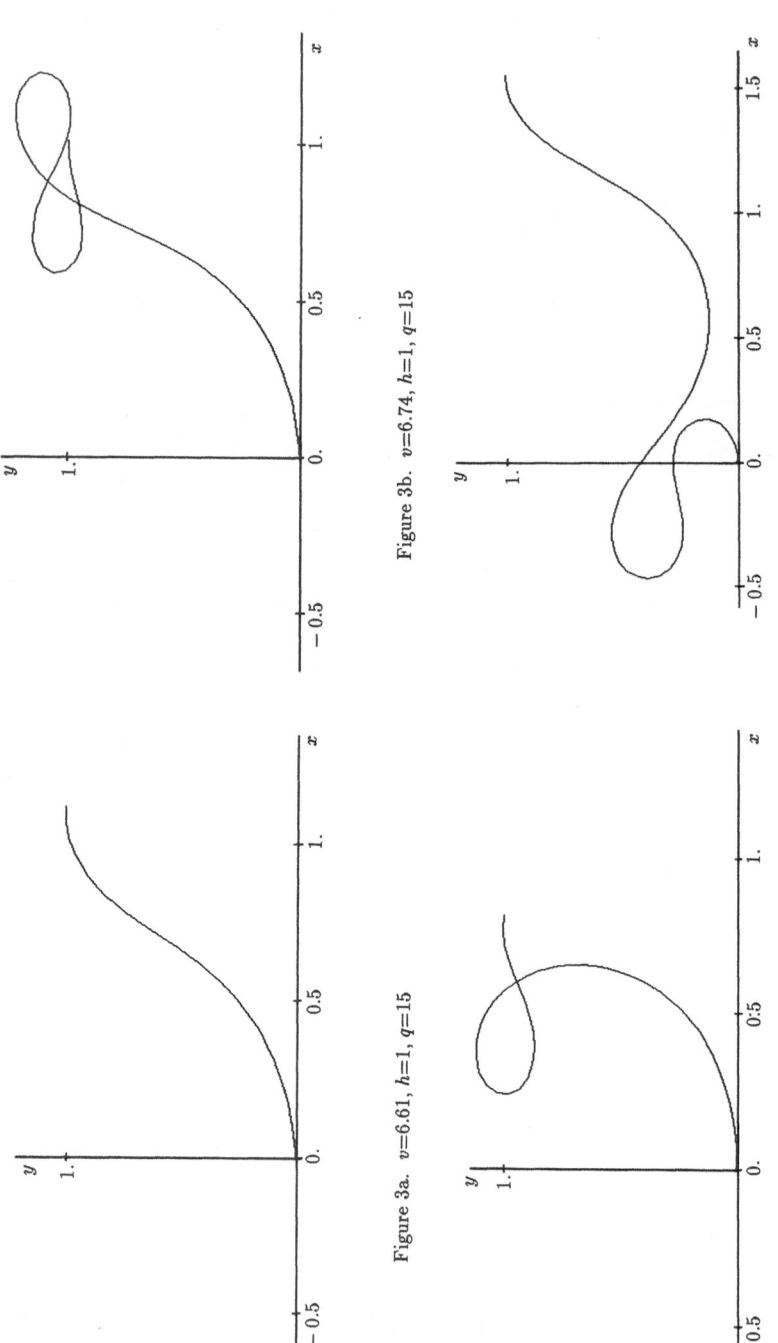

Figure 3a. $v=6.61$, $h=1$, $q=15$

Figure 3b. $v=6.74$, $h=1$, $q=15$

Figure 3c. $v=6.98$, $h=1$, $q=15$

Figure 3d. $v=35.74$, $h=1$, $q=15$

53

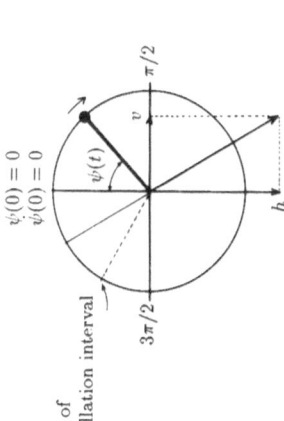

$\psi(0) = 0$
$\dot{\psi}(0) = 0$

end of
oscillation interval

$3\pi/2$

$\psi(t)$

v

h

$\pi/2$

Figure 4a. Kirchhoff's pendulum analogy for $q=0$.

Constant gravity field $(-h, v)$

$\psi(0) = 0$
$\dot{\psi}(0) = 0$

$3\pi/2$

$\psi(t)$

$v - qt$

h

$\pi/2$

Figure 4b. Generalized pendulum analogy.

Time dependent gravity field $(-h, v - qt)$.

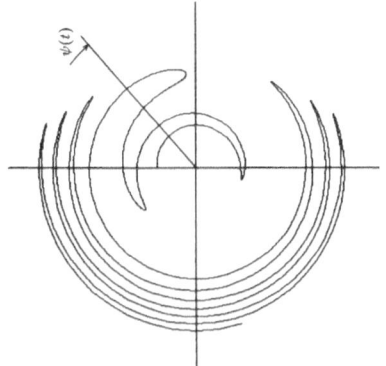

$\psi(t)$

Figure 5b. $v=35.63$, $h=1$, $q=15$

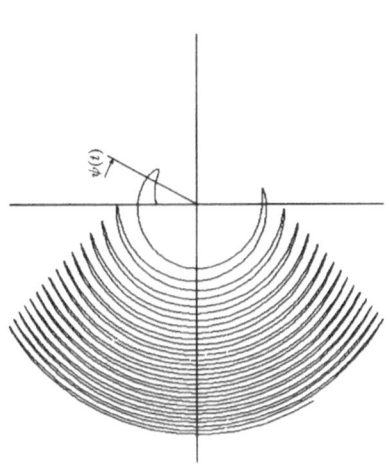

$\psi(t)$

Figure 5a. Pendulum analogy (radial coordinate grows

linearly in time) for $v=6$, $h=1$, $q=15$.

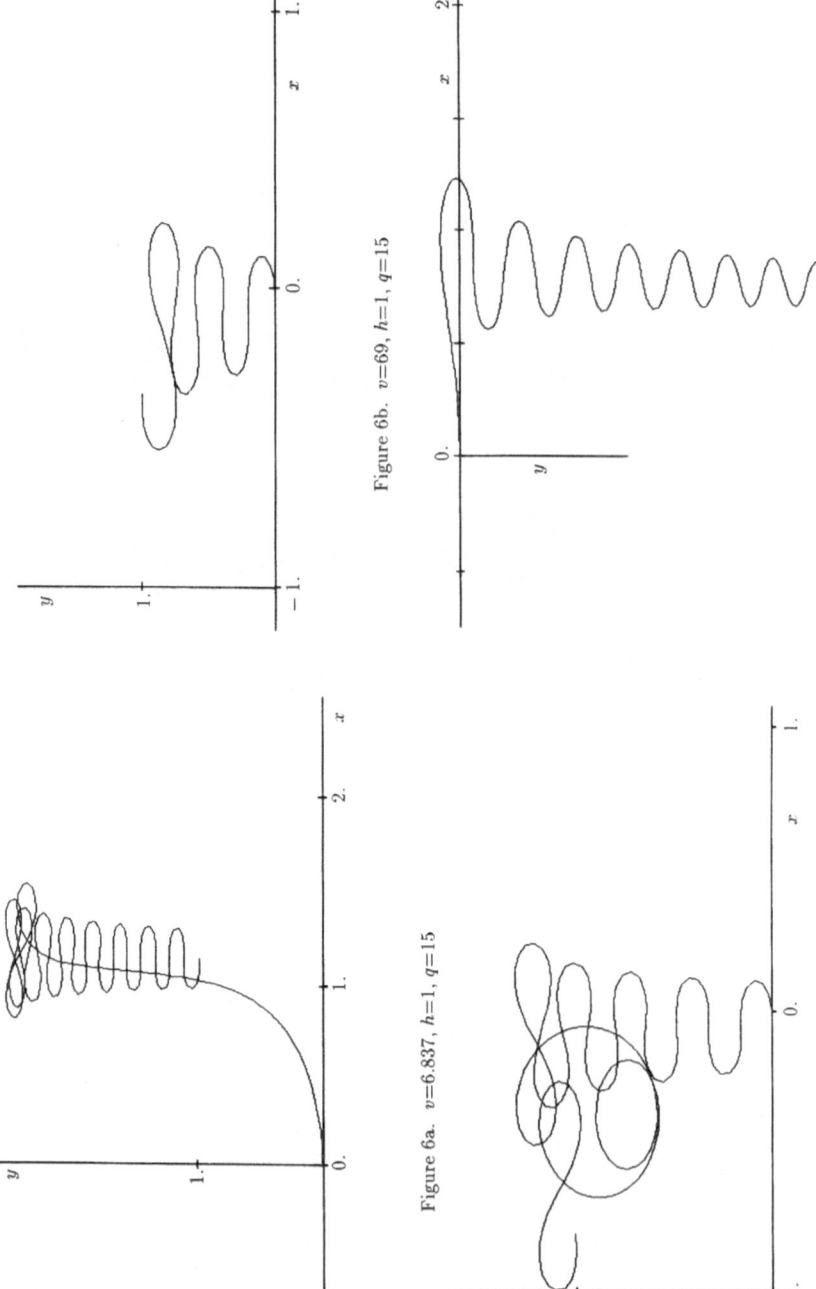

Figure 6b. $v=69$, $h=1$, $q=15$

Figure 6d. $v=5$, $h=1$, $q=15$

Figure 6a. $v=6.837$, $h=1$, $q=15$

Figure 6c. $v=100$, $h=1$, $q=15$

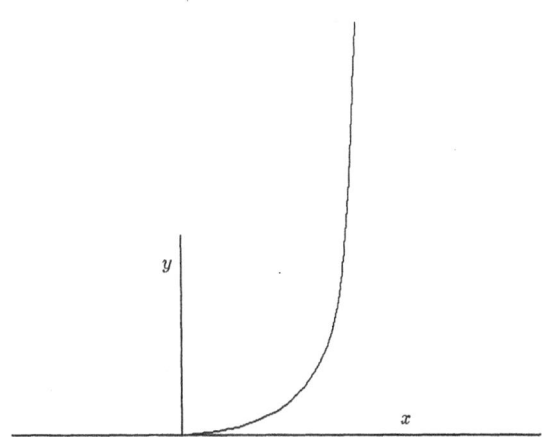

Figure 7a. (First) critical solution in cartesian

coordinates (v=6.837, h=1, q=15)

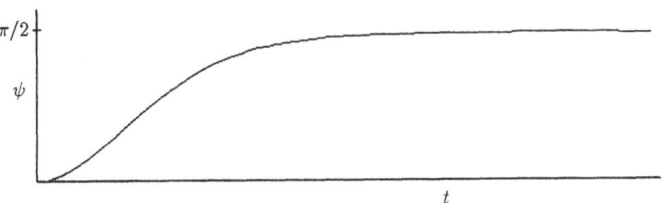

Figure 7b. (First) critical solution in (ψ, t) coordinates

(v=6.837, h=1, q=15).

Mathematische Probleme des rechnergestützten Entwurfs höchstintegrierter Schaltungen

K. Merten
Siemens AG
Otto-Hahn-Ring 6
D-8000 München 83

0 Einleitung

Die Herstellung von VLSI-Schaltungen ist ohne den Einsatz von CAD-Werkzeugen und -Systemen nicht mehr denkbar. Sie haben auf allen Entwurfsebenen Eingang gefunden. Erst durch die Unterstützung der Entwicklung mit solchen hochspezialisierten Programmen war das Design komplexer integrierter Schaltkreise mit bis zu 4 Mill. Transistoren möglich. Mit Hilfe von Prozeß- und Bauelementsimulation konnte die benötigte Technologie bereitgestellt werden. Die Erfahrung und das Wissen von Ingenieuren, Technologen und Schaltungsdesignern stehen heute abrufbar in Form von Regeln, Modellen, Algorithmen etc. in mächtigen Programmsystemen zur Verfügung. An der Verfeinerung solcher Systeme wird unter großem Wettbewerbsdruck weltweit gearbeitet. Die Programmentwicklung bedient sich abgesehen von den relevanten Ingenieurdisziplinen zweier Hilfswissenschaften, der Informatik und der (angewandten) Mathematik. Im zweiten Fall ist die Behauptung von Edward David (Präsident von Exxon Research), daß die heute so sehr gefeierte Hochtechnologie in ihrem Wesen "mathematische" Technologie ist, sicherlich nicht übertrieben. Leider ist dies von der angewandten Mathematik lange nicht oder nur teilweise erkannt worden.

Folgende Übersicht ordnet beispielhaft den wesentlichen Entwurfsebenen des rechnergestützten IC-Entwicklungsprozesses die dort benötigten Disziplinen der angewandten Mathematik zu.

58

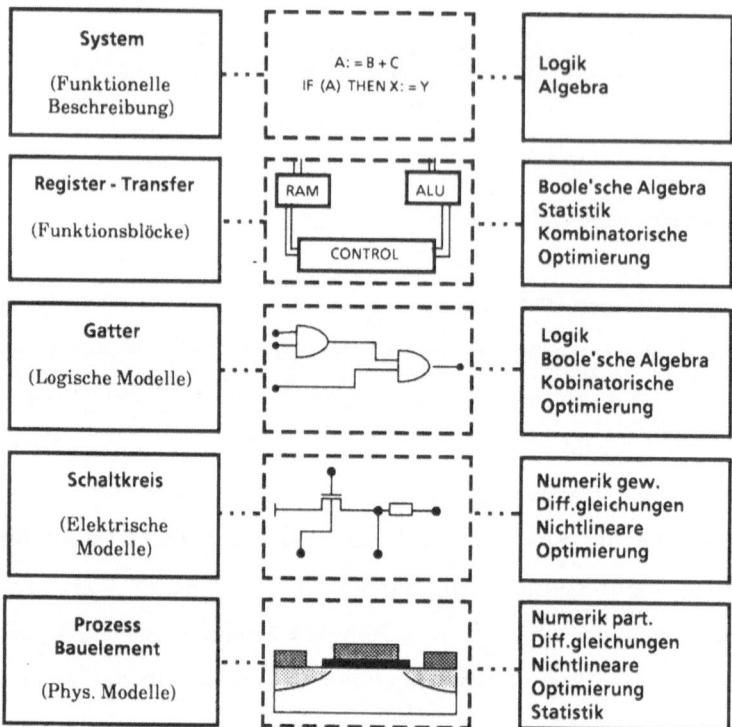

Abb. 1: Zuordnung mathematischer Disziplinen zu Entwurfsebenen

Die Aufstellung ist sicher nicht vollständig. Wir wollen in dieser Arbeit einige typische Probleme aufgreifen. An diesen soll exemplarisch gezeigt werden, mit welcher hochgradig nichttrivialen mathematischen Fragestellung die CAD-Entwicklung konfrontiert wird.

1   Designverfahren

Wir unterscheiden bei den Designverfahren heute zwischen den zellenorientierten und den transistororientierten Designverfahren. Zellenorientierte Designverfahren basieren auf dem Entwurf mit sogenannten Zellen. Unter Zellen versteht
man hier vorgefertigte und verifizierte Schaltungsteile, die in verschiedenen
Representationen aus der Systembibliothek abrufbar sind. Eine Zelle kann z. B.
ein einfaches Gatter, ein Flip-flop oder ein Zähler sein. Diese Methode erlaubt
eine vollständige CAD-Unterstützung. Bei transistororientierten Designverfahren
wird das Layout einer Schaltung technologienah auf Transistorebene entworfen.
Dies geschieht mit Unterstützung einzelner CAD-Werkzeuge. Dieses Verfahren
ist fehleranfällig und zeitaufwendig, führt jedoch zu flächen- und laufzeitopti-
mierten Schaltungen. Für Bausteine, die in großen Stückzahlen (wie z. B. Mikro-
prozessoren und Speicherbausteine) gefertigt werden sollen, wird diese
Designmethode bevorzugt. Sie wird außerdem für das Design von Zellen
verwandt. Zellenorientierte Designverfahren sind den optimierten transistororientierten Designverfahren insbesondere bzgl. Entwurfssicherheit und Ent-
wicklungszeit deutlich überlegen. Sie werden vorwiegend für das Design soge-
nannter kundenspezifischer Bausteine, die nicht in großen Stückzahlen herge-
stellt werden, herangezogen. Mit steigender Komplexität treten diese Verfahren
immer mehr in den Vordergrund. Mit dem CAD-System VENUS® kann z. B. die
Logik einer ganzen Flachbaugruppe mit Hilfe von zellenorientierten Designverfahren in einen einzigen Chip umgesetzt werden Eine umfassende Darstellung
des Themas mit vielen Literaturhinweisen ist in [1] zu finden.

® Eingetragenes Warenzeichen der Siemens AG, München

2. Das Plazierungsproblem

Auf der Ebene des physikalischen Entwurfs wird bei zellenorientierten Design-verfahren aus der Schaltungsstruktur ein volles Layout des Chips erstellt. Die Zellen- und Verknüpfungsinformation wird umgesetzt in eine Zellplazierung und Zellverdrahtung unter Berücksichtigung der Rand- und Rahmengenerierung des Chips. Vorgelegt ist eine komplexe Optimierungsaufgabe, die folgendermaßen formuliert werden kann: Ordne k Zellen derart an, daß die Gesamtfläche minimal wird und (oder) das "Ergebnis" unter gewissen lokalen Nebenbedingungen verdrahtbar ist. Die Zellenanzahl k liegt hierbei typischerweise zwischen 10^1 und 10^4. Wesentliche Randbedingungen sind:

- Layoutstil (Zellenkonzept)
- Designregeln
- Bibliotheksdaten
- Anwendervorgaben

Neben dem Optimierungsziel Chipfläche (Ausbeute, Fertigungskosten) und der Vollständigkeit des Entwurfs sind noch Leitungslaufzeit (Taktfrequenz) und Chipformat (Gehäuse) zu nennen. Die eigentliche Schwierigkeit dieses Problems der kombinatorischen Optimierung liegt in der Definition der Zielfunktion und der Komplexität der Aufgabe (NP-vollständiges Problem).

Bei der sogenannten Gate-Array-Designmethode wird nach Vorfertigung der Transistoren in Reihen die Plazierung einer logischen Funktion durch die in der Bibliothek abgelegte Verdrahtungsinformation auf einen "optimalen Einbauplatz" vorgenommen. Im zweiten Schritt erfolgt dann die Verdrahtung der Logikfunktionen nach Vorgabe der Netzliste. Mit Gate-Arrays sind kosten- und zeitgünstige Entwürfe möglich.

61

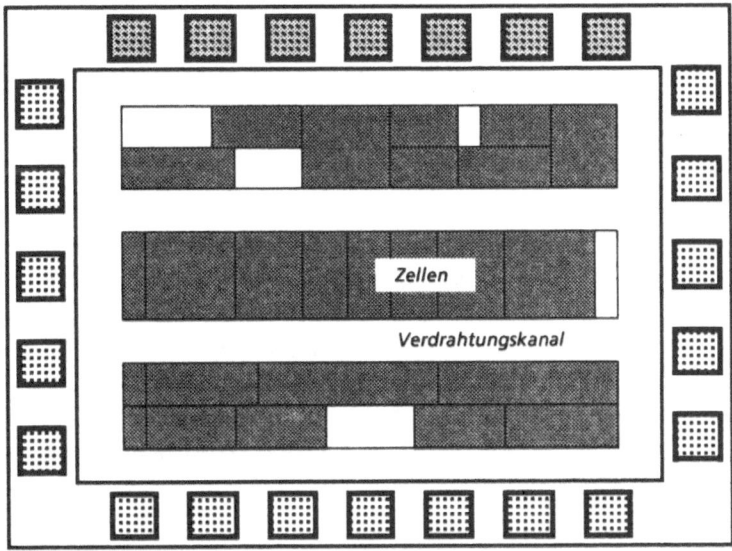

Abb. 2: Anordnungsschema für Gate-Arrays mit Reihenarchitektur

Für diese Designmethode kann das Plazierungsproblem wie folgt mathematisch annähernd korrekt formuliert werden (vgl. [2]):

Gegeben sei ein Master mit p Einbauplätzen und z Zellen, die darauf plaziert werden sollen. Gesucht sind Unbekannte $x_{i,j}$ mit $x_{i,j} \in (0,1)$ und $i = 1, ..., p$, $j = 1, ..., z$. $x_{i,j} = 1$ heißt, Zelle j wird auf Position i eingebaut, $x_{i,j} = 0$ heißt, Zelle j wird nicht auf Position i eingebaut.

Die Gleichungen

(2.1) $$\sum_{i=1}^{p} x_{i,j} = 1, \qquad j = 1, ..., z$$

stellen sicher, daß jeder Zelle tatsächlich ein Einbauplatz zugewiesen wird.

Die Ungleichungen

(2.2)
$$\sum_{j=1}^{z} x_{i,j} \leq 1, \qquad i = 1, ..., p$$

stellen sicher, daß auf jedem Einbauplatz höchstens eine Zelle plaziert wird. Haben zwei Variable $x_{i,j}$ und $x_{m,n}$ den Wert 1, heißt dies, daß die Zelle j auf Einbauplatz i und Zelle n auf Einbauplatz m positioniert werden. Diese Vorgabe verursacht "Kosten" $c(i, j, m, n)$ (etwa: Manhattan-Abstand zwischen Einbauplatz i und Einbauplatz m multipliziert mit der Anzahl der Netze/Verbindungen, die sowohl Zelle j als auch Zelle n enthalten). Diese Kosten treten nur auf, wenn $x_{i,j} \cdot x_{m,n} = 1$ gilt. Daraus ergibt sich folgendes quadratisches Zuordnungsproblem:

$$\text{Min} \sum_{i=1}^{p} \sum_{j=1}^{z} \sum_{m=1}^{p} \sum_{n=1}^{z} c(i,j,m,n) \, x_{i,j} \cdot x_{m,n}$$

$$\sum_{i=1}^{p} x_{i,j} = 1 \qquad , \quad j = 1, ..., z$$

(2.3)

$$\sum_{j=1}^{z} x_{i,j} \leq 1 \qquad , \quad i = 1, ..., p$$

$$x_{i,j} \in (0,1) \; .$$

Die Lösung von Problemen dieser Art ist heute erst bis zu einer Größenordnung von etwa 15 möglich. Bei den vorgelegten Komplexitäten müssen auf das Problem zugeschnittene Heuristiken herangezogen werden (etwa Lösung von (2.3) durch Block- und Hierarchiebildung). Bei zellenorientierten Designmethoden ohne Vorfertigung (Standard- und Makrozellen) wird das sogenannte Min-Cut Verfahren [3] mit recht gutem Erfolg eingesetzt. Ausgehend von der Vereinigung aller Zellen wird durch Partitionierung nach dem Prinzip "minimaler Schnitt von Verbindungen" eine schnittoptimale Lösung gefunden. Diese heuristische Vorgehensweise ist in Abb. 3 veranschaulicht.

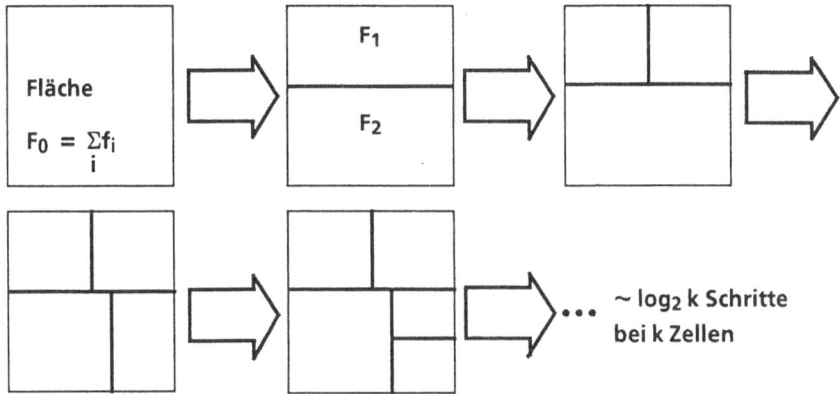

Abb. 3: Min-Cut Algorithmus / Anordnung durch fortgesetzte Aufteilung

Der fortschreitende Einsatz von zellenorientierten Designmethoden fordert aus Kostengründen weiterhin dringend besser optimierende Verfahren.

3. Laufzeitberechnung

Nach Plazierung und Verdrahtung erwartet der Designer auf der Basis des fertigen Layouts Angaben über Pfadlaufzeiten. Diese geben ihm Auskunft über zeitkritische Signale und ermöglichen eine Logiksimulation mit realen Laufzeiten. Die Berechnung der Verzögerungen in einem Signalnetz geschieht durch Abbildung des Netzes in einen (linearen) RC-Baum und Lösung des diesen RC-Baum beschreibenden gewöhnlichen Differentialgleichungssystems. Eingehende Parameter sind Ausgangswiderstände (High → Low / Low → High) der sendenden Zelle und die Eingangskapazitäten der Empfänger. Darüber hinaus

64

werden natürlich die Geometrie des Netzes und entsprechende Technologie-
parameter (zur Berechnung der Leitungswiderstände und -kapazitäten)
benötigt. Diese Werte stehen in der Systembibliothek bzw. in der
schaltungsspezifischen Netzliste zur Verfügung.

Jedes Leitungsnetz wird durch einen RC-Baum mit K_i, $i = 1, ..., n$, Knoten
modelliert. Dem Knoten K_i wird der (näherungsweise) konstante Widerstand R_i
des entsprechenden Leitungsstückes und die zugehörige Kapazität C_i
zugeordnet. Speziell modelliert R_1 die Summe aus dem Innenwiderstand der
treibenden Zelle und dem Gesamtwiderstand des ersten Leitungsstückes. Die
Eingangskapazitäten der Empfänger werden dem Knoten des letzten Leitungsstückes des Pfades dorthin zugerechnet.

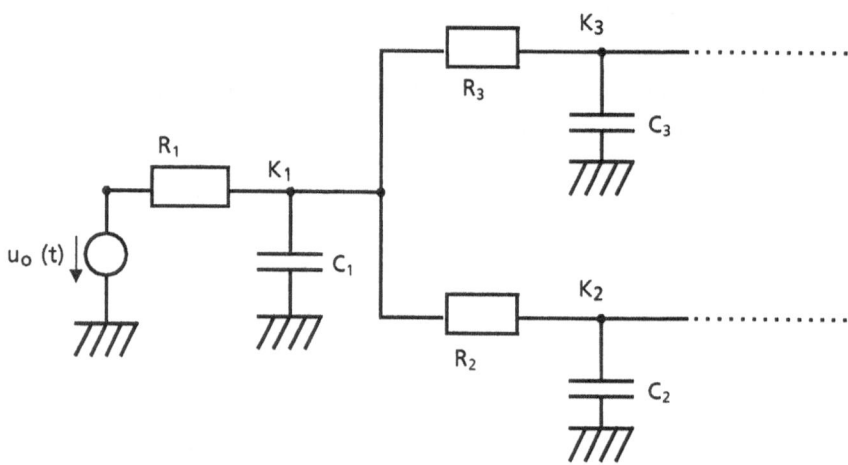

Abb.1 Modellierung eines Leitungsnetzes

Die Anwendung der Knotengleichungen führt auf ein System gewöhnlicher
linearer Differentialgleichungen erster Ordnung der Form

(3.1) $C\,\dot{v}(t) + M\,v(t) = R_1{}^{\text{-}1}\!\cdot\!e_1$

mit den Anfangsbedingungen

(3.2) $v(0) = 0$.

Hierbei gilt

$C = \text{diag}(C_1, ..., C_n)$ (Kapazitätsmatrix)

$M = (m_{i,j})$ (Admittanzmatrix)

$v(t) = (v_1(t), ..., v_n(t))^T$

$e_1 = (1, 0, ..., 0)^T$.

Man kann zeigen, daß das System (3.1) ein sogenanntes "steifes" Differentialgleichungssystem ist. Typischerweise sind auf einem Semi-Custom-IC im Mittel mehr als 5000 RC-Bäume mit zwischen 20 und 100 Knoten abzuarbeiten. Eine numerische Berechnung der Lösung $v(t)$ von (3.1) ist also mit hohem Rechenaufwand verbunden. Aus diesem Grund sind eine Reihe von schnelleren Verfahren vorgeschlagen worden (z. B. Penfield'sche Formeln [5]), die aber nicht ausreichend genau sind. Das System (3.1) hat für die Anwendung sehr gute Struktureigenschaften. M ist nämlich STIELTJES-Matrix, also $M^{\text{-}1} \geqq 0$ und M positiv definit. Dies erlaubt die Anwendung eines Monotoniesatzes [6]. In [7, 8] wird gezeigt, wie mit Hilfe des Ansatzes

(3.3) $$u(t) = e - \sum_{i=1}^{m} a_i\, x_i\, e^{-\frac{1}{\lambda_i} t}, \qquad m < n$$

durch Variation der Gewichte a_i leicht scharfe untere und obere Schranken u^* und u^{**} zu beschaffen sind. Hierbei sind x_i und λ_i die Eigenvektoren bzw.

Eigenwerte der Matrix $M^{-1}C$. In [9] wird basierend auf dem Ansatz (3.3) eine weitere Möglichkeit gezeigt, in akzeptabler Rechenzeit genaue Näherungen zu bestimmen. Das dort vorgestellte Verfahren minimiert den Fehler in den Anfangswerten bzgl. der l_2-Norm, liefert i. a. also keine Einschließungen.Diese Verfahren ermöglichen eine effiziente näherungsweise Berechnung der Laufzeiten in linearer Zeit (bzgl. der Anzahl der Knoten). Sie sind an realen Schaltungen, die mit dem CAD-System VENUS entworfen wurden, erprobt worden. Die Wahl $m \leqq 2$ erweist sich als ausreichend.

4 Schaltwerksimulation

Programme der Schaltwerksimulation sind ein unverzichtbares Werkzeug für transistororientierte Designverfahren. Mit ihnen wird das elektrische Verhalten der Schaltung auf Transistorebene simuliert. Die Anwendung der Kirchhoffschen Regel auf ein Netzwerk aus elektrischen Bauelementen (Transistoren, Widerstände, Kapazitäten) liefert ein implizites, nichtlineares, steifes, unregelmäßiges und schwach gekoppeltes System von Algebrodifferentialgleichungen der Form

$$(4.1) \qquad\qquad F(t, y(t), \dot{y}(t)) = 0 \ , \qquad 0 < t < T \ .$$

Die gesuchten Lösungen $y(t) = (y_1(t), ..., y_n(t))^T$ können je nach Formulierung die Spannungen an den Knoten des Netzes oder die Ströme in den Zweigen sein. Der Nachweis der Existenz von Lösungen von (4.1) gelingt i. a. nicht, elektrische Plausibilitätsbetrachtungen müssen hier bemüht werden. Eingangsparameter für die Gewinnung der Darstellung (4.1) sind Technologiedaten der Prozeß- und Bauelementcharakterisierung und sogenannte Transistormodelle oder -kennlinien. Dies sind aus numerischer Sicht Näherungskurven für das nichtlineare Verhalten eines Transistors (z. B. Drain-Source Strom I_{DS} in Abhängigkeit der

Gate-Source Spannung U_{GS}). Die Bestimmung korrekter Modelle aus Meßdaten ist ein nichtriviales numerisches Teilproblem der Schaltwerksimulation. Man geht mittlerweile mehr und mehr dazu über, sich die Eingangsdaten für (4.1) mit Hilfe von Prozeß- und Bauelementsimulatoren zu beschaffen. In Kapitel 5 wird diese Vorgehensweise ausführlicher diskutiert.

Ganz konkret wird das in Abb. 5 dargestellte Schieberegisters durch das nicht-lineare System

$$(C_1+C_2+C_3)\,\dot{u}_1(t) - I_1\,(u_1(t)) + I_2(u_1(t),\, v_1\,(t)) + I_3\,(u_1(t),\, v_2(t),\, u_2(t)) - C_1\,\dot{v}_1(t) - C_3\dot{v}_2(t) \quad = 0$$

$$(C_4+C_5+C_6)\,\dot{u}_2(t) - C_6\dot{u}_3(t) - I_3\,(u_1(t),\, v_2(t),\, u_2(t)) - C_4\,\dot{v}_2(t) \quad = 0$$

$$(C_6 - C_7)\,\dot{u}_3(t) - C_6\dot{u}_2(t) - I_4(u_3(t)) + I_5(u_3(t),\, u_2(t)) \quad = 0$$

modelliert.

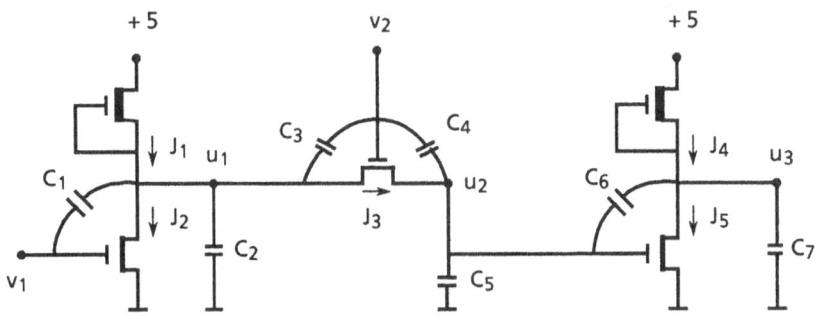

Abb. 5: **Dynamisches MOS-Schieberegister**

Das Problem der effizienten numerischen Lösung von (4.1) ist bis heute nicht befriedigend gelöst [10]. Dies gilt insbesondere vor dem Hintergrund großer Komplexitäten (bis zu 10^4 Transistoren) und dem häufigen Einsatz dieser Programme. Jede deutliche Verbesserung in Richtung Rechenzeit erbringt der Halbleiterindustrie sofort eine Ersparnis an Rechenkosten in Millionenhöhe. Der klassische numerische Lösungsansatz wie er etwa in [11] dargestellt ist, findet sich in dem weitverbreiteten Programmpaket SPICE wieder. Neuere Ansätze versuchen die natürliche Partitionierung von VLSI-Schaltungen auszunutzen. Die Gleichungen werden gemäß den Teilschaltungen angeordnet und blockweise mit den üblichen Techniken abgearbeitet. Diskrete Varianten des Newtonverfahrens zur Lösung der nichtlinearen Gleichungssysteme wie sie in [12] zu finden sind, werden hier (teilweise unter anderem Namen) in allen Variationen verwandt. Ein durchschlagender allgemeiner Erfolg war allen Ansätzen nicht beschieden. Implementierungen auf Vektorrechnern ermöglichten jedoch zwischenzeitlich die Simulation auch komplexer Schaltungsteile mit über 10^4 Transistoren. Eine recht erfolgreiche Realisierung auf dem Vektorrechner VP200 mit einem Beschleunigungsfaktor bis zu 10 ist in [13] beschrieben. Diese schnellen Simulationsmöglichkeiten wurden intensiv für das Design des 4M Speichers im MEGA-Projekt der SIEMENS AG genutzt. Vielversprechend scheinen neuere numerische Forschungsergebnisse über verallgemeinerte Runge-Kutta-Fehlberg Verfahren [14] zu sein.

5 Halbleitersimulation

Der Einsatz von CAD-Werkzeugen hat auch auf den unteren physikalischen Ebenen (Prozeß, Bauelement (Device)) in den letzten Jahren in immer stärkerem Maße Eingang gefunden. Kostspielige und zeitintensive Messungen werden in bestem CAD-Verständnis durch Simulation der physikalisch/elektrischen Vorgänge abgelöst. Insbesondere die japanische Halbleiterindustrie scheint diese Möglichkeiten im Sinne von Zeitersparnis bei neuen Technologieentwicklungen erkannt zu haben. In Entwicklung sind für diese Ebenen übergreifende CAD-Systeme, die die Simulation und Optimierung durchgängig von den Herstellungsprozeßschritten über die physikalische Charakterisierung einzelner integrierter Bauelemente bis zu den elektrischen Eigenschaften ganzer Schaltungen ermöglichen.

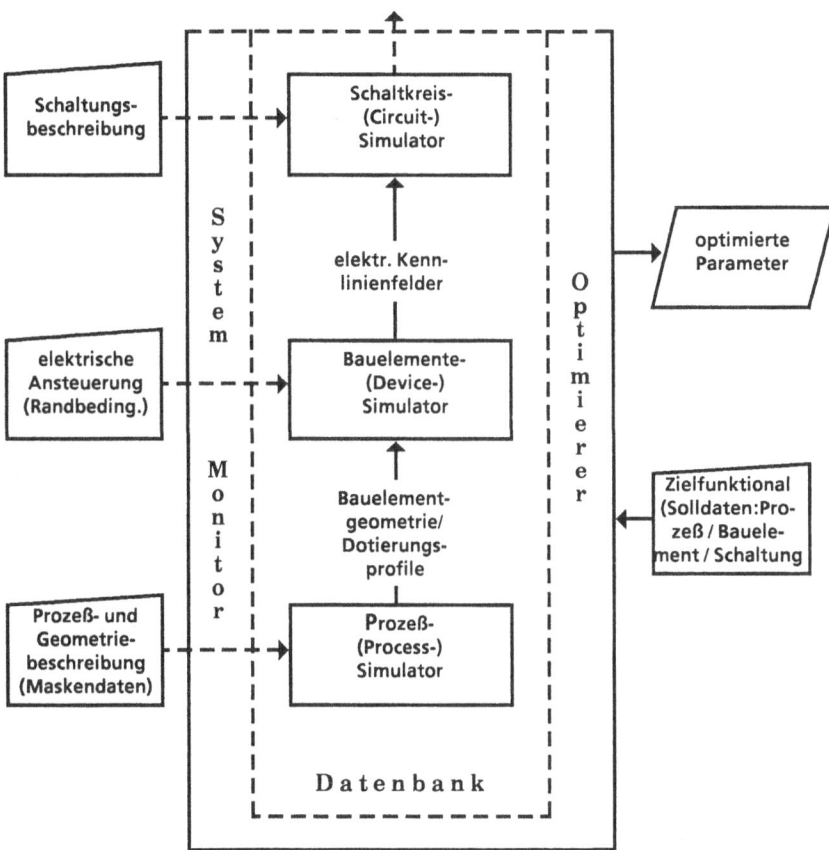

Abb. 6: SATURN: Prozeß-, Bauelement- und Schaltwerksimulation

Am Beispiel des CAD-System SATURN der SIEMENS AG sind in Abb. 6 die Großfunktionen und Abläufe skizziert. Aus der Fülle der mathematischen Problemstellungen greifen wir einige wesentliche heraus.

Durch Aufheizung der Siliziumwafer (Temperung) werden die Dotierungsprofile (gezielt) verändert. Diese Änderung der Dotierstoffkonzentration C (x, t) wird durch dieDiffusionsgleichung

(5.1)
$$\frac{dC}{dt} = div\left(D(C) \cdot \ grad\, C + H(C) \cdot \ grad\, u\right)$$

modelliert, wobei (5.1) über das Potential u (x, t) an

(5.2)
$$\Delta u - f(C) = 0$$

gekoppelt ist.

Numerisch zu lösen ist also ein Verbund von parabolischer und elliptischer Differentialgleichung bei Vorgabe entsprechender Anfangsrandwerte und Randbedingungen.

Die Isolation integrierter Bauelemente erfolgt durch Oxidation des Siliziums. Das Verständnis und die Stabilisierung dieses Prozeßschrittes bereitet allen Halbleiterherstellern größte Schwierigkeiten. Aus mathematischer Sicht handelt es sich hier um ein sogenanntes "freies" Randwertproblem; die Grenze des Oxydwachstums kann erst bei der (numerischen) Lösung des Differentialgleichungssystems berechnet werden. Für die Sauerstoffkonzentration C (x, t) gilt

(5.3)
$$\frac{dC}{dt} = div(D \cdot \ grad\, v) - v \cdot \ grad\, C \qquad ,$$

wobei der Oxydgeschwindigkeitsvektor v (x, t) und der Druck P (x,t) noch

(5.4)
$$\mu\Delta v \ = \ grad\, P \qquad und$$
$$div\, v \ = \ 0$$

genügen.

Diese Problemklasse ist Gegenstand neuerer numerischer Forschungen. Erfolg-
versprechende Lösungen für obiges Problem zeichnen sich ab. Sie würden dazu
beitragen, einen schwierigen Prozeßschritt per Simulation richtig zu verstehen
und in den Griff zu kriegen.

Die Bauelementsimulation beruht auf den Maxwell'schen Gleichungen in ihrer
reduzierten Halbleiterform. Sie werden ausführlich etwa in [15, 16] erläutert.
Das Ergebnis dieser Simulation sind z. B. Parametersätze für die Schaltkreis-
berechnung. Sie werden diskret in Tabellenform der Schaltwerksimulation über-
geben. Diese setzen sie dann intern in "Kennlinien" per Spline-Approximation
um. Ein frühzeitiger Designbeginn mit Parametern einer neuen Technologie, die
durch Simulation gewonnen wurden, ist somit möglich.

Simulationssysteme werden aber auch eingesetzt, um die Ausbeute in den
Prozeßlinien zu sichern und zu erhöhen.

Programme der Prozeß- und Bauelementsimulation beherrschen heute
standardmäßig planare Strukturen und zwei Raumdimensionen. Die Strukturver-
kleinerung in den Submikron-Bereich erfordert die Einbeziehung der dritten
Dimension und allgemeiner nichtplanarer Oberflächen. Damit konzentriert sich
die eigentliche Problematik weiter verstärkt auf die Numerik. Nichtlineare
diskrete Probleme mit bis zu über $5 \cdot 10^5$ Unbekannten erfordern z. B. den
Einsatz von Vektorrechnern mit speziell angepaßten Algorithmen.

Eine weitere Herausforderung an Techniken der nichtlinearen Optimierung ist
durch das sogenannte "inverse" Problem gegeben (wie sind bei vorgegebener
Schaltung und Spezifikation die elektrischen Charakteristika des Bauelements
und daraus resultierend das Dotierungsprofil und die Prozeßdaten abzuändern,
um die Anforderungen optimal zu erfüllen?).

Simulatoren sind nicht nur Werkzeuge, die verstandene physikalische Modelle
berechnen, sie dienen auch der Weiter- und Neuentwicklung von Modellen und
ermöglichen so erst das richtige Erfassen komplizierter physikalisch/elektrischer

Vorgänge. Die Entwicklung neuer numerischer Verfahren darf also nicht nachhinken, sie muß immer "einen Schritt voraus" sein.

Literaturverzeichnis

[1]   E. Hörbst, M. Nett, H. Schwärtzel: VENUS Entwurf von VLSI Schaltungen,
        Springer Verlag Berlin Heidelberg New York, 1986

[2]   M. Grötschel: Interner Bericht Uni Augsburg, 1986

[3]   M. A. Breuer: A class of min-cut placement algorithms, Proc. 14th Design
        Autom. Conf., 1977, pp. 284 - 290

[4]   U. Lauther: Channelrouting in a general cell environment, Proc. VLSI 85,
        Tokyo 1985

[5]   J. Rubinstein, P. Penfield and M. A. Horowitz: Signal Delay in RC Tree
        Networks, IEEE Transactions on CAD, Vol. CAD-2, No. 3, July 1983

[6]   N. Rouche, P. Habets, M. Laloy: Stability Theory by Liapunov's Direct
        Method, Applied Mathem. Sciences 22, Springer-Verlag, New York
        Heidelberg Berlin, 1977

[7]   K. Glashoff, K. Merten: Neue Verfahren zur Laufzeitberechnung bei
        Semi-Custom-Schaltungen, Informatik in der Praxis, Herausg. H.
        Schwärtzel, Springer Verlag Berlin Heidelberg New York, 1986

[8]   K. Glashoff, K. Merten: Accurate Bounds for Signal Delay in RC Tree
        Networks, erscheint demnächst

[9]   K. Glashoff, K. Merten: Effiziente numerische Verfahren für die Laufzeit-
        berechnung bei Semi-Custom-Schaltungen, erscheint demnächst

[10]   R. Bulirsch, A. Gilg: Effiziente numerische Verfahren für die Simulation
        elektrischer Schaltwerke, Informatik in der Praxis, Herausg. H.
        Schwärtzel, Springer Verlag Berlin Heidelberg New York, 1986

[11]   L. W. Nagel: SPICE2: A computer program to simulate semiconductor
        circuits, Mem. ERLM520, University of California, Berkeley, 1975

[12]   J. M. Ortega, W. C. Rheinboldt: Iterative Solution of Nonlinear Equations
        in Several Variables, Academic Press New York London, 1970

[13]   U. Feldmann, K.-G. Rauh, K. Steger: Circuit Simulation on
        Vectorprocessors, Proc. COMP EURO 87, Herausg. W. E. Proebster,
        H. Reiner, Hamburg, May 11-15, 1987, S. 246-249.

[14]   P. Kaps, P. Rentrop: Generalized Runge-Kutta of order four with step size
        control for stiff ordinary differential equations, Numer. Math. 33, pp. 55 -
        68, 1979

[15]   S. Selberherr: Analysis and Simulation of Semiconductor Devices, Wien,
        Springer Verlag 1984

[16]   P. A. Markowich: The Stationary Semiconductor Device Equations, Wien,
        Springer Verlag 1986

STABILITY PROBLEMS IN OFFSHORE INSTALLATION OPERATIONS

P.S. Teigen, STATOIL, Norway *

ABSTRACT

The paper considers a specific problem in relation to off-shore installation operations. The treatment is purely theoretical, and a considerable effort is made to expose the general structure of the solution, with particular emphasis on stability problems.

INTRODUCTION

Following the rapid development of offshore technology in the past decade, much attention has been given to extreme loads and fatigue life of installed offshore structures. Perhaps due to the variety of techniques and operational strategies involved, less effort has been devoted to theoretical problems related to the installation of a structure, although, not infrequently, this phase is rather critical.

In some cases sub-sea structures will have to be emplaced with great precision, and it is therefore of importance to know exactly under what conditions these precision requirements can be met.

Although not always possible, very often the simplest and least expensive way of conducting an installation operation, is to lower the object from a single surface vessel, without the use of guidelines or other positioning devices.

---

* This paper is in a final form and no version of it will be submitted for publication elsewhere.

In the present paper this situation is treated rather extensively, the main objective being to analyse the motion of a submerged structure due to environmental action and to the motion of the surface vessel.

PRELIMINARY RESULTS

We first consider a relatively simpel, 2-dimensional situation in which an unspecified object is suspended freely from a surface vessel moving in a circular orbit with the waves (fig. 1). We ignore the possibility of the connecting wire going slack, and in effect we are considering a rigid connection. Furthermore we neglect the influence of damping on the motion of the submerged object.

With reference to fig. 1, letting 1 denote the fixed length of the wire and r the orbital radius for the motion of the surface vessel, we have for the potential energy of the load

$$V = -(r \cos(\omega t) + l \cos(\theta)) \, m\zeta g \qquad (2.1)$$

Here ω is the frequency of oscillation for the vessel (i.e. the wave frequency), m is the total mass of the load, and ζ is defined by

$$\zeta = (1 - \frac{\varrho_w}{\varrho_L}) \qquad (2.2)$$

where ϱ_w is the water density and ϱ_L the mean density of the submerged body. With no loss of generality, in 2.1 an arbitrary constant and a phase angle are both put equal to zero.

Letting m_a denote added mass, the relevant kinetic energy of the system is

$$T = \frac{1}{2}(m + m_a)[(-\omega r \sin(\omega t) - l\dot\theta \sin(\theta))^2$$

$$+ (\omega r \cos(\omega t) + l\dot\theta \cos(\theta))^2] \tag{2.3}$$

With

$$L = T - V \tag{2.4}$$

we have

$$\frac{d}{dt}\left(\frac{\partial L}{\partial \dot\theta}\right) - \frac{\partial L}{\partial \theta} = 0 \tag{2.5}$$

which after some computational work yields the following equation

$$(m + m_a)[l\ddot\theta + \omega^2 r \sin(\theta - \omega t)]$$

$$+ m\zeta g \sin(\theta) = 0 \tag{2.6}$$

2.6 is a rather complex, non-linear, ordinary differential equation.

Introducing

$$\alpha = \frac{m\zeta g}{\omega^2 l(m + m_a)} \quad , \quad \beta = \frac{r}{l} \quad , \quad \tau = \omega t \tag{2.7}$$

2.6 simplifies to

$$\frac{d^2\theta}{d\tau^2} + \alpha \sin(\theta) + \beta \sin(\theta - \tau) = 0 \tag{2.8}$$

The linear counterpart to 2.8, assuming θ small, is

$$\frac{d^2\theta}{d\tau^2} + (\alpha + \beta \cos(\tau))\theta = \beta \sin(\tau) \tag{2.9}$$

2.9 is a non-homogeneous Mathieu type equation.

A detailed treatment of 2.9 and the associated homogeneous equation is given in [1]. There it is shown how extremely accurate analytic solutions may be obtained in a constructive manner. Here we content ourselves by pointing out some general properties of 2.9.

In all cases of practical interest a general solution to 2.9 will be available in the form

$$\theta = \theta_h + \theta_p \tag{2.10}$$

where

$$\theta_h(\tau) = Ae^{\nu\tau} \sum_{n=-\infty}^{+\infty} c_n e^{in\tau}$$

$$+ Be^{-\nu\tau} \sum_{n=-\infty}^{+\infty} c_n e^{-in\tau} \tag{2.11}$$

and

$$\theta_p(\tau) = \sum_{n=1}^{\infty} d_n \sin(n\tau) \tag{2.12}$$

Here θ_p is an odd, 2π-periodic solution to the inhomogeneous equation 2.9, and θ_h is a solution to Mathieu's equation:

$$\frac{d^2\theta}{d\tau^2} + (\alpha + \beta \cos(\tau))\theta = 0 \tag{2.13}$$

A and B are constants depending on the initial conditons, while the c_n's are fixed, non-arbitrary constants ([3]).

The behaviour of the total solution is decisively influenced by the parameter v appearing in 2.11. In general v is a function of α and β, and there are three typical cases ([3]):

i) $v = i\delta$, $0 < \delta < 1$

ii) $v = \frac{i}{2} + \delta$, δ-real

iii) v-real

In the first case, which is the one most likely to occur in practice, the solution is stable (bounded), and for normal inital conditons the original non-linear problem is adequately handled by 2.9.

Case ii) and iii) both yield exponentially growing, unbounded solutions, and the linearised equation 2.9 is no longer a valid approximation to 2.8. However, the instability of the solution to 2.9 in these cases will serve as an indicator for unacceptably large angular excursions ($\theta \approx 1$) in the original equation 2.8 (observe that 2.9 is a reasonable approximation of 2.8 at least for θ less than 0.5).

In practice, for Re $v \neq 0$ and "normal" initial conditions (e.g. $\theta(0)$, $\theta'(0)$ "small"), 2.8 will in contrast to 2.9, give rise to bounded, oscillatory solutions with distinctive beats (see fig. 3).

The present physical problem requires β to be small.

Although no precise upper bound can be given, a reasonable condition to be enforced is

$$\beta < 0.1 \qquad\qquad (2.14)$$

Similarily, it can be argued (see [1]) that

$$1.6 \ 1^{-1} < \alpha < 50 \ 1^{-1} \qquad\qquad (2.15)$$

It must, however, be pointed out that the above estimates are only tentative, and accordingly should not be used in any strict sense.

We now illustrate the theory by means of two numerical examples:

In fig. 2 we have assumed $\alpha = 0.1$ and $\beta = 0.02$. The solutions are bounded, but aperiodic. Two sets of inital conditons are chosen, and as predicted, solutions of the linear equation are in excellent agreement with those based on the non-linear equation.

In fig. 3 $\alpha = 0.25$ and $\beta = 0.05$. These parameter values represent an unstable situation for eq. 2.13 (or 2.9). Linearisation is only valid up to a certain point (for which $\theta \approx 1$), and the solutions of 2.8 and 2.9 are seen to split apart for τ around 130.

AN EXTENDED MODEL, INCLUDING HEAVE COMPENSATION AND DAMPING

We now proceed to consider a more general physical situation. To this end we assume that the submerged load is subjected to a linear type of hydrodynamic damping force, i.e. the damping force is proportional to the velocity of the load. We further assume that the suspension point for the wire moves in an elliptical orbit with arbitrary eccentricity. This appears to be a valid approximation in regular waves provided the suspension point is close to the center of gravity for the surface vessel, or provided that the roll (pitch) motion of the surface vessel is not too large.

An analysis similar to the one preceding equation 2.6, this time yields (see [1]),

$$(m + m_a)\{l^2 \ddot{\theta} - \tfrac{1}{2} r\omega^2 l[(1 + a) \sin(\omega t - \theta)$$

$$+ (1 - a) \sin(\omega t + \theta)]\} + lm\zeta g \sin(\theta)$$

$$= -k\{\dot{\theta}l^2 + \tfrac{1}{2} \omega r l[(1+a)\cos(\omega t-\theta)+(1-a)\cos(\omega t+\theta)]\} \qquad (3.1)$$

Here the quantities l (fixed), r, ω, g, m, ζ etc. are given the same meaning as before, k is the damping coefficient (dimension kgs^{-1}) and a is the degree of heave compensation accomplished.

Thus, in the above notation, a = 1 corresponds to a circular motion for the suspension point of the wire, while a = 0 corresponds to perfect heave compensation (horizontal motion). In practical applications heave reduction of more than 90% of the original heave motion is very hard to achieve.

By introducing α, β, τ as in equation 2.6, and defining

$$\kappa = k[\omega(m + m_a)]^{-1} \qquad (3.2)$$

3.1 is transformed to the following dimensionless form

$$\frac{d^2\theta}{d\tau^2} + \kappa \frac{d\theta}{d\tau} + \{\alpha + \beta a(\cos(\tau) + \kappa \sin(\tau))\} \sin(\theta)$$

$$= \beta \{\sin (\tau) - \kappa \cos (\tau)\} \cos (\theta) \tag{3.3}$$

A linear equation corresponding to 3.3 is obtained replacing sin θ by θ and cos θ by 1. Hence

$$\frac{d^2\theta}{d\tau^2} + \kappa \frac{d\theta}{d\tau} + \{\alpha + \beta a(\cos (\tau) + \kappa \sin (\tau))\} \theta$$

$$= \beta \{\sin (\tau) - \kappa \cos (\tau)\} \tag{3.4}$$

3.4 has the distinctive traits of an inhomogeneous Mathieu type equation, and it is in fact possible to transform it into a genuine Mathieu equation (see [1]).

As with 2.8, disregarding some exceptional cases ([1]), the general solution θ(τ) to 3.4 has the structure

$$\theta = \theta_h + \theta_p \tag{3.5}$$

where θ_p is 2 π-periodic, and θ_h can be written in the form

$$\theta_h = Ae^{(\nu-\frac{1}{2}\kappa)\tau} \sum_{n=-\infty}^{+\infty} a_n e^{in\tau} + Be^{-(\nu+\frac{1}{2}\kappa)\tau} \sum_{n=-\infty}^{+\infty} b_n e^{-in\tau} \tag{3.6}$$

where A, B are arbitrary constants, and a_n, b_n are fixed, non-arbitrary (related) constants ([1]).

An interesting feature of eq. 3.6 is that instabilities may theoretically still occur, provided
$$| \text{Re}\nu | > \frac{\kappa}{2} \tag{3.7}$$

It can be proved that |Reν| has an overall extremum for

$\alpha = \frac{1}{4} (1 + \kappa)$, ([1]). For this particular value of α, assuming $\beta \ll 1$,

$$\kappa > a\beta \tag{3.8}$$

provides an excellent criterium for the stability of the function Θ, ([1]). Consequently, if 3.8 is satisfied, the stability of Θ is secured for all values of α.

In principle, the relationship bewtween 3.3 and 3.4 is similar to the one that exists between 2.7 and 2.8. However, it must be strongly emphasized that situations which could cause exponentially growing solutions to 3.4 are comparatively rare. Therefore, in almost all cases of practical interest, 3.4 will give a sufficiently accurate representation of 3.3.

As in the previous case we conclude this section with some numerical examples: Figure 4 presents a numerical integration of equation 3.4 with initial conditions ($\Theta(0) = \Theta'(0) = 0$, and with $\alpha = 0.25$, $\beta = 0.04$, $a = 0.5$ and $\kappa = 0.01$. Since $\kappa < \beta a$ the solution is unstable. In the first part of the time trace the influence of the 2π-periodic "steady state" solution is clearly visible. However, for $\tau > 200$ the solution is totally dominated by the transient (which has an oscillatory "period" of 4π).

Fig. 5 shows another simulation which differs from fig. 4 in the choice of κ only. Here $\kappa = 0.03$, so $\kappa > a\beta$ and consequently the solution becomes stable. The 4π-"periodic" transient decays slowly, and eventually we are left with a 2π-periodic "steady state" solution.

For even larger values of κ the picture will remain essentially similar to fig. 5, the main difference being increased attenuation of the transient part of the solution.

The case $\alpha \approx 0.25$ is definitely realistic, but not really typical. For $\beta \ll 1$, we will find that $iv \in R$ for most values of α. This points to a solution composed of a 2π-periodic steady state part, as before, together with an irregular transient, containing a multitude of frequencies (see [1]), and decaying exponentially at a rate of $-\frac{\kappa}{2}$ An example with $\alpha = 0.5$, $\beta = 0.04$, $a = 0.5$ and $\kappa = 0.02$ is given in fig. 6.

Two numerical simulations of the non-linear equation 3.3 are presented in figures 7 and 8, corresponding to fig. 4 and 6 respectively. Fig. 6 and fig. 8 bear a close resemblance, and in this case the linear equation is perfectly adequate. The situation with fig. 4 and fig. 7 is quite revers; the two graphs differ in a way essentially similar to the two examples given in fig. 3.

CONCLUDING REMARKS

Some noticeable progress is achieved in analysing the motion behaviour of a submerged structure, suspended from a single surface vessel. However, one should bear in mind the limitations imposed through the mathematical modelling. Essentially, we are considering a 2-dimensional, deep water operation in regular waves.

The motion stability problem is discussed and characterized. Some simple and useful criteria pertaining to this question are explicitly given. Both heave compensation and, in particular, damping have a favourable influence on the behaviour of the submerged body, but large excursions may still occur.

The dynamical equations are presented in a form well suited for numerical integration. For a more comprehensive treatment, with analytic solutions included, the reader is referred to [1]. A more general problem, comprising variable wire length l(t) and irregular wave excitation, is discussed in [2].

REFERENCES

[1]   Teigen, P.S.: Some mathematical problems related to offshore installation operations, STATOIL-report, Trondheim, 1987.

[2]   Teigen, P.S.: A general approach to offshore installation operations, STATOIL-report, Trondheim, 1987. (To appear).

[3]   Whittaker, E.T. & Watson, G.N.: A course of modern analysis, Cambridge University Press, 4th ed., 1927.

The authors present adress:

        Statoil
        Postuttak
        N-7004 TRONDHEIM
        NORWAY

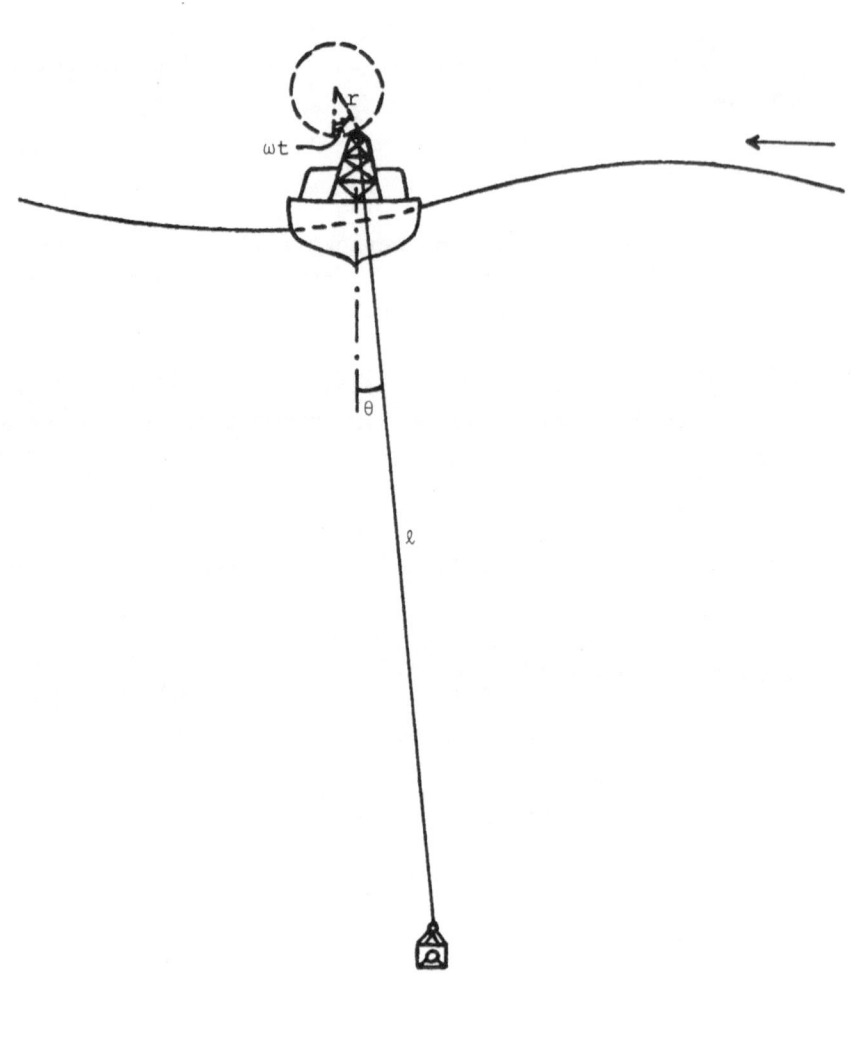

Fig. 1: Definition sketch for the physical model.

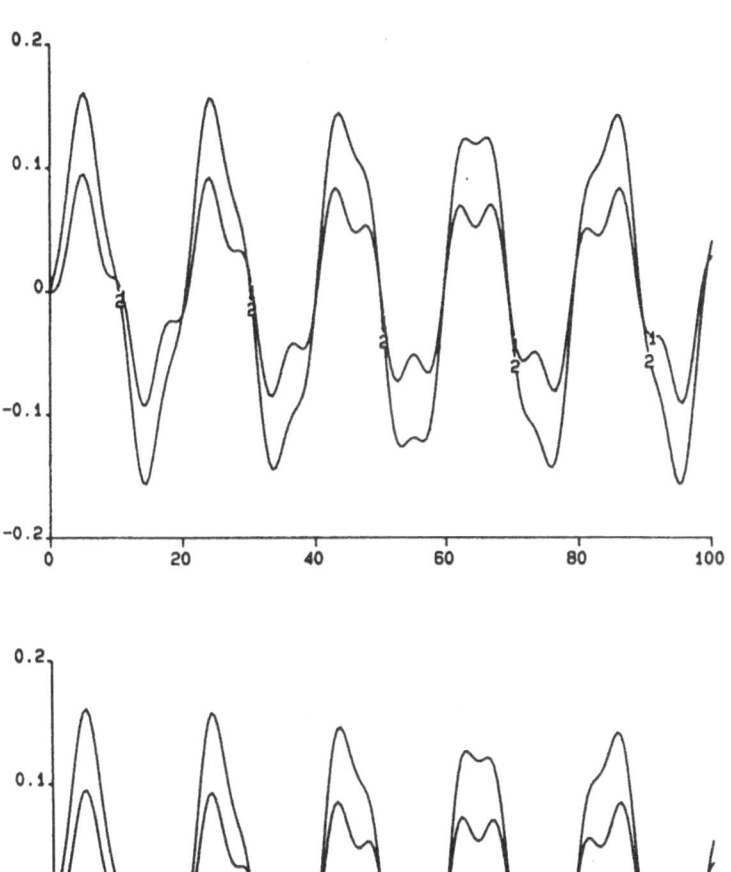

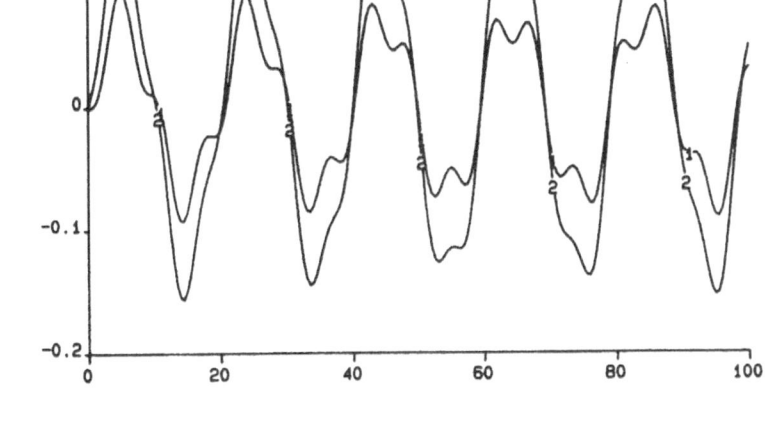

Fig. 2: Numerical integration of 2.8 (top) and 2.9 (bottom)
for $\alpha = 0.1$, $\beta = 0.02$
1: $\theta(0) = \theta'(0) = 0$ 2: $\theta(0) = 0$, $\theta'(0) = 0.02$

88

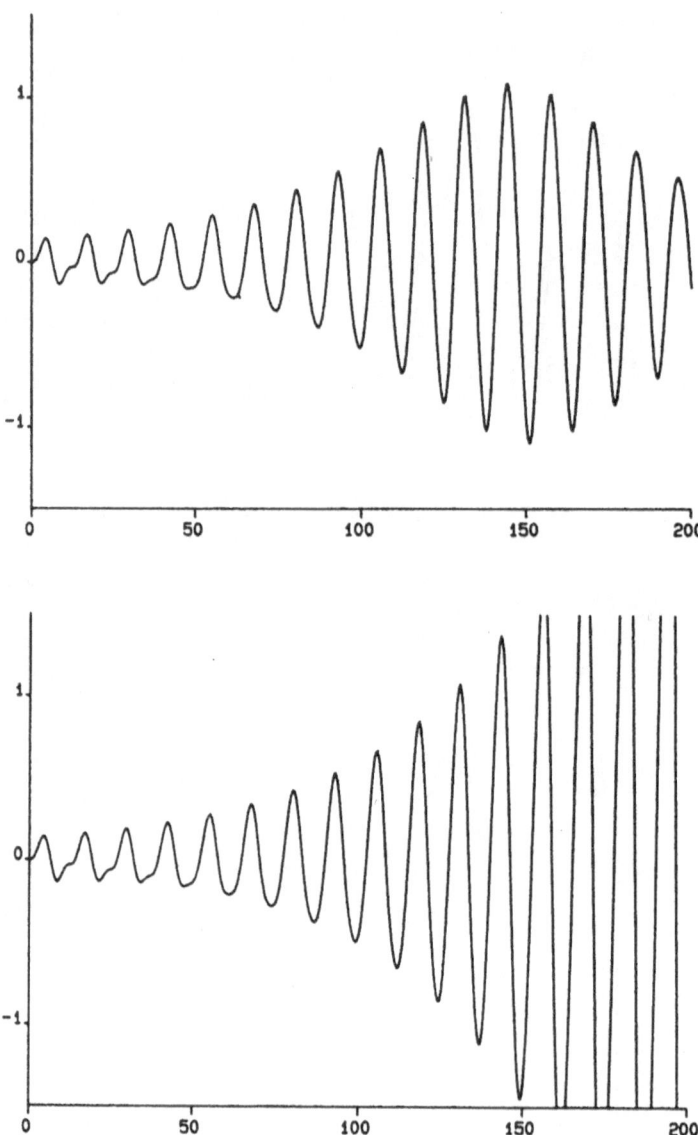

Fig. 3: Numerical integration of 2.8 (top) and 2.9
(bottom) for α = 0.25, β = 0.05.
θ(0) = θ'(0) = 0

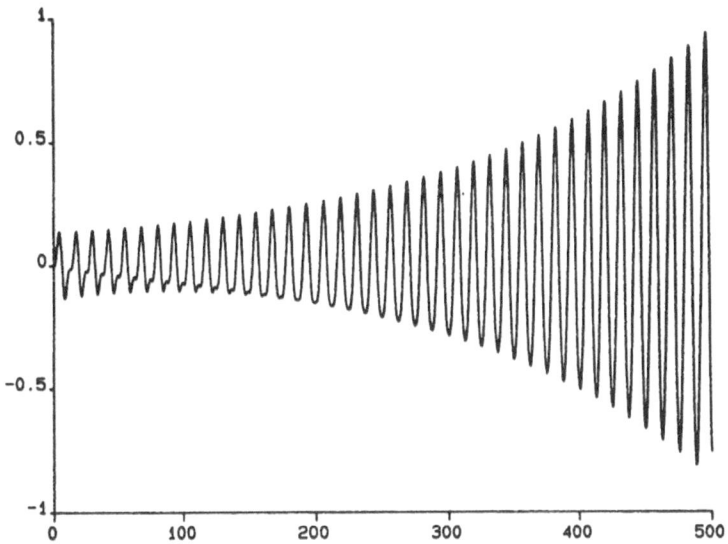

Fig. 4: Numerical simulation of eq. 3.4
$\alpha = 0.25$, $\beta = 0.04$, $\kappa = 0.01$, $a = 0.5$
$\theta(0) = \theta'(0) = 0$

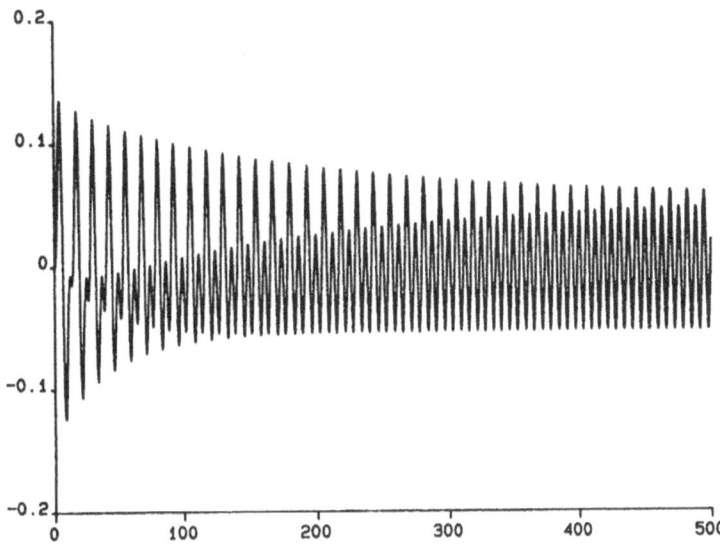

Fig. 5: Numerical simulation of eq. 3.4
$\alpha = 0.25$, $\beta = 0.04$, $\kappa = 0.03$, $a = 0.5$
$\theta(0) = \theta'(0) = 0$

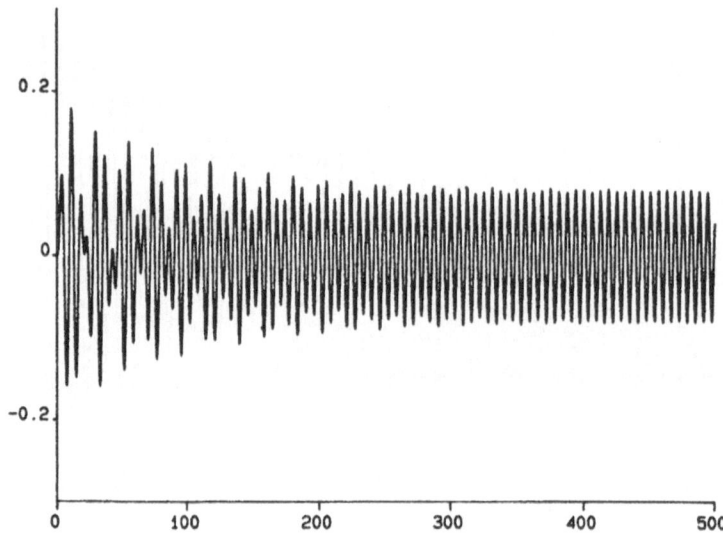

Fig. 6: Numerical simulation of eq. 3.4
$\alpha = 0.5$, $\beta = 0.04$, $\kappa = 0.02$, $a = 0.5$
$\theta(0) = \theta'(0) = 0$

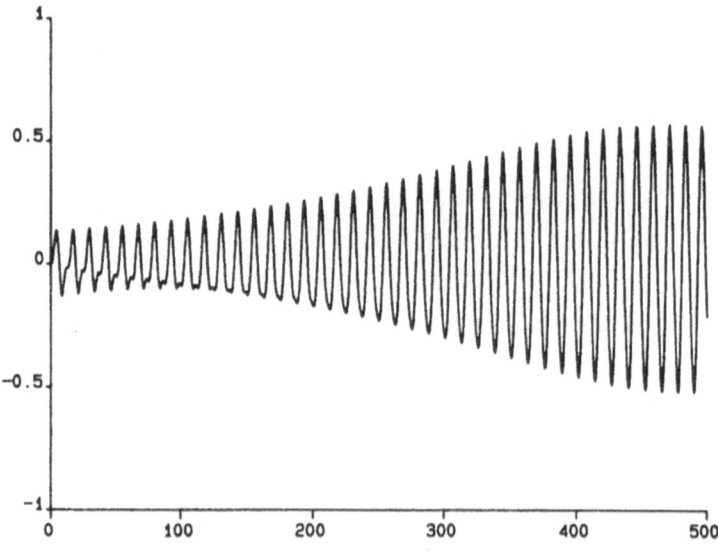

Fig. 7: Numeric. simulation of eq. 3.3
$\alpha = 0.25$, $\beta = 0.04$, $\kappa = 0.01$, $a = 0.5$
$\theta(0) = \theta'(0) = 0$ (cf. fig. 4)

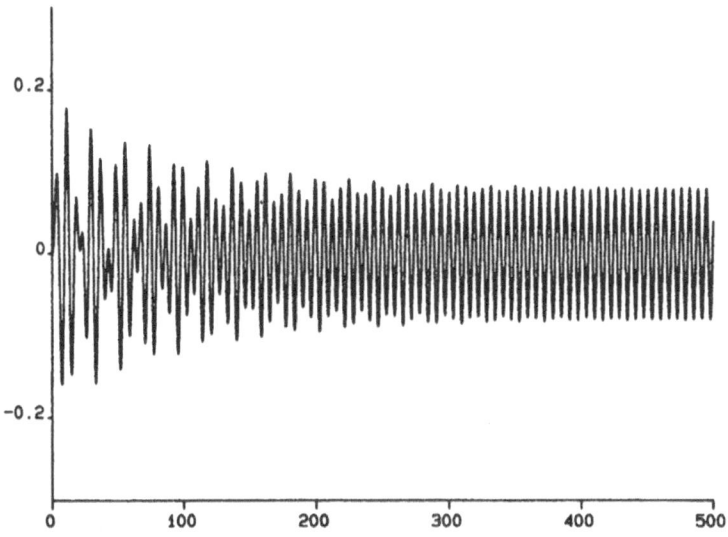

Fig. 8: Numerical simulation of eq. 3.3
         α = 0.5, β = 0.04, κ = 0.02, a = 0.5
         θ(0) = θ'(0) = 0   (cf. fig. 6)

IDENTIFICATION OF DETERMINISTIC NONLINEAR DYNAMICAL SYSTEMS

J. Struckmeier/F.-J. Pfreundt, Kaiserslautern

Abstract: We propose an identification procedure for state-space systems with nonlinearities in the state. The theoretical background and the algorithm are described. To show the efficiency of the algorithm we apply it to two real world examples.

1   Introduction

The general problem of identification may be described by the following picture:

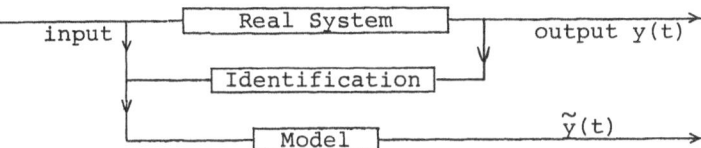

The measured data of a physical system are used to generate a mathematical model of the system. In many cases the best way to do is to build a physical model based for example on the principles of mechanics. But you may get into the following difficulties:

- the physical model is good but the evaluation is extremely time consuming or

- the physics itself is complicated, not all relevant effects are known, parameters are to estimate.

In those cases numerical identification is a way to get a model which is easy to evaluate and may give a reasonable description

---

"This paper is in final form and no version of it will be submitted for publication elsewhere."

of the system. The disadvantage of that problem is the loss of physical insight.

At our group in Kaiserslautern we have studied the following two problems:

Computersimulation of car driving in order to check the critical safety components.

The parts of the car which cause the greatest trouble are the tires. They behave extremely nonlinear and good three-dimensional models use finite elements with more than 10 000 knots. For our purpose we need a nonlinear model with less than 20 degrees of freedom.

Computersimulation of the water temperature in a solar heated swimming pool.

The aim of this project is to find out the usefulness of a solar heating equipment and to give hints for the design of such a system.

In our case the system has as inputs variables: Solar radiation, air humidity, windvelocity, air temperature, incomming heat from the solar collectors, status messages a.s.o. Essentially the weather is the input and all possible convection, radiation and diffusion effects occur in the system. A physical model will be complicated and will contain rather vage parameters.

The identification results, which you will find at the end, give in the linear case no useful results, so that we had to develop a nonlinear identification procedure, which contains an identification of the nonlinear structure.

In the following chapters we will present first the theoretical background and then the algorithm we use to solve the mentioned problems.

2   Theory

     a) Model

     Our aim is to study models in the state space which contain nonlinear interactions within the system:

$$x_{t+1} = Ax_t + By_t + \sum_{l=2}^{p} D_1 x_t^{[1]}$$

(2.1)

$$y_t = Cx_t$$

where

$$p \in \mathbb{N}, \quad x_t \in \mathbb{R}^n, \quad y_t \in \mathbb{R}^m, \quad u_t \in \mathbb{R}^r$$

(2.2)

$$x_t^{[1]} := (x_1^1, x_1^{1-1} x_2, \ldots, x_1^{1-1} x_n, \ldots, x_n^1)^t$$

A, B, C and D_1 are matrices with appropriate dimension. The set of (p+2)-tuples $(A,B,C,D_2,\ldots,D_1) =: (A,B,C,D_1)$ is called P_p.

     b) Observability

     For the purpose of identification it is necessary to work with observable systems. The following conditions guarantee observability for our model-systems:

(2.3)
$$rg(C) = m$$
$$rg(C^t \ A^t C^t \ \ldots \ A^{t(n-m)} C^t) = n$$

(vgl. Casti [1])

     c) Equivalence classes

     The next step is to find an equivalence relation on the set of systems: The equivalence classes are the sets of systems with the same input-output-behaviour. With these equivalence classes of systems one can obtain a complete set of invariants and finally some kind of "normal forms" for the specified systems.

Definition 1

Let $T \in GL(n)$. Then we define the mapping $T^{[1]}$ by
$$T^{[1]} x_t^{[1]} := (Tx)^{[1]}.$$
The mapping $T^{[1]}$ has the following properties:

Lemma 1

i) $\quad (T^{[1]})^{-1} = (T^{-1})^{[1]}$

ii) $\quad (T_1 T_2)^{[1]} = (T_1^{[1]})(T_2^{[1]})$

With this definitions of $T^{[1]}$ and Lemma 1 we get the following theorem referring to the input-output-equivalence for our non-linear systems.

Theorem 1

Consider $\Sigma, \Sigma' \in P_p$. If some $T \in GL(n)$ exist with

$$A' = TAT^{-1}$$

(2.4) $\qquad B' = TB$

$$C' = CT^{-1}$$

$$D_1' = TD_1 (T^{-1})^{[1]} \qquad 1 = 2, \ldots, p$$

then Σ and Σ' are input-output-equivalent.

The transformation property in theorem 1 leads to the following lemma:

Lemma 2

The transformation property in theorem 1 forms an equivalence relation on P_p.

d) Normal forms

Now we want to obtain a special set of invariants which

leads to a special canonical form in each equivalence class. Through the linear parts of every system have the same transformation behaviour as a linear system, we follow the paper of Guidorzi [2]:

Consider $(A,B,C,D_1) \in P_p$

$$(2.5) \qquad
\begin{matrix}
c_1^t & A^t c_1^t & \cdots & A^{t(n-m+1)} c_1^t \\
\vdots & \vdots & & \vdots \\
c_m^t & A^t c_m^t & \cdots & A^{t(n-m+1)} c_m^t
\end{matrix}$$

and the sequence

$$(2.6) \qquad c_1^t \; \cdots \; c_m^t \; A^t c_1^t \; \cdots \; A^{t(n-m+1)} c_m^t \; .$$

Let $A^{t\nu_i} c_i^t$ be the first vector in the ith row of (2.5) which is linear dependent on its predecessors in (2.6). It follows that

$$(2.7) \qquad A^{t\nu_i} c_i^t = \sum_{j=1}^{m} \sum_{k=1}^{\nu_{ij}} \alpha_{ijk} A^{t(k-1)} c_j^t$$

with

$$(2.8) \qquad \nu_{ij} = \begin{cases}
\nu_i & i = j \\
\min(\nu_i + 1, \nu_j) & i > j \\
\min(\nu_i, \nu_j) & i < j
\end{cases}$$

Lemma 3

The coefficients ν_i have the following properties:

i) If $A^{t\nu_i} c_i^t$ is linear dependent on the predecessors in (2.6) then the same is true of the vectors $A^{t(\nu_i + k)} c_i^t$, $k \geq 1$

ii) $\nu_i > 0$, $\forall_i = 1, \ldots, m$

iii) $\sum_{i=1}^{m} \nu_i = n$

The linear independent vectors $A^{t(k-1)}c_j^t$ in (2.6) are called regular vectors.

Let $R_v := (c_1^t \ldots A^{t(\nu_1-1)}c_1^t \ldots c_m^t \ldots A^{t(\nu_m-1)}c_m^t)$.

If we now define $(b_{ijk}, d_{ijk}^{(1)})$ as follows:

Definition 2

(2.9) (i) $(b_{ijk}) := R_v^t B$

with

$$
(2.10) \qquad (b_{ijk}) = \begin{pmatrix} b_{111} & \cdots & b_{r11} \\ \vdots & & \vdots \\ b_{11\nu_1} & \cdots & b_{r1\nu_1} \\ \vdots & & \vdots \\ b_{1m1} & \cdots & b_{rm1} \\ \vdots & & \vdots \\ b_{1m\nu_m} & \cdots & b_{rm\nu_m} \end{pmatrix}
$$

(2.11) (ii) $(d_{ijk}^{(1)}) := R_v^t D_1 (R_v^{t[1]})^{-1} \qquad \forall\, 1 = 2,\ldots,p$

with

$$
(2.12) \qquad (d_{ijk}^{(1)}) = \begin{pmatrix} d_{111}^{(1)} & \cdots & d_{n_111}^{(1)} \\ \vdots & & \vdots \\ d_{11\nu_1}^{(1)} & \cdots & d_{n_11\nu_1}^{(1)} \\ \vdots & & \vdots \\ d_{1m1}^{(1)} & \cdots & d_{n_1m1}^{(1)} \\ \vdots & & \vdots \\ d_{1m\nu_m}^{(1)} & & d_{n_1m\nu_m}^{(1)} \end{pmatrix}
$$

then Lemma 4 holds:

Lemma 4

The function $f = (f_i, f^\alpha_{ijk}, f^b_{ijk}, f^{d(1)}_{ijk})$ constitutes a basis for the specified equivalence relation.

If we choose the unit vectors as basis of the state space, we obtain "normal forms":

$$(2.13\ a))\qquad C = \begin{pmatrix} 1 & 0 & & & & & & & & \\ 0 & 0 & \cdots & 1 & 0 & & & & & \\ & \vdots & & & & & & & & \\ & \vdots & & & & & & & & \\ & \vdots & & & & & & & & \\ 0 & & & & & 0 & 1 & 0 & \cdots & 0 \end{pmatrix}$$

$$\underset{\text{1. column}}{\uparrow} \qquad \underset{(\nu_1+1)\text{-column}}{\uparrow} \qquad \underset{(\nu_1+\ldots+\nu_{m-1}+1)\text{-column}}{\uparrow}$$

$A = (A_{ij}), \quad i,j = 1,\ldots,m$

$$A_{ii} = \begin{pmatrix} 0 & 1 & & & \\ \vdots & & 0 & 1 & \\ \vdots & & & & \\ \vdots & & & & \\ 0 & & & 0 & 1 \\ \alpha_{ii1} & \cdots & & & \alpha_{ii\nu_i} \end{pmatrix}$$

$$(2.13\ b))$$

$$A_{ij} = \begin{pmatrix} 0 & & \cdots & & & 0 \\ \vdots & & & & & \vdots \\ \vdots & & & & & \vdots \\ \vdots & & & & & \vdots \\ 0 & & & & & \vdots \\ \alpha_{ij1} & \cdots & \alpha_{ij\nu_{ij}} & 0 & \cdots & 0 \end{pmatrix}$$

$$(2.13\ c))\quad B = \begin{pmatrix} B_1 \\ \vdots \\ \vdots \\ \vdots \\ B_m \end{pmatrix}, \quad B_i = \begin{pmatrix} b^t_{i1} \\ \vdots \\ \vdots \\ \vdots \\ b^t_{i\nu_i} \end{pmatrix} = \begin{pmatrix} b_{1i1} & \cdots & b_{ri1} \\ \vdots & & \vdots \\ \vdots & & \vdots \\ b_{1i\nu_i} & \cdots & b_{ri\nu_i} \end{pmatrix}$$

$$(2.13 \text{ d})) \quad D_1 = \begin{pmatrix} D_1^{(1)} \\ \vdots \\ D_m^{(1)} \end{pmatrix}, \quad D_i^{(1)} = \begin{pmatrix} d_{i1}^{(1)\,t} \\ \vdots \\ d_{iv_i}^{(1)\,t} \end{pmatrix} = \begin{pmatrix} d_{1i1}^{(1)} & & d_{n_1 i1}^{(1)} \\ \vdots & & \vdots \\ d_{1iv_i}^{(1)} & \cdots & d_{n_1 iv_i}^{(1)} \end{pmatrix}$$

$$1 = 2,\ldots,p$$

e) Transformation to input-output-models

Now we want to consider the transition to input-output-models. In the linear case, Guidorzi had shown the existence of an input-output-model of the form

$$P(z)y(t) = Q(z)u(t)$$

which forms the identical input-output-behaviour. ($P(z)$ and $Q(z)$ are polynomial matrices in the shift operator z: $zy(t) = y(t+1)$.) The transition for systems $(A,B,C,D_1) \in P_p$ is much more difficult:

The existence of a finite input-output-model for a system from P_p is not guaranteed in every case. Consider the system of nonlinear equations

$$(2.14) \quad \begin{pmatrix} 1 & & & O \\ -z & \ddots & & \\ & \ddots & \ddots & \\ O & & -z & 1 \end{pmatrix} \begin{pmatrix} x_{v_1+\ldots+v_{j-1}+1}(t) \\ \vdots \\ x_{v_1+\ldots+v_j}(t) \end{pmatrix}$$

$$= \begin{pmatrix} y_j(t) \\ -b_{j1}^t u(t) \\ \vdots \\ -b_{j(v_j-1)}^t u(t) \end{pmatrix} + \begin{pmatrix} O \\ -\sum\limits_{l=2}^{P} d_{j1}^{(1)} x_t^{[1]} \\ \vdots \\ -\sum\limits_{l=2}^{P} d_{j(v_j-1)}^{(1)} x_t^{[1]} \end{pmatrix}$$

and

$$x_1(t) = y_1(t)$$

$$x_{\nu_1+1}(t) = y_2(t)$$

(2.15)

$$\vdots$$

$$x_{\nu_1+\ldots+\nu_{m-1}+1}(t) = y_m(t)$$

Splitting the nonlinear terms on the right hand side in terms
depending on $x_1(t), x_{\nu_1+1}(t), \ldots, x_{\nu_1+\ldots+\nu_{m-1}+1}(t)$ and
$x_2(t), \ldots, x_{\nu_1}(t), x_{\nu_1+2}(t), \ldots, x_n(t)$, we obtain a nonlinear
system of equations

$$f(\bar{x}(t)) + Fx(t) = r + g(y(t))$$

where

$$\bar{x}(t) = (x_2(t), \ldots, x_{\nu_1}(t), x_{\nu_1+2}(t), \ldots, x_n(t)).$$

This is an implicit system of nonlinear equations which can be
locally inverted around a solution. The inverse function can be
described by an infinite power serie. More over the existence of
a solution of the system is not guaranteed in the real vector
space. So we have to restrict ourselves to systems, where the
nonlinear terms of the model can be directly observed on the
output of the system: $f(\bar{x}) \equiv 0$.
We call such a system direct-polynomial-model.
Then it follows

$$P(z)y(t) = Q(z)u(t) + \sum_{l=2}^{P} R_l(z)y^{[l]}(t)$$

with $P(z)$ and $Q(z)$ like in the linear case and

$$R_1(z) = \begin{pmatrix} r_{11}^{(1)}(z) & \cdots & r_{1m_1}^{(1)}(z) \\ \vdots & & \\ r_{m1}^{(1)}(z) & \cdots & r_{mm_1}^{(1)}(z) \end{pmatrix}$$

(2.16)
$$m_1 = \binom{m+1-1}{1}, \quad 1 = 2,\ldots,p$$

$$r_{ij}^{(1)}(z) = \delta_{ij\nu_i} z^{\nu_i-1} + \ldots + \delta_{ij2} z + \delta_{ij1} .$$

If we consider the vectors $u(t)$ and $y^{[1]}(t)$ formally as one vector $w(t)$, we obtain an input-output-model of the form
$$P(t)y(t) = R(z)w(t)$$
and

(2.17)
$$y_i(t+\nu_i) = \sum_{j=1}^{m} \sum_{n=1}^{\nu_{ij}} \alpha_{ijk} y_j(t+k-1)$$
$$+ \sum_{j=1}^{s} \sum_{k=1}^{\nu_i} \eta_{ijk} w_j(t+k-1)$$

We call this model direct polynomial input-output-model and the components of the vector $w(t)$ model-inputs.

3   Identification algorithm

The following flow chart shows the steps of the identification algorithm:

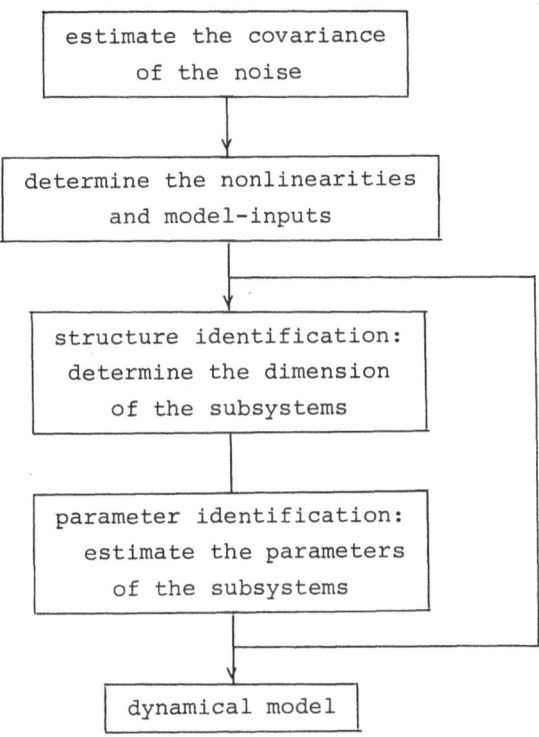

Flow chart of the identification algorithm.

a) The idea of the range error

We consider the described direct-polynomial input-output-model of the form

(3.1)

$$y_i(t+\nu_i) = \sum_{j=1}^{m} \sum_{k=1}^{\nu_{ij}} \alpha_{ijk} y_j(t+k-1)$$

$$+ \sum_{j=1}^{s} \sum_{k=1}^{\nu_i} \eta_{ijk} w_j(t+k-1)$$

Remember that the dimension of the state space is determined by the sum over the ν_i's and that the model-inputs consist of a subset of the available inputs, some nonlinear terms of these

inputs and some nonlinear terms of the measured outputs. If we
want to predict the vector $y^i(t+1)$ by a linear combination of
the $y^j(t-k)$, $w^j(t-k)$, then the error vector ε, which originates
from a L^2-approximation, has the norm $\lambda_i = [(y^i(t+1) -$
$P_i y^i(t+1))^t(y^i(t+1) - P_i y^i(t+1))]^{1/2}$ where P_i denotes the projection onto the subspace spanned by the vectors of the linear
combination. The measured data are corrupted by additive noise.
Our identification procedure should minimize the projection
error λ_i in relation to the noise variance.

So if we write

$$y^i(t+k) := (y_i(t+k),\ldots,y_i(t+k+L-1))^t, \quad i = 1,\ldots,m$$

(3.2)

$$w^i(t+k) := (w_i(t+k),\ldots,w_i(t+k+L-1))^t, \quad i = 1,\ldots,s$$

we obtain

(3.3) $\quad y^i(t+\nu_i) = \sum\limits_{j=1}^{m} \sum\limits_{k=1}^{\nu_{ij}} \alpha_{ijk} y^j(t+k-1) + \sum\limits_{j=1}^{s} \sum\limits_{k=1}^{\nu_i} \eta_{ijk} w^j(t+k-1)$

for

$$L \geq \sum\limits_{j=1}^{m} \nu_{M_j} + s\nu_M + 1, \quad \nu_M = \max_i \nu_i .$$

If we define

(3.4) $\quad \begin{aligned} L_k(y_i) &:= (y^i(t) y^i(t+1) \ldots y^i(t+k-1)) \\ L_k(w_i) &:= (w^i(t) w^i(t+1) \ldots w^i(t+k-1)) \end{aligned}$

and

$$R(\delta_1,\ldots,\delta_{m+s}) := [L_{\delta_1}(y_1)|\ldots|L_{\delta_m}(y_m)|L_{\delta_{m+1}}(w_1) \ldots$$

(3.5)

$$L_{\delta_{m+s}}(w_s)]$$

$$S(\delta_1,\ldots,\delta_{m+s}) := R^t(\delta_1,\ldots,\delta_{m+s})R(\delta_1,\ldots,\delta_{m+s})$$

we can define the so-called Predicted-Percent-Reconstruction-
Error (PPCRE) π_i, with

$$\pi_i = \pi(v_{i1}, \ldots, v_{i(i-1)}, v_i^{+1}, \ldots)$$
$$= 100 \cdot \sqrt{\lambda_i / (L-1) \sigma_i^2}$$

(3.6)
$$\lambda_i := \frac{\det S(v_{i1}, \ldots, v_{i(i-1)}, v_{i+1}, \ldots)}{\det S(v_{i1}, \ldots, v_{i(i-1)}, v_i, \ldots)}$$

σ_i variance of the noise at the i-th output

as an indicator for the quality of a one-step-ahead prediction
by a model with the structure above.

b) The estimation of the noise variance

Denote the noise corrupted measurements by

(3.7)
$$y_i^*(t) := y_i(t) + d(y_i(t)), \quad i = 1, \ldots, m$$
$$w_i^*(t) := w_i(t) + d(w_i(t)), \quad i = 1, \ldots, s$$

Assume that the noise has mean value zero and that the noise of
the input and output is uncorrelated. Then it follows that

(3.8)
$$\plim_{L \to \infty} \frac{\lambda_{min}}{L} = \lim_{L \to \infty} \frac{1}{L} S_i + N(d_i)$$

where $N(d_i)$ the covariance matrix of the noise vector.
If we also assume that the noise of the different inputs and
outputs is identical distributed one finds

(3.9)
$$\plim_{L \to \infty} \frac{\lambda_{min}}{L} = \sigma^2$$

λ_{min} smallest eigenvalue of $S^*(v, \ldots, v)$, $v \geq v_M$.
Therefore we obtain the following algorithm for the estimation
of the noise variance:

Compute the smallest eigenvalue of the matrices

$S^*(1, \ldots, 1), S^*(2, \ldots, 2), S^*(3, \ldots, 3)$ etc.

until a stabilization of the eigenvalue can be

established. Then

$$\sigma^2 = \lambda_{min}/L$$

is an estimation for the noise covariance.

c) Algorithm for the determination of the model-inputs
(dynamic correlation)

Choose two integers N and M.

Let N the dimension of the single subsystems. M the bound for
the polynomial order in the state space.

Estimate the noise covariance at which the vectors $w_i(t)$ are
replaced by the measured input-vectors $u_j(t)$.

1st step:

(determination of the first model-input $w_1(t)$)

α) Compute the reconstruction error of S*(N,...,N,N-1) by the
formula (3.6) successively for

(3.10) $\qquad u_i, u_i^2, \ldots, u_i^M, \qquad i = 1, \ldots, r$
$\qquad\qquad y_j^2, \ldots, y_j^M, \qquad j = 1, \ldots, m$

where just one element in (3.10) is taken as input.

β) Choose the smallest error and fix the first model-input.

γ) Cancel the chosen model-input from the list (3.10).

ith step:

(determination of the ith model-input $w_i(t)$)

α) Compute the reconstruction error of S*(N,...,N,N-1,...,N-1)
by the formula (3.6) successively for the already existing
elements of (3.10), where the first (i-1)th model-input are
determined by the steps 1 to (i-1).

β) Choose the smallest error and fix the ith model-input.

107

γ) Cancel this model-input from the list (3.10).

Follow this step until no significant change in the errors can be determined, the error becomes smaller than a chosen error bound or all elements of (3.10) are chosen.

d) Structure- and parameter identification

Once the model-inputs are fixed by dynamic correlation, the next step is to determine the dimension of the state space (the dimension of the single subsystems) and to compute the parameters needed to determine the model-system.

1st step:

Construct the matrices

(3.11)
$$S*(2,1,\ldots,1),S*(2,2,1,\ldots,1),\ldots,S*(2,2,\ldots,2)$$
$$S*(3,2,\ldots,2),S*(3,3,2,\ldots,2),\ldots,S*(3,3,\ldots,3)$$

etc.

and compute the corresponding reconstruction errors

(3.12)
$$\pi(2,1,\ldots,1),\pi(2,2,1,\ldots,1),\ldots,\pi(2,\ldots,2)$$
$$\pi(3,2,\ldots,2),\pi(3,3,2,\ldots,2),\ldots,\pi(3,\ldots,3)$$

etc.

If there is no significant difference between the errors
$$\pi(\mu_1-1,\ldots,\mu_i-1,\ldots) \quad \text{and} \quad \pi(\mu_1,\ldots,\mu_i,\ldots)$$
from the sequence (3.12) (i.e. by increasing the dimension of the ith subsystem there is no improvement of the projection error mentioned), then one can fix the dimension of the current subsystem by $v_i = \mu_i-2$. Let γ_i the vector of the model-parameter of the current subsystem

(3.13) $\gamma_i = (\alpha_{i11},\ldots,\alpha_{i1v_i},\ldots,\alpha_{imv_m},\eta_{i11},\ldots,\eta_{i1v_i},\ldots,\eta_{isv_i})^t$

Let

(3.14)
$$R_i = R(\nu_{i1}, \ldots, \nu_{im}, \nu_i, \ldots, \nu_i)$$
$$S_i = S(\nu_{i1}, \ldots, \nu_{im}, \nu_i, \ldots, \nu_i),$$

then it follows from (3.1) that

(3.15)
$$\gamma_i = (R_i^T R_i)^{-1} R_i^{*t} y^{*i}(t+\nu_i)$$

$$\hat{\gamma}_i = \hat{S}_i^{-1} R_i^{*t} y^{*i}(t+\nu_i)$$

where $\hat{S}_i := S_i^* - LN(d_i)$.

$\hat{\gamma}_i$ is an estimation of the model-parameter by a least-square-formula.

In this way the first subsystem is determined.

(i+1)th step:

Construct the matrix

(3.16)
$$S^*(\nu_{i1}, \ldots, \nu_i, \nu_{i(i+1)}, \ldots)$$

by cancelling the matrix S* of the ith step. Test (beginning with the matrix (3.16)) the respective reconstruction errors of the matrices which arise by increasing the dimension of the single subsystems (the dimension of the subsystems that have already been identified are not modified) by the formula of the 1st step. If the dimension of a further subsystem is fixed (analogous to step 1), one can estimate the parameter of this subsystem by formula (3.15).

Repeat this step, until every subsystem is identified.

Thus all steps of the flow chart are specified and the algorithm is completely described.

3   Examples

   a) Nonlinear tire model

       To develop a simple tire model we consider a car
driving over a speedbump. For this system we only have the
geometry of the speedbump as input. The outputs are some measured acceleration and force time series. As a simple example we
choose the x- and the z-acceleration, measured on the wheel
suspension, as outputs. (Pictures 1.1, 1.2, 2.1, 2.2)

   b) Simulation of the water temperature in a solar-heated
      swimmingpool

       The used data are measured in the swimmingcenter of the
German city UNNA during August 1985. Picture 3 shows (some of)
the input time series. In picture 4 the linear and nonlinear
results are presented. It's remarkable that we get in the nonlinear case such a good reconstruction for the whole dataset,
although we only use 200 points for the identification of the
model.

   Final remarks

       As the examples show our identification procedure works
very well, but there remain some things to improve. At first we
have to include real stochastic inputs. Then there is a need
for on-line procedures. This can certainly not be managed by the
existing algorithm but some of the underlying ideas should be
useful. Concerning the theoretical part we want to look for
better normal forms. That mean to develop models with less
parameters.

4   References

[1]   Casti, John L.: Nonlinear System Theory, Orlando, 1985

[2]   Guidorzi, Roberto P.: Invariants and Canonical Forms for
       Systems Structural and Parametric Identification, in:
       Automatica, 17, 1981, p. 117-133

[3]   Guidorzi, Roberto P.; Losito, Maria P.; Muratori, Tiziana:
       The Range Error Test in the Structural Identification of
       Linear Multivariable Systems, in: IEEE Tr. on Automatic
       Control, AC-27, 1982, p. 1044-1053

[4]   Struckmeier, Jens: Ein einfacher Identifikationsalgorith-
       mus für nichtlineare Systeme, Diplomarbeit, Kaiserslautern,
       1987

J. Struckmeier, F.-J. Pfreundt

Department of Mathematics

University of Kaiserslautern

Erwin-Schrödinger-Strasse

D-6750 Kaiserslautern

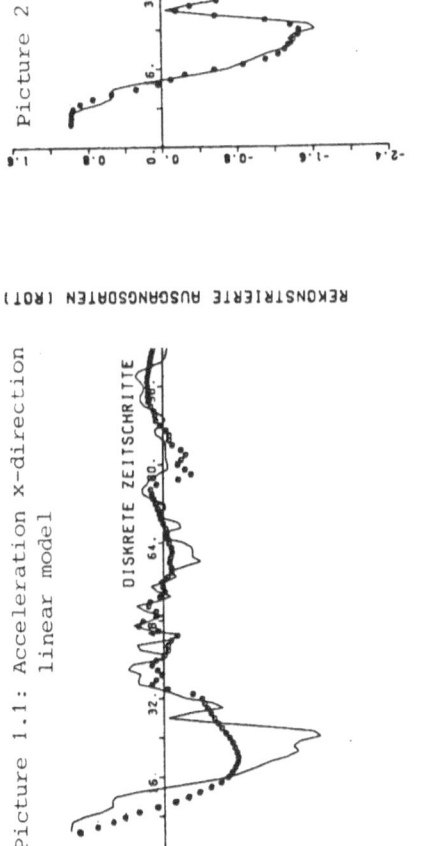

Picture 2.1: Acceleration x-direction nonlinear model

Picture 1.1: Acceleration x-direction linear model

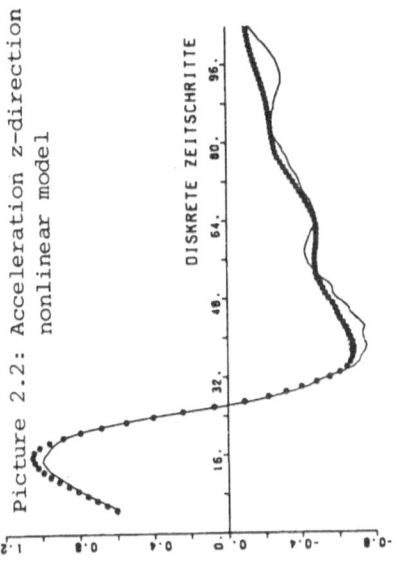

Picture 2.2: Acceleration z-direction nonlinear model

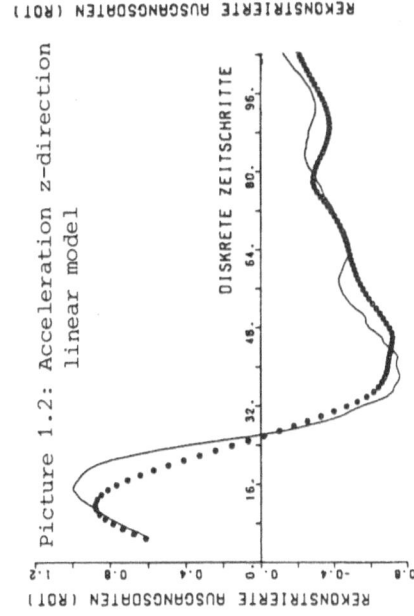

Picture 1.2: Acceleration z-direction linear model

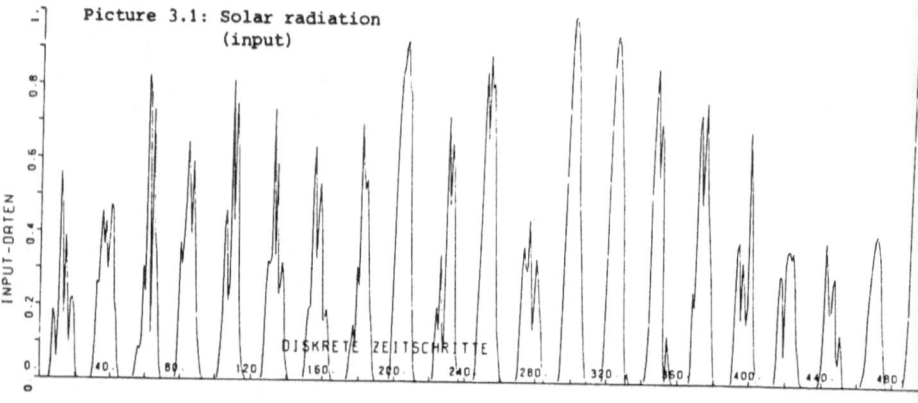

Picture 3.1: Solar radiation (input)

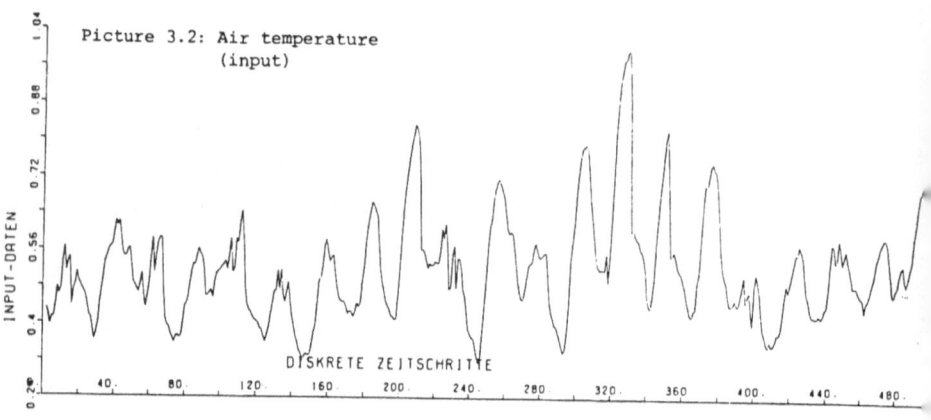

Picture 3.2: Air temperature (input)

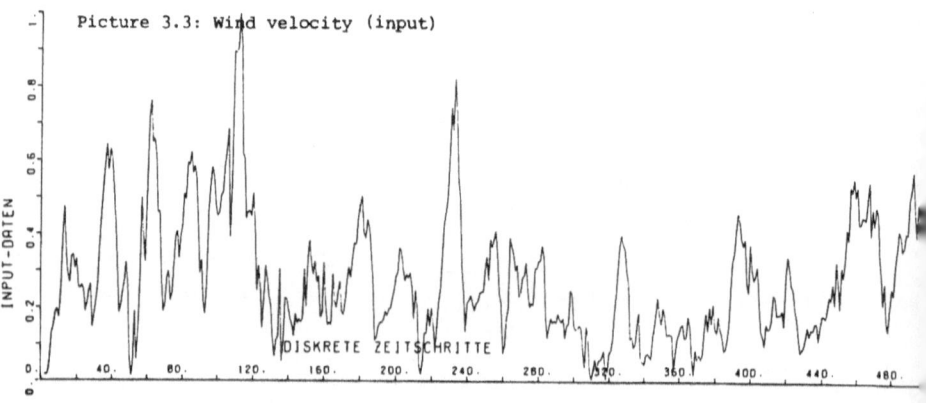

Picture 3.3: Wind velocity (input)

113

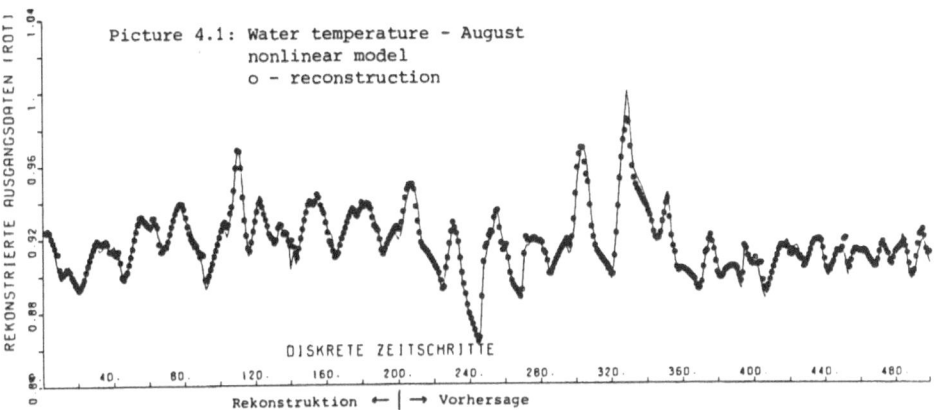

Picture 4.1: Water temperature - August
          nonlinear model
          o - reconstruction

REKONSTRIERTE AUSGANGSDATEN (ROT)

DISKRETE ZEITSCHRITTE

40.   80.   120.   160.   200.   240.   280.   320.   360.   400.   440.   480.

Rekonstruktion ←— | —→ Vorhersage

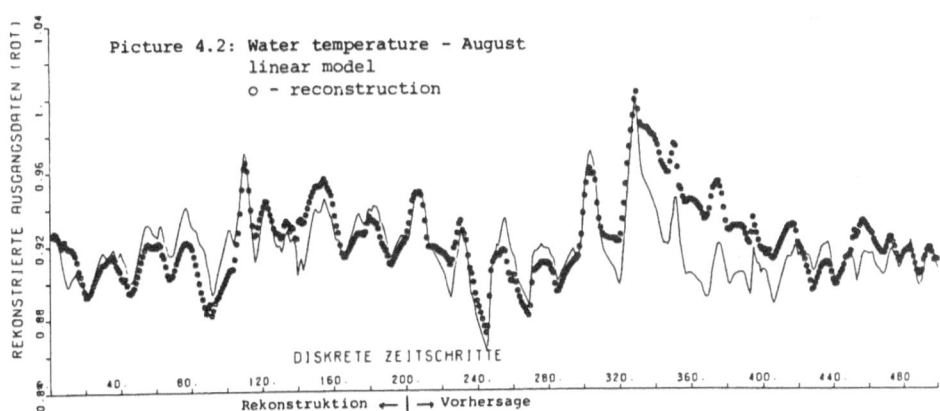

Picture 4.2: Water temperature - August
          linear model
          o - reconstruction

REKONSTRIERTE AUSGANGSDATEN (ROT)

DISKRETE ZEITSCHRITTE

40.   80.   120.   160.   200.   240.   280.   320.   360.   400.   440.   480.

Rekonstruktion ←— | —→ Vorhersage

INTRODUCTION TO NILPOTENT APPROXIMATION FILTERING

Michiel Hazewinkel
Centre for Mathematics and Computer Science
P.O. Box 4079, 1009 AB Amsterdam, The Netherlands

The socalled reference probability of unnormalized probability method for nonlinear filtering problems leads to a (robust) infinite dimensional filter of bilinear type. If the associated Lie algebra is topologically solvable or nilpotent an infinite dimensional version of Wei-Norman theory applies. If not then ideas of nilpotent approximation lead to (potential) approximation filters. This note is not so much a definite report on results as on outline of a research program.

1. STATEMENT OF THE PROBLEM

In full generality *filtering* is concerned with obtaining estimates concerning a stochastic process $\{x_t\}$, the *signal process*, on the basis of another related process $\{y_t\}$, the *observation process*. In this paper we have the following realization of this situation in terms of stochastic differential equa-

$$dx_t = f(x_t)dt + G(x_t)dw_t \quad , \quad x_t \in \mathbf{R}^n, \ w_t \in \mathbf{R}^m \tag{1.1}$$

$$dy_t = h(x_t)dt + dv_t, \quad y_t \in \mathbf{R}^p, \ v_t \in \mathbf{R}^p \tag{1.2}$$

where f, G, h are vector and matrix valued functions of the right dimensions and w_t and v_t are independent Wiener noise processes also independant of the initial state x_0. The problem is the following. For a given (interesting) function $\phi(x)$ of the state x, give a calculation procedure for the best estimate $\widehat{\phi(x_t)}$ at time t given the observations y_s, $0 \leqslant s \leqslant t$. More generally one also considers finding $\widehat{\phi(x_t)}$ given y_s, $0 \leqslant s \leqslant t_1$, $t_1 < t$ *(prediction)* and finding $\widehat{\phi(x_t)}$ given y_s, $0 \leqslant s \leqslant t_2$, $t < t_2$ *(smoothing)*. Of particular importance is finding $\hat{x}_t$ *(state estimation)*.

Ideally one would like the calculation procedure to be *finite dimensional, exact, recursive,* and *robust*. The first three adjectives here mean (more or less by definition) that the calculation procedure, the *filter*, should be of the form

$$dm_t = \tilde{\alpha}(m_t)dt + \sum_{j=1}^{r} \tilde{\beta}_j(m_t)d\tilde{\zeta}_j(y_t) \tag{1.3}$$

$$\widehat{\phi(x_t)} = \gamma(m_t, y_{1t}, \ldots, y_{pt}) \tag{1.4}$$

Here $\tilde{\alpha}, \tilde{\beta}_j, \tilde{\zeta}_j, \gamma$ are known functions and vectorfields and m_t evolves over a finite dimensional manifold (finite dimensionality); recursiveness is embodied by the fact that (1.3) is directly driven by the observations and that $\widehat{\phi(x_t)}$ only depends on the filter state m_t; and the current observations; (1.4) of course also reflects exactness. For robustness one requires that the filter equations be driven by y_t itself instead of also involving the dy_t. I.e. one requires (1.3) to be replaced by an equation

$$\frac{dm_t}{dt} = \alpha(m_t) + \sum_{j=1}^{r} \beta_j(m_t)\zeta_j(y_{1t}, \ldots, y_{pt}). \tag{1.5}$$

Thus while (1.3) is a stochastic differential equation its robust version (if it exists) (1.3) can be treated pathwise and makes sense as a family of differential equations, one for each possible observation path $\{y_t\}$.

The problem now is: given a system (1.1), (1.2) and a function ϕ how to find a filter (1.4), (1.5); i.e. how to determine the functions γ and ζ_j and vectorfields α and β_j occurring in (1.4), (1.5).

2. THE DMZ FILTER

Under mild regularity assumptions on f, G, h and reachability and observability conditions on the system (1.1), (1.2) the conditional state $x_t = E[x_t | y_s, 0 \leqslant s \leqslant t]$ has a density $\pi(x,t)$.

2.1. (Duncan [2], Mortensen [6], Zakai [9]). Under appropriate regularity conditions there exists an unnormalized version $\rho(x,t)$ of $\pi(x,t)$ (i.e. $\rho(x,t) = \sigma(t)\pi(x,t)$ for some unknown function $\sigma(t)$) which satisfies the stochastic partial differential equation

$$d\rho = \mathcal{L}\rho dt + \sum_{i=1}^{p} h_i(x) dy_{it}. \tag{2.2}$$

Here $\mathcal{L}$ is the second order partial differential operator defined by

$$\mathcal{L}\psi = \frac{1}{2} \sum_{i,j=1}^{n} \frac{\partial^2}{\partial x_i \partial x_j} ((GG^T)_{ij}\psi) - \sum_{i=1}^{n} \frac{\partial}{\partial x_i} (f_i\psi) - \frac{1}{2} \sum_{j=1}^{p} h_j^2 \psi. \tag{2.3}$$

Here G^T is the transpose of the matrix valued function G and $(GG^T)_{ij}$ is the (ij)-th entry of the matrix GG^T, f_i is the i-th component of the function f and h_j the j-th component of the function h.

The stochastic PDE (2.2) is to be regarded as a Fisk-Stratonovic stochastic PDE. To obtain the equivalent Ito version remove the term $-\frac{1}{2}\sum h_j^2 \psi$ in (2.3).

Consider the time dependant gauge transformation

$$\tilde{\rho}(x,t) = \exp(-h_1(x)y_{1t} - \ldots - h_p(x)y_{pt})\rho(x,t). \tag{2.4}$$

Substituting this into (2.2) yields an equation

$$\frac{\partial \tilde{\rho}(x,t)}{\partial t} = \mathcal{L}\tilde{\rho} - \sum_{i=1}^{p} y_i(t)\mathcal{L}_i\tilde{\rho} - \sum_{i,j=1}^{p} y_i(t)y_j(t)\mathcal{L}_{ij}\tilde{\rho} \tag{2.5}$$

where

$$\mathcal{L}_i = [h_i, \mathcal{L}] := h_i\mathcal{L} - \mathcal{L}h_i, \quad \mathcal{L}_{ij} = \mathcal{L}_{ji} = \frac{1}{2}[h_i, [h_j, \mathcal{L}]]. \tag{2.6}$$

Given $\phi(x)$ and $\tilde{\rho}(x,t)$ the best estimate $\hat{\phi(x_t)}$ can be calculated by

$$\rho(x,t) = \exp(h_1(x)y_{1t} + \ldots + h_p(x)y_{pt}) \tag{2.7}$$

$$\hat{\phi(x_t)} = (\int \rho(x,t)dx)^{-1} \int \phi(x)\rho(x,t)dx. \tag{2.8}$$

Note that (2.5) together with the output map (2.7), (2.8) is a recursive, exact and robust filter. The only trouble with it (from the calculation point of view) is that it is infinite dimensional.

3. WEI-NORMAN THEORY [8]

For the moment let us consider control systems of the form

$$\dot{x} = u_1 A_1 x + \ldots + u_k A_k x, \quad x \in \mathbf{R}^n \tag{3.1}$$

where the A_i are $n \times n$ matrices and the u_i are inputs (known functions of time). Adding a few more terms (with $u_j = 0$, $j > k$) we may as well assume that $A_1, \ldots, A_k$ are a basis of a Lie algebra of $n \times n$ matrices (under the commutator difference product $[A,B] = AB - BA$). Let us look for solutions of the form

$$x(t) = e^{g_1 A_1} \ldots e^{g_k A_k} x(0) \tag{3.2}$$

where the $g_i(t)$ are still to be determined functions of time. By differentiating (3.2), inserting $\exp(-g_1 A_1) \cdots \exp(-g_i) \exp(g_i A) \cdots \exp(g_1 A)$ just after $\dot{g}_{i+1}A_{i+1}$ in the result, using the Baker-Cambell-Hausdorff formula, using (3.1) and collecting terms, one finds a set of equations

$$\dot{g}_i + \sum_{j=1}^{k} \dot{g}_j h_{ji}(g_1,...,g_k) = u_i, \quad i = 1,...,k \tag{3.3}$$

with $h_{ij}(0,...,0)=0$ and the following properties of the $h_{ij}(g_1,...,g_k)$:

$$h_{ij} \text{ only involves } g_1,...,g_{i-1} \tag{3.4}$$

and if $A_{l+1},...,A_k$ are a basis of an ideal or $\mathfrak{a} \subset \mathfrak{g}$ (so that $[A_i,\mathfrak{a}] \subset \mathfrak{g}$ for all i) then

$$h_{ji} = 0 \text{ for } i = i,...,l; \ j = l+1,...,k \tag{3.5}$$

so that the equations for $g_1,...,g_l$ do not involve $g_{l+1},...,g_k$ at all. It is also important to note that the h_{ij} are universal functions depending only on the Lie algebra $\mathfrak{g}$ and the chosen basis and totally independent of the particular matrix realization (representation) we may be dealing with. In particular if $\mathfrak{a}$ is an ideal of $\mathfrak{g}$ and $A_1,...,A_k$ is a basis as above then

$$\text{equations for } g_1,...,g_l \text{ only depend on } \mathfrak{g} / \mathfrak{a}. \tag{3.6}$$

In case that $\mathfrak{g}$ is nilpotent (or more generally solvable) equations (3.3) therefore take a particularly pleasant triangular form which can be solved just using quadratures. Indeed if L is nilpotent, so that

$$L \underset{\neq}{\supset} [L,L] = L_2 \underset{\neq}{\supset} [L,L_2] = L_3 \underset{\neq}{\supset} \cdots \underset{\neq}{\supset} [L,L_r] = L_{r+1} = 0$$

and if we choose a basis

$$A_1,...,A_{k_1}, A_{k_1+1},...,A_{k_2},...,A_{k_{r-1}+1},...,A_{k_r}, \quad k_r = k$$

such that

$$A_{k_{i-1}+1},...,A_{k_i}, \quad k_0 = 0$$

is a basis for L_i, $i = 1,...,r$, then the equations take the form

$$\dot{g}_1 = u$$

...

$$\dot{g}_{k_1} = u k_1$$
$$\dot{g}_{k_1+1} = u_{k_1+1} + \alpha_{k_1+1}(u_1,...,u_{k_1}; g_1, \ldots, g_{k_1})$$

...

$$\dot{g}_{k_2} = u_{k_2} + \alpha_{k_2}(u_1,..,u_{k_1}; g_1, \ldots, g_{k_1}) \tag{3.6}$$
$$\dot{g}_{k_2+1} = u_{k_2+1} + \alpha_{k_2+1}(u_1,...,u_{k_2}; g_1,...,g_{k_2})$$

...

$$\dot{g}_{k_3} = u_{k_3} + \alpha_{k_3}(u_1,...,u_{k_3}; g_1,...,g_{k_2}).$$

...

Now note that the robust DMZ filter equation (2.5) is of the form (3.1) except that it takes place in a function space. So in particular if the Lie algebra generated by the operators $\mathcal{L},\mathcal{L}_i,\mathcal{L}_{ij}$ in (2.5) is nilpotent (solvable) and finite dimensional with basis $A_1,...,A_k$ and we have given an initial density $\rho_0(x)$ and function ϕ then equations (3.6) together with the output equation

$$(g_1,...,g_k) \mapsto \tilde{\rho}(x,t) = \exp(g_1 A_1)\cdots\exp(g_k A_k)\rho_0(x)$$
$$\tilde{\rho}(x,t) \mapsto \rho(x,t) = \exp(h_1(x)u_1)\cdots\exp(h_p(x)y_p)\tilde{\rho}(x,t)$$
$$\hat{\phi}(x) = \left(\int \rho(x,t)dx\right)^{-1} \int \phi(x)\rho(x,t)dx$$

constitute a recursive exact robust filter for $\hat{\phi}(x_t)$. It is not really finite dimensional because the A_i here are operators and calculating $\exp(g_i A_i)$ (for known $g_i(t)$) amounts to solving $\frac{d}{dt} B_i = \dot{g}_i A_i B_i$, $B_0 = id$ which is again a partial differential equation.

4. THE IDENTIFICATION CASE

The problem of identifying a linear system

$$dx_t = Ax_t dt + B dw_t, \quad dy_t = Cx_t + dv_t \tag{4.1}$$

i.e. the problem of determining the unknown matrices A, B, C on the basis of the observations, can be viewed as a nonlinear filtering problem for the system with state vector (x, A, B, C) obtained by adding the equations $dA = 0$, $dB = 0$, $dC = 0$ to (4.1). It can be proved that the Lie algebra generated by the $\mathcal{L}, \mathcal{L}_i, \mathcal{L}_{ij}$ in this case is topologically solvable. I.e. there is a sequence of ideals $\mathfrak{a}_i$ such that $\mathfrak{g}/\mathfrak{a}_i$ is finite dimensional solvable for all i and $\bigcap_i \mathfrak{a}_i = \{0\}$. Because of (3.6) this yields a sequence of approximate filters via

$$e^{g_1 A_1} \cdots e^{g_1 A_{k_1}} p_0, \quad e^{g_1 A_1} \cdots e^{g_{k_2} A_{k_2}} p_0, \cdots$$

where $A_1, ..., A_{k_1}, A_{k_1+1}, ..., A_{k_2}, \cdots$ are such that the equivalence classes of $A_1, ..., A_{k_r}$ mod $\mathfrak{a}_r$ are a basis for $\mathfrak{g}/\mathfrak{a}_r$. Cf [5] for more details.

5. NILPOTENT AND SOLVABLE APPROXIMATIONS

However, in many cases, the Lie algebra generated by $\mathcal{L}, \mathcal{L}_i, \mathcal{L}_{ij}$ will not be topologically solvable. For instance in the case of perturbed linear systems

$$dx = (Ax + \epsilon P_A(x))dt + (B + \epsilon P_B(x))dw_t, \quad dy = (C + \epsilon P_C(x))dt + dv_t \tag{5.1}$$

where the $P_A(x), P_B(x), P_C(x)$ are polynomial higher order disturbances. In this case the Lie algebra tends to be $W_n = \mathbb{R} < x_1, ..., x_n; \frac{\partial}{\partial x_1}, ..., \frac{\partial}{\partial x_n} >$, the Lie algebra of all differential operators (any order) with polynomial coefficients. In this case the higher order operations come with higher powers of ϵ in the sense that

$$\text{Lie}(\mathcal{L}, \mathcal{L}_i, \mathcal{L}_{ij}) \bmod \epsilon^n \text{ is finite dimensional for all } n \tag{5.2}$$

(and these algebras are solvable). Again there result approximate filters and they seem to perform well [3,4]. Still more generally there is no small parameter at all, but there still is a natural gradation structure on the Lie algebra. To see why this might be the case and why this will give us possibilities for constructing approximate filters observe that the operators $\mathcal{L}, \mathcal{L}_i, \mathcal{L}_{ij}$ are of the general forms

$$\mathcal{L} = \sum a_{ij} \frac{\partial^2}{\partial x_i \partial x_j} + \sum b_j \frac{\partial}{\partial x_j} + c$$

$$\mathcal{L}_i = \sum d_{ij} \frac{\partial}{\partial x_j} + e_i$$

$$\mathcal{L}_{ij} = f_{ij} .$$

where the $a_{ij}, b_{ij}, f_{ij}, e_i, c$ are explicit functions of the G_{ij}, f_i, h_j and their derivatives. Commuting various $\mathcal{L}$'s brings at least one derivative of the G_{ij}, f_i, h_j in each term, third order brackets bring second order derivatives or products of first order derivatives, ...

Now if the system described by the f_i, h_j, G_{ij} is supposed to model some real world phenomenon then we can not assume that we know these functions perfectly. In general one would expect that the values of the functions would be known very well, their derivatives less so, their second derivatives still less, etc., and by the time r-th derivatives come into play their values are almost totally unknown.

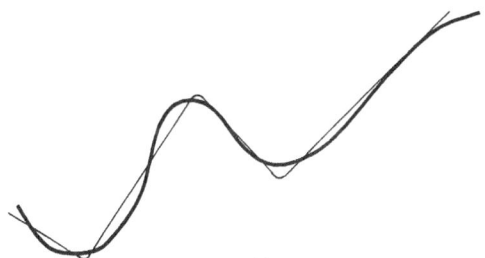

For $r = 2$ the kind of approximation involved is somewhat like illustrated on the above, i.e. something like a piecewise linear approximation with rounded corners. One expects a system close to real one in this sense of diminishing importance of higher derivatives (globally) to behave much like the true one. The comulative effect of small inaccuracies in first derivatives, larger ones in second derivatives, ..., very large ones in r-th derivatives will be such that r order brackets are almost totally unknown. And thus a system approximation which just happened to have all these zero would perform much as the original one but that one would have a filter as in section 3 above and this filter should also give reasonable results for the true system by considering the stability properties of the composed system

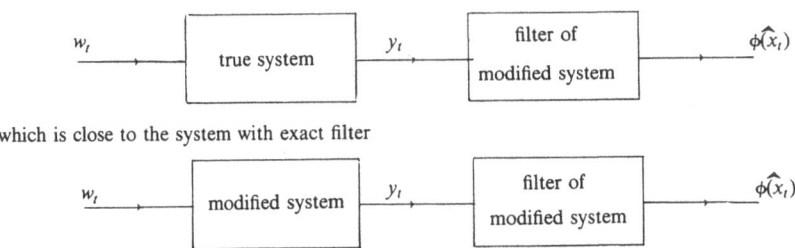

which is close to the system with exact filter

Now such a modified system which just happens to have all terms in r-th order brackets of the ℓ, ℓ_i, ℓ_{ij} equal to zero will probably not as a rule exist. But the corresponding filters can certainly be constructed. It suffices to introduce a counting mechanism and to consider the Lie algebra generated by the operator $z\ell, z\ell_i, z\ell_{ij}$. This one is topologically nilpotent and so Wei-Norman theory can be applied to Lie $(z\ell, z\ell_i, z\ell_{ij})$ mod z^n for all n (after which one sets $z = 1$.) Here z is an extra parameter.

The argument above indicates that such a procedure could work well. Another not unrelated argument can be based on Volterra series expansions. These ideas have of course a good deal to do with nilpotent and solvable approximation ideas [1], [7].

REFERENCES

1. *Solvable approximation to control systems,* SIAM J. Control and Opt. 32 (1984), 40-54.

2. *Probability densities for diffusion processes with applications to nonlinear filtering,* Ph.D. thesis, Standord, 1967.

3. *On deformations, approximations and nonlinear filtering,* Systems and Control Lett. 1 (1982), 32-36.

4. *Lie algebraic methods in filtering and identification,* Report PM-R86-6, November 1986, CWI, Amsterdam; to appear in Proc. 8-th Int. Symp. IAMP (Luminy, 1986), World Scientific and in Proc. 1-st World Congress Bernouilli Society (Taskent, 1986), VNU Science Press.

5. *Current algebras and the identification problem,* Stochastics 11 (1983), 65-101.

6. *Optimal control of continuous time stochastic differential equations,* Ph.D. thesis, Berkeley, 1966.

7. *Intrinsic nilpotent approximation,* preprint MIT, LIDS-R-1482, 1985; to appear Acta Appl. Math., 1987.

8. *On the global representation of the solutions of linear differential equations as products of exponentials,* Proc. AMS 15 (1964), 327-334.

9. *On the optimal filtering of diffusion processes,* Z. Wahrsch. und verw. Gebiete 11 (1969), 230-243.

DYNAMICS OF "SLURRIES"

M. Primicerio, Firenze

1. Introduction

   This is a report on a work in progress in which our research group is currently involved, in the framework of a contract awarded by Snamprogetti to the University of Florence (Department of Chemistry and Mathematical Institute) and Eniricerche. The members of the group are: E. Comparini, A. Fasano, P. Paoli, A. Papini, S. Paveri-Vontana, G. Pettini, M. Primicerio, R. Ricci and M. Ughi. The aim of our study is to develop a mathematical model for the rheological behaviour of coal slurries, i.e. of stable concentrated coal-water suspensions (C.W.S.). The oil crisis of the Seventies made it economically interesting to explore the possibilities offered by coal as energy source, trying to reduce its cost and in particular transportation costs which influence the final price up to 60% (of course this is a rough average estimate). Using pipelines to transport suspensions of solid particles in a carrier fluid is a rather old technique. In case of coal, the production of highly concentrated slurries could result in cutting transportation costs, especially because the CWS can be fired directly in burners, without a preliminary dewatering process, if the concentration of coal is large enough.

   The growing interest of industries and of scientists in these fields is testified by the annual international meetings in which a very large number of papers have been presented in different areas such as preparations of CWS, characterization and rheology, hydraulics and pipelining, atomization, combustion, environmental effects, economic analysis and so on. The proceedings of these meetings and in particular [1], [2], [3], [4] are very good sources of information.
───────
"This paper is in final form and no version of it will be submitted for publication elsewhere."

High concentrations (up to about 70% in weight) have been obtained by means of an optimized grain size distribution: approximately bi-modal distributions have proved to give the best results (typical values of the peaks of the distribution function are 10 and 100 microns).

Moreover, suitable fluidizing additives are studied to form a sort of protective film around coal particles so that most of the water is prevented to enter the coal grains and is available as a lubricant to reduce friction forces between adjacent particles. At these concentrations slurry can be burned without drying, although several technical problems have still to be solved for the construction of engines using CWS as fuel.

Concerning stability, static and dynamic stability have to be considered separately. Static stability, against sedimentation, is typically measured as follows: slurry is left for 5 days in a cylindrical container (radius 10 cm, height 50 cm). Then the sample is frozen and two slices of it are taken at the top and at the bottom of the cylinder and the ratio r = concentration at the top/concentration at the bottom is taken as the measure of the static stability. The usual requirement for stability is $r_5 > 0.95$. A stability ratio after 20 days can also be defined. Typical requirement is $r_{20} > 0.90$.

The most delicate point is dynamical stability: experiments show that the rheological properties of CWS may exhibit dramatic changes after the slurry has been kept in motion for several hours in test circuits. Indeed, the pressure gradient needed to maintain a given flow rate in the pipe loop increases during the motion and there maybe a progressive thickening of the slurry up to a level in correspondence of which it is no longer pumpable.

The mechanism of this degradation phenomenon is the main object of our research. The goal is to identify the relevant factors affecting it, to give a possible mathematical description of the process in order to suggest, accordingly, laboratory tests to predict the "degradation time" of the slurry.

2. The rheology of slurries

The rheology of suspensions is a widely studied field (see e.g. [5], [6]). Confining to the case of suspensions of solid spherical particles in a newtonian fluid (water), experiments show that there exists a limit relative concentration c_S beyond which the rheology of the suspension is no longer newtonian.

Roughly speaking, if the concentration c is less than c_S, then the interaction between the particles and the liquid is the main mechanism of momentum transfer, while if $c > c_S$ the interactions among particles play the major role and the behaviour becomes nonlinear.

In the case $c < c_S$ the viscosity η can be defined as the ratio between the (tangential) stress τ and the strain rate σ.

The way η depends on c has been investigated by several Authors, starting from the classical paper by Einstein [7] in which the simple relationship

$$(2.1) \qquad \eta = \eta_0 (1+0.25c)$$

is theoretically established (η_0 is the viscosity of the pure liquid). Law (2.1) is experimentally verified for c up to 10%. For higher concentration (still below c_S) nonlinear relation-ship has been proposed, mainly on empirical basis.
For the case of CWS, we are essentially in the non newtonian range. Most Authors postulate a power-law relationship between τ and $\dot\sigma$.

$$(2.2) \qquad \tau = K \dot\sigma^n,$$

where K and n are to be determined experimentally via curve fitting.

Alternatively, one can define the apparent viscosity η_A to be the ratio between τ and $\dot\sigma$ at a standard reference shear rate $\dot\sigma_0$.

It is immediately found that

(2.3) $\eta_A = K \dot{\sigma}_0^{n-1}.$

Of course, the values of K and n (or of η_A), depend on several factors included in the adopted formulation for the mixture. Moreover they will also depend on the concentration of the slurry and of the additive.

As we said above, our main concern will be their dependence on the history of the slurry.

Here we list some basic experimental facts:

1) No ageing phenomena are relevant. A CWS left at rest for weeks has essentially the same values of K and n as initially.

2) Let the slurry flow in a test circuit. Measuring its rheological properties at different times, one finds that a law like (2.2) can be assumed to hold but that K increases and n decreases as time passes. The same occurs if the CWS is stirred in a blender.

3) The material has a non-fading memory: let the slurry be kept circulating for a time Δ_1, then at rest for a time Δ_0, and again circulating for a time Δ_2. The effect on K and n is the same irrespectively of the value of Δ_0, from 0 to several days.

Two more effects have been pointed out:

a) in many cases, there is a short-time effect due to stirring (or circulation in test loops): an initial reduction of K, before the long-time effect starts and K begins to increase;

b) temperature history can also have relevant effects on the rheology of CWS.

Effect a) seems to be essentially of thixotropic type, although it is difficult to decide whether it presents hysteresis loops or not. In this first approach we will ignore it, also because the phenomenon, when it occurs, has an influence of few percents.

On the contrary, effect b) seems to be very important and it is currently under theoretical and experimental investigation. Here we will assume that the suspension is maintained at constant temperature, according to the experimental conditions in laboratory.

3. The model

We refer to the simplest symmetrical case, i.e. a one-dimensional laminar flow such that

(3.1) $\underline{v} = v(x,t)\ \underline{j}$

($\underline{v}$ is the velocity, $\underline{j}$ the unit vector in the direction of motion and x is a coordinate in a direction normal to $\underline{j}$). We also assume that the slurry is incompressible with density ρ_0.

The conservation of momentum implies (in this geometry the Lagrange and Euler derivatives coincide)

(3.2) $\rho_0\ \dfrac{\partial v}{\partial t} = \dfrac{\partial T}{\partial x}$,

if body forces are negligible. Assuming (2.2) to hold, one has

(3.3) $\rho_0\ \dfrac{\partial v}{\partial t} = \dfrac{\partial}{\partial x}\left[K\left(\dfrac{\partial v}{\partial x}\right)^n\right]$.

Of course, (3.3) reduces to the linear Navier-Stokes equation - written in this simple symmetrical case - when n = 1 and K is a constant or a prescribed function of x and t. When n ≠ 1, even in the case of constant K equation (3.3) may degenerate or become singular at points where $v_x = 0$.

As a first step, it seems reasonable to isolate what we consider to be the main feature of the behaviour of slurries and to focus our attention on the possible law governing the evolution of K, keeping n = 1 throughout. Thus, K and η will obviously coincide.

According to the experimental facts described in Sec. 2,

we assume that the rate of change of K at a given point P and time t is a function of the rate at which energy E is dissipated in the fluid at this same position and time. This rate is given in general by

(3.4) $\qquad \dot{E}(P,t) = 2\ K\ \nabla\ \underline{v} : \nabla\ \underline{v} - (\text{curl } \underline{v})^2,$

where $\nabla\ \underline{v} : \nabla\ \underline{v}$ is the saturated tensor product.

Thus, in the simple geometry we are dealing with it is

(3.5) $\qquad \dot{E}(x,t) = K\ \left[\dfrac{\partial v}{\partial x}\right]^2,$

and our assumption reads

(3.6) $\qquad \dfrac{\partial K}{\partial t} = f\ \left[K\left[\dfrac{\partial v}{\partial x}\right]^2\right],$

where f is a function to be identified. We decided to consider the case of a linear relationship and, denoting by α a constant to be determined, we have the following coupled equations

(3.7) $\qquad \rho_0\ \dfrac{\partial v}{\partial t} = \dfrac{\partial}{\partial x}\ \left[K\dfrac{\partial v}{\partial x}\right],$

(3.8) $\qquad \dfrac{\partial K}{\partial t} = \alpha\ K\ \left[\dfrac{\partial v}{\partial x}\right]^2,$

which are to be supplemented with initial conditions

(3.9) $\qquad v(x,0) = v_0(x),$

(3.10) $\qquad K(x,0) = K_0(x),$

and with boundary conditions, e.g. of the form

(3.11) $\qquad v(0,t) = 0,$

(3.12) $\qquad v(L,t) = F(t).$

Here, v_0 and K_0 are given functions with domain $[0,L]$ and $F(t)$ is defined for $t \geqslant 0$ (we assume for simplicity $v_0(0) = 0$, $v_0(L) = F(0)$). Equations (3.7), (3.8) are assumed to hold in in $(0,L) \times (0,T)$ for a given $T \geq 0$.

Let us begin by investigating the possibility of solutions such that

$$(3.13) \qquad K(x,t) = \chi(x)\psi(t).$$

It is immediately seen that (3.8) and (3.13) imply that v_x does not depend on x. Thus

$$(3.14) \qquad v(x,t) = F(t)\ x/L.$$

Consequently, (3.8) yields

$$(3.15) \qquad K(x,t) = K_0(x)\ \exp[\alpha L^{-2}\int_0^t F^2(\tau)d\tau].$$

Thus, we obtain the following necessary conditions on the data v_0, K_0, F:

$$(3.16) \qquad v_0(x) = F(0)\ x/L,$$

$$(3.17) \qquad K_0'(x) = C\rho_0 x,$$

$$(3.18) \qquad \dot{F}(t) = C\ F(t)\ \exp[\alpha L^{-2}\int_0^t F^2(\tau)d\tau]$$

for some real constant C.

In particular, if $K(x,0)$ is constant, the only possibility of having a solution of the form (3.13) is $F(t) = F_0$ constant and

$$(3.19) \qquad K = K_0\ \exp(\alpha\ F_0^2\ t/L^2).$$

This form can be useful to give a first estimate of the order of magnitude of the parameters. In particular, one finds values of the order 10^{-5} sec^{-1} for the quantity $\alpha\ F_0^2\ /L^2$.

4. A local well-posedness result.

There is no need to outline that proving well-posedness

results (existence, uniqueness, continuous dependence on the data) is a first important check for a mathematical model. In this sense, even a local theorem can be interesting although the final goal remains the proof of global results, i.e. for arbitrary values of T.

Define

$$(4.1) \qquad a[v] \equiv \rho_0^{-1} \, K_0(x) \, \exp[\alpha \int_0^t v_x^2(x,\tau) \, d\tau] \, ,$$

$$(4.2) \qquad b[v] \equiv \rho_0^{-1} \left\{ K_0'(x) + 2\alpha K_0(x) \int_0^t v_x(x,\tau) v_{xx}(x,\tau) \, d\tau \right\} \cdot$$

$$\cdot \exp[\alpha \int_0^t v_x^2(x,\tau) \, d\tau] \, .$$

Thus, solving (3.7) - (3.12) means to find $v(x,t)$ such that

$$(4.3) \qquad v_t = a[v] \, v_{xx} + b[v] \, v_x, \quad \text{in } D_T \equiv (0,L) \times (0,T)$$

and that (3.9), (3.11), (3.12) are fulfilled.

Using the notation of [8], let

$$X(M,N) \equiv \left\{ u \in \bar{C}_{2+\delta}(D_T) : \ |\bar{u}|_{1+\delta} < M, \ |\bar{u}|_{2+\delta} < N \right\}$$

for some $\delta \in (0,1)$, and M, N, positive constants to be chosen later.

For a given $u \in X(M,N)$, solve the linear problem

$$(4.4) \qquad v_t = a[u] \, v_{xx} + b[u] \, v_x,$$

with the same initial and boundary conditions. The existence of $v \in \bar{C}_{2+\delta}$ is a classical result, provided that the data are regular enough, as we will assume.

In the sequel, we will denote by μ any constant depending on M and possibly on the data, while we will use ν if the constant depends on N as well. Finally, λ will denote constants depending on the data only.
We have

(4.5) $a \geqslant 1, \qquad |\bar{a}|_{\delta} \prec \mu,$

(4.6) $|b| \prec \mu + \nu T.$

If we define

(4.7) $V(x,t) = v(x,t) - F(t)\, v_o(x)/v_o(L),$

we have

(4.8) $V_t - a[u]\, V_{xx} = b[u]\, V_x - \left\{ \dot{F}v_o + a[u]\, Fv_o'' + b[u]\, Fv_o' \right\} v_o^{-}$

in D_T. Moreover

(4.9) $V(x,0) = V(0,t) = V(L,t) = 0.$

Using estimate (3.23) page 200 of [8] we obtain

(4.10) $|\bar{V}|_{1+\delta} \prec (\lambda + \mu + \nu T)\, T^{\gamma}$

for some $\gamma \succ 0$.

Now, we apply Schauder estimates to problem (4.8), (4.9) in which the terms on the r.h.s. of (4.8) are considered as given functions.

(4.11) $|\bar{V}|_{2+\delta} \prec \mu \left\{ \lambda + \mu + \nu T^{\gamma} \right\}.$

From (4.10) and (4.11) it is clear that it is possible to choose M,N, and T so that the transformation defined via (4.4), (3.9), (3.11), (3.12) maps X(M,N) into itself.

Now consider u_1 and u_2 and the corresponding solutions v_1 and v_2. It is clear that $w = v_1 - v_2$ solves the equation

(4.12) $w_t = a[u_1]\, w_{xx} + b[u_1]\, w_x + v_{2xx}\Delta a + v_{2x}\Delta b$

with zero data. In (4.12) $\Delta a = a[u_1] - a[u_2]$ and $\Delta b = b[u_1] - b[u_2]$.

Schauder estimates on w give

(4.13) $|\overline{w}|_{2+\delta} \ll \mu(|\overline{v_2}|_{1+\delta}|\overline{\Delta b}|_{\delta}+|\overline{v_2}|_{2+\delta}|\overline{\Delta a}|_{\delta})$.

Now we use (4.10), (4.11) and the fact that

(4.14) $|\overline{\Delta b}|_{\delta} \ll \nu|\overline{w}|_{2+\delta}$,

(4.15) $|\overline{\Delta a}|_{\delta} \ll \nu T^{\hat{\gamma}}|\overline{w}|_{2+\delta}$

to conlude that there exists T_0 such that for $T \ll T_0$ the transformation is a contraction. Hence, applying Banach's fixed point theorem concludes the proof.

5. Discussing time scales.

Let us come back to equations (3.7), (3.8) and give them non dimensional form. Introduce

(5.1) $K = K/K_0$, $y = x/L$, $U = v/v_0$, $\tau = v_0^2 \alpha t/L^2$,

where K_0 and v_0 are reference values for the viscosity and the velocity, respectively: e.g. $K_0 = K(0,0)$, $v_0 = v(L,0) = F(0)$. One gets, in $(0,1) \times (0,\overline{T})$

(5.2) $\epsilon U_\tau = (kU_y)_y$,

(5.3) $k_\tau = k\ U_y^2$,

where

(5.4) $\epsilon = \rho_0 \alpha v_0^2/k_0$.

We have already seen that the order of magnitude of $\alpha\ v_0^2/L^2$ is 10^{-s} sec^{-1}. Typical values for ρ_0, K_0, L in C.G.S. units are such that ϵ can be estimated to be less that 10^{-3}.

In view of this fact it is reasonable to substitute to (5.2) its stationary version

(5.5) $(k\ U_y)_y = 0$.

From a physical point of view, this fact simply means that the time needed by the fluid to reach a stationary velocity profile is much less than the time needed for a significant modification of its viscosity by means of the postulated mechanism. This is what happens in reality and, by the way, if the contrary would be true the actual measure of K with a viscometer could be very difficult.

In the spirit of the argument above, we will consider the problem (5.3), (5.5) with data

(5.6) $k(y,0) = k_0(y)$, $y \in (0,1)$

(5.7) $U(0,t) = 0$, $t \in (0,\overline{T})$

(5.8) $U(1,t) = g(t)$ $t \in (0,\overline{T})$.

We will prove that the problem has a unique solution $\{U(y,\tau), k(y,\tau)\}$ for any $\overline{T}$.

From (5.5) we get

(5.9) $k(y,\tau) \, U_y(y,\tau) = H(\tau)$,

where H is a function to be determined. Therefore, (5.3) and (5.6) give

(5.10) $k^2(y,y) = k_0^2(y) + 2 \int_0^\tau H^2(s) \, ds$,

whence

(5.11) $U_y(y,\tau) = H(\tau) \left[k_0^2(y) + 2 \int_0^\tau H(s) \, ds \right]^{-1/2}$.

Using (5.7) and (5.8) we obtain the equation for the unknown function $H(\tau)$:

(5.12) $g(\tau) = H(\tau) \int_0^1 \left\{ k_0^2(y) + 2 \int_0^\tau H^2(s) \, ds \right\}^{-1/2} dy$.

We can give (5.12) a rather simpler form setting

(5.13) $G(\tau) = 2 \int_0^\tau H^2(s) \, ds$.

Now the new unknown function G solves

(5.14) $\dot{G}(\tau) = 2\ g^2(\tau)\ \left\{\int_0^1 [k_0^2(y) + G(\tau)]^{-1/2}dy\right\}^{-2}.$

We will show that (5.14), with the initial condition
$G(0) = 0$, determines $G(\tau)$ uniquely and independently of $\bar{T}$.
But, before studying (5.14) in the general case, we will
briefly discuss the special situation $k_0(y) = k_0$ constant: in
this case it is

(5.15) $\dot{G}(\tau) = 2g^2(\tau)\ [k_0^2 + G(\tau)].$

Therefore

(5.16) $G(\tau) = k_0^2\ \left\{\exp[2\int_0^T g^2(s)\ ds] - 1\right\},$

i.e.

(5.17) $k(\tau) = k_0\ \exp[\int_0^T g^2(s)\ ds],$

which is a particular case of (3.15). Of course this solution
exists only if g^2 is integrable in any interval, as we will
assume (from a practical point of view the solution is meaningful only until the exponential factor reaches the value 2
or 3).

Let us come back to (5.14). The Cauchy problem for this
O.D.E. is locally solvable since the r.h.s. is Lipschitz continuous w.r.t. G and since g is assumed to be smooth. Moreover, (5.16) provide an upper bound for the solution if the
constant k_0 is an upper bound for $k_0(y)$. Consequently the
solution can be continued over any time interval.

6. Conclusions.

The validation of the model requires the identification
of the constant α for different experimental situations. This
part of the research is still in progress since practical

experiments with the simple geometry we sketched (or in a cylindrical symmetry, which is essentially equivalent) are practically very difficult because they require extremely long times.

The way we are following is to investigate the dependence of $\dot{K}$ on $\dot{E}$ in situations in which the latter quantity is larger (turbulent motion). We remark that the proof of our local existence theorem is valid even if this dependence is nonlinear.
Another point to be clarified is whether or not the form of this dependence is influenced by the Reynolds number.

Finally, it will be necessary to study the case n $\neq$ 1 (non newtonian behaviour) and a law of evolution for n.

Acknowledgements

We express our gratitude to Snamprogetti for stimulating and supporting this research, to the researchers of Eniricerche for the fruitful cooperation and discussions. We are particularly indebted to Drs. Carniani, Ercolani and Donati for their help and particularly cooperative attitude.

References

1   Coal Slurry Combustion and Technology, Proceedings of the 6. International Symposium, Orlando (USA). U.S. Dept. of Energy, Pittsburgh 1984.

2   Coal Slurry Fuels Preparation and Utilization, Proceedings of the 7. International Symposium, New Orleans (USA). U.S. Dept. of Energy, Pittburgh 1985.

3   Coal Slurry Fuels Preparation and Utilization, Proceedings of the 8. International Symposium, Orlando (USA). U.S. Dept. of Energy, Pittsburgh 1986.

4   Coal & Slurry Technology, Proceedings of the 12. International Conference, New Orleans (USA). U.S. Dept. of Energy, Pittsburgh 1987.

5   F.R. Eirich (ed.), Rheology: Theory and Applications,
     Academic Press, New York 1956.

6   R.E. Meyer (ed), Theory of Dispersed Multiphase Flows,
     Academic Press, New York 1983.

7   A. Einstein, Ann. Physik 19 (1906) pp. 289-306.

8   A. Friedman, Partial Differential Equations of Parabolic
     Type, Prentice Hall, Englewood Cliffs, 1964.

Prof. Dr. Mario Primicerio
Università degli Studi di Firenze
Facoltà di Scienze Matematiche Fisiche e Naturale
I - 50134 Firenze

PRACTICAL APPLICATIONS OF THE NOTIONS
OF PROPAGATION OF SINGULARITIES

C. Bardos

I.   Introduction

The purpose of this talk is to show that deep theoretical approach of linear partial differential equations turns out to be a compulsory tool in understanding and solving some practical problems. Most of the material of this talk is joint work with G. Lebeau and J. Rauch and detailed versions or extensions will appear in the near future.

The problem is the stabilization of large structures. It comes from robots or space stations technology. It has attracted the attention of many applied mathematicians working in control theory; it has been suggested to us by J.L. Lions in a meeting in Corsica in 1976, and the writer of this note takes this opportunity for acknowledging the fantastic influence that J.L. Lions had on the development of applied mathematic and in particular on his own understanding of the subject. Related informations and complementary references can be found in the Von Neumann lecture of J.L. Lions [9]. The problem was already proposed to us in 1976 and most of the intuitive ideas to attack it had been described in two papers of J. Rauch and M. Taylor [15], [16]; however by that time it was not possible to handle realistic problems, in fact domain with boundaries, due to the lack of sharp enough results concerning propagation of singularities near the boundary. This gap was filled by the work of Melrose and Sjostrand [11], in 1978 and we use both their results and their method. The reader will notice many connections with the theory of scattering waves by an exterior obstacle as it was initiated by Lax and Phillips [7] and further developed by Ralston [14], Melrose [10] and others (cf. also Morawetz, Strauss and Ralston [12]).

Some results had been already obtained by specialists of control theory like Chen [3] or Lagnese [6]. Their method relies on the use of multipliers it gives shorter proof, better evaluation of the constants which appears in the computation but it is restricted to very special geometry and bring no light for the understanding of the problem. In the present description we restrict ourselves to the genuine wave equation which is the model problem of the theory. However the reader will understand that the method can be applied to the other hyperbolic problems of mathematical physic in particular to Maxwell equation and linear elasticity equations.

II.   Description of the problems

We denote by Ω a bounded domain of R^n with smooth boundary $\partial\Omega$ and by $\Box$ the wave operator.

(1) $\Box u = \partial_t^2 u - \Delta u$

Γ is an non empty open subset of $\partial\Omega$ and T a positive number. Γ is the region where the control device will act and T the time during which it has to act. However the discussion concerns only the definition of these two objects.

The observability is the "applied version" of the Holmgren theorem it address the following question: find Γ and T large enough such that any solution of the problem

(2) $\Box u = 0$ in $\Omega \times [0, T]$, $u\Big|_{\partial\Omega \times [0, T]} = 0$

is identically zero when it satisfies the relation

(3) $\dfrac{\partial u}{\partial n}\Big|_{\Gamma \times [0, T]} = 0$

In (3) $\partial/\partial n$ denotes the derivative of   u according to the direction of the outward normal to $\partial\Omega$.

The problem of the stable observation (s.o.) is to find Γ and T large enough to ensure the existence of a finite constant such that one has

$$(4) \qquad \int_{\Omega} (|\partial_t u|^2 + |\nabla_x u|^2)\,dx \leq C \int_{\Gamma \times [0, T]} \left|\frac{\partial u}{\partial n}\right|^2 d\sigma dt$$

for any solution of (2) with finite energy.

The problem of the exact controlability is to find for any pair of initial data (u_0, u_1) a function $g(u_0, u_1)$ defined on $\Gamma \times [0, T]$ such that the corresponding solution of:

(5) $\Box u = 0$ in $\Omega \times]0, T[$, $u(x, 0) = u_0$, $\partial_t u(x, 0) = u_1$

(6) $u\big|_{(\partial\Omega\backslash\Gamma)\times]0, T[} = 0$

(7) $u\big|_{\Gamma\times]0, T[} = g(u_0, u_1)$

satisfies at time T the relation

(8) $u(x, T) \equiv \partial_t u(x, T) \equiv 0$, $\forall x \in \Omega$

g is the control, it brings back to zero the initial vibration.

To introduce the stabilization, we consider for instance the following problem.

(9) $\Box u = 0$, $u(x, 0) = u_0$, $\partial_t u(x, 0) = u_1$

(10) $\partial_t u = -\lambda(x)\partial_n u$, $\lambda(x) > 0$ on $\Gamma \times R_t^+$

(11) $u\big|_{(\partial\Omega\backslash\Gamma)\times R_t^+} = 0$

Multiplying the equation (9) by $\partial_t u$ gives after integration by part the relation:

(12) $$\frac{1}{2}\,d_t \int_\Omega (|\partial_t u|^2 + |\nabla_x u|^2)\,dx + \int_\Gamma \lambda(x)|\partial_n u|^2 d\sigma = 0$$

Therefore one introduces the space E_0 of pairs

$$f = (u, v) \; \{u \in H^1(\Omega)/u\big|_{\partial\Omega\backslash\Gamma} = 0,\; v \in L^2(\Omega)\}$$

with the corresponding energy norm.

The mapping $(u_0, u_1) \longrightarrow (u(.,t), \partial_t u(.,t))$ is described by a strongly continuous contraction semi group denoted $Z(t)$. Iwasaki [5] has shown that any solution of (9), (10), (11) goes to zero when t goes to infinity however this does not implies the exponential decay of the semi group $Z(t)$ and the problem of the stabilization is to find sufficient conditions concerning Γ such that this exponential decay is true.

III. The broken hamiltonian flow

The broken hamiltonian decsribes the basic geometric structure of the propagation of singularities. We denote by M the manifold $\Omega \times R_t$ and by $\Sigma_b \subset T^*(\partial M) \cup T^*(M)$ the set of points:

$$(p, q) = \{ (x, t), (\tau, \zeta) \}$$

which satisfies one of the two following conditions:

(i) $x \in \Omega$ and $|\tau|^2 = |\zeta|^2$

(ii) $x \in \partial\Omega$ and $|\tau|^2 \geq |\zeta|^2$

   The set of C^∞ rays (cf. Melrose and Sjostrand [11] for details) defines a foliation of Σ_b. It contains the rays propagating in Ω and reflecting on $\partial\Omega$ according to the rule of optical geometry, the gliding and glancing rays and any C^1 combination of the above.

Definition 1: *We say that $\Gamma \times]0, T[$ geometrically controls Ω (in short G C (M)) if for any C^∞ ray $\gamma \subset \Sigma_b$, $\gamma \cap (\Gamma \times]0, T[)$ contains at least one non diffractive point.*

IV. The control of the Dirichlet problem

   We denote by E_s the space of solution u of the wave equation with Dirichlet boundary condition which satisfy:

(13) $u \in H^{s+1} (\Omega \times]0, T[)$

   E_s is isomorphic to the space of initial data in $H^{s+1} (\Omega) \cap H_0^1 (\Omega) \times H^s (\Omega)$ with compatibility conditions (for $s > 0$).

Theorem 1: *If for any C^∞ ray γ, $\gamma \cap \Gamma \times]0, T[$ contains at least one non diffractive point then the following statements are true:*

(i) Any extensive distribution solution of

$$\Box u = 0, \quad u\Big|_{\partial\Omega x]0,T[} = 0$$

with derivative normal $\partial_n u$ in $L^2(\Gamma x]0,T[)$ belongs to E_0 and there exists a finite constant C such that:

(14)
$$||u||_0^2 \leq C \int_{\Gamma x]0,T[} |\partial_n u|^2 d\sigma dt$$

(ii) Any solution corresponding to an initial data in the space E_{-1} can be exactly controlled by at least one distribution $g \in L^2(\Gamma x]0,T[)$, i.e. the solution u of:

(15) $\Box u = 0, \quad u(x,0) = u_0, \quad \partial_t u(x,0) = u_1$

(16)
$$u(x,t)\Big|_{\partial\Omega\backslash\Gamma} = 0, \quad u(x,t)\Big|_{\Gamma x]0,T[} = 0$$

satisfies

(17) $(u(x,t) \equiv 0$ for $t \geq T$

(iii) Any extendible distribution solution of $\Box u = 0$, $u\Big|_{\partial\Omega} = 0$ can be exactly controlled by a distribution g with support in $\Gamma x]0,T[$, i.e. there exists g with support contained in $\Gamma x]0,T[$ such that the solution $v(x,t)$ of the problem:

(18) $\Box v = 0, \quad v(x,0) = \partial_t v(x,0) \equiv 0$

(19)
$$v(x,t)\Big|_{\partial\Omega\backslash\Gamma} = 0, \quad v(x,t)\Big|_{\Gamma x]0,T[} = g(x,t)$$

satisfies the relation:

(20) $(u+v)(x,t) \equiv 0$ for $t \geq T$

Theorem 2: *If any extendible distribution can be exactly controlled then for any C^∞ ray γ we have*:

$$\bar{\gamma} \cap \Gamma x[0,T] \neq \{\emptyset\}$$

The proof of the theorem 2 is easy because, according to the classical result of Hormander [4], one can construct a solution u_γ of $\Box u_\gamma = 0$, $u_\gamma|_{\partial\Omega} = 0$, which is singular everywhere on g and C^∞ on $M\backslash\gamma$. By inspection one sees that it is impossible to bring this solution to zero with a control acting outside γ.

Remark 1. The points (i), (ii), (iii) of the theorem 1 are almost equivalent to the statement $\gamma \cap \Gamma x]0,T[=\{\emptyset\}$ of the theorem 2. All these statements turns out to be completely equivalent in the absence of some pathological situations created by the diffractive points.

The point (ii) is deduced from the point (i) by a duality argument (cf. Lions [9]). This type of argument is very classical in control theory and leads indeed to a constructive proof which can be used for the computation of g. To deduce (iii) from (ii) one uses simply the fact that any distribution is locally of finite order.

Remark 2. With a very simple multiplier one can show that any solution u of finite energy of $\Box u = 0$, $u|_{\partial\Omega} = 0$ satisfies the relation:

$$(21) \quad \int_{\Gamma \times]0, T[} |\partial_n u|^2 d\sigma dt \leq \int_{\partial\Omega \times]0, T[} |\partial_n u|^2 d\sigma dt \leq C \|u\|_0^2$$

The point (i) of the theorem 1 which is the essential of our contribution turns out to be the converse of the relation (2) and shows that:

$$\int_{\Gamma \times]0, T[} |\partial_n u|^2 \, d\sigma dt$$

defines on the space E_0 an equivalent norm.

The proof goes as follow: first we prove a microlocal result near the boundary, second we propagate this result and finally we use some functional analysis; we extend u by zero outside Ω and we denote this function by $\underline{u}$; of course we have:

$$(22) \quad \Box \underline{u} = \partial_n u \otimes \delta_{\partial\Omega}$$

Lemma 1. Let $p \in \partial M$ and U be a neighbourhood of p in $R_x^n \times R_t$ assume that u is an extendible distribution which satisfies

(23) $\Box u = 0$ in $U \cap M$, $u\big|_{\partial\Omega \cap U} = 0$

and

(24) $\partial_n u\big|_{\partial M \cap U} \in L^2 (\partial M \cap U)$

then for any $m = (p, q) \in T^* (R_x^n \times R_t)$ with q non diffractive, we have microlocally $\underline{u} \in H_m^1$.

Proof. First one shows that $\partial_n u \otimes \delta_{\partial\Omega}$ belongs to $H-\frac{1}{2}(U)$, near the characteristic cone, then by integration along any incoming bicharacteristic we obtain

(25) $\underline{u} \in H_m^{\frac{1}{2}}$ (near the characteristic cone)

Finally we introduce a first order operator P with the following properties:

(26) Π_x (ess sup P) $\cap$ M C U

(27) $[\Box, P] = C$

with C microlocally elliptic of order two near M. The computation of $(\underline{u}, [\Box, P]\underline{u})$, taking in account (25) and (24) gives the result. Next using the result of Melrose and Sjostrand [11] we prove the:

Lemma 2. let u be an extendible distribution solution in M of the problem

$$\Box u = 0 \text{ in } M, \quad u\big|_{\partial M} = 0$$

let $\gamma \subset \Sigma_b$ be any C^∞ ray then if there exists one point $m_0 \in \gamma$ such that $\underline{u} \in H_{m_0}^1$ we have $\underline{u} \subset H_m^1$ for any $m \in \gamma$.

With the lemmas 1 and 2 we obtain easily the:

Corollary 1. Assume that $\Gamma \times]0, T[$ has the property G C (M) then any extendible distribution u with the following properties:

(28) $\Box u = 0$ in M, $\nabla_{x', t} u \in L^2 (\partial \Omega \times]0, T[)$

(29) $\partial_n u \in L^2 (\Gamma \times]0, T[)$

belongs to the space $H^1 (\Omega \times]0, T[$.

In (28) $\nabla_{x', t}$ denotes the derivatives along ∂M. Using the classical results (cf. Lions and Magenes [9], or Chazarain and Pirion [2]) concerning the mixed Cauchy problem we reduce the above situation to the case where u = 0 on ∂M and then we use the Lemmas 1 and 2.

Lemma 3. If we assume the condition C G(M) the expression

(30) $$|||u|||_G = \int_{\Gamma \times]0, T[} |\partial_n u|^2 d\sigma dt$$

defines an equivalent norm on the set on extendible distribution of finite energy solution of $\Box u = 0$, $u\big|_{\partial M} = 0$.

Proof. First we prove that this is indeed a norm. In fact

∂_t is an homomorphism of the space N of solutions u of:

(31) $\Box u = 0, \quad u\big|_{\partial M} = 0, \quad \partial_n u\big|_{\Gamma \times]0,T[} = 0$

Therefore for any $u \in N$ we have also $\partial_t u \in N$, this implies that N is both closed and compact in E_0, therefore N is of finite dimension and for any eigenvector w of the automorphism ∂_t we have the relation:

(32) $\lambda^2 w + \Delta w = 0, \quad w\big|_{\partial \Omega} = 0, \quad \partial_n w\big|_\Gamma = 0$

This implies that w is identically zero; therefore we have $N = \{0\}$ and $|||\cdot|||$ defines a norm. Finally one shows (this is not trivial and the hypothesis C G(M) has to be used once more) that with this norm we have a Banach space. With the closed grap theorem the proof of the point (i) of the theorem 1 is complete.

IV. The stabilization

To avoid technical difficulties we consider only the following situation. We assume that $\partial \Omega$ is the union of two disjoints families of connected smooth manifolds $\partial \Omega_i$ and $\partial \Omega_j$ we denote $\Gamma = \bigcup_j \partial \Omega_j$ and we consider the following problem:

(33) $\Box u = 0$ in $\Omega \times R_t^+$, $(u(x,0), \partial_t u(x,0) = (u_0, u_1)$

(34) $u \Big|_{(\partial\Omega \backslash \Gamma) \times R_t^+} = 0$

(35) $\dfrac{\partial u}{\partial t} + \lambda(x) \dfrac{\partial u}{\partial n} = 0$ on $\Gamma \times R_t^+$

where $\lambda(x)$ is a measurable function which satisfies the relations:

(36) $0 < \alpha \leq \lambda(x) \leq \beta < +\infty$ $\forall x \in \Gamma$

and we consider the related semi group $Z(t)$ described in the introduction, acting on the space:

$$E_0^{\Gamma} = \{(u,v) \in H^1(\Omega) \times L^2(\Omega) \mid u\Big|_{\partial\Omega \backslash \Gamma} = 0\}$$

with the usual norm.

Theorem 3: *Assume that for every T>0 there exists a C^∞ ray γ with*

(37) $\gamma \cap \overline{\Gamma} \times [0,T] = \{\emptyset\}$

then:

(i) $\forall \epsilon > 0$ and T>0 there exists a Cauchy data f with energy norm 1 such that one has:

(38) $(1 - \epsilon) \leq ||Z(t)f||$, $\forall t$, $0 \leq t \leq T$

(ii)   The   spectral radius of Z(t) is exactly 1   and   the part of   this spectra   contained in   the circle $|z|=1$ is continuous spectra.

Theorem 4: *We assume the existence of a finite T such that $\Gamma \times]0,T[$ has the property C G(M), then there exists a finite strictly positive α with*

(39) $||Z(t)|| \leq Ce^{-\alpha t}$ $\forall t>0$

The proof of the theorem 3 is done by construction of a   counter example.   Once again one introduce a solution u of   the problem

$$\Box u = 0 \text{ in } M, \quad u\Big|_{\partial M} = 0$$

singular   on γ and smooth in $M\backslash\gamma$.   Let $\rho_\epsilon(t)$ be a   family of   smooth functions converging   to   the Dirac   mass δ_t. We introduce $u_\epsilon = \rho_\epsilon * u / ||\rho_\epsilon * u||_0$ and v_ϵ given by:

(40) $\Box v_\epsilon = 0 \text{ in } \Omega \times R_t^+, \quad v_\epsilon(x,0) = \partial_t v_\epsilon(x,0) \equiv 0$

(41) $\partial_t v_\epsilon + \lambda(x)\partial_n v_\epsilon = \partial_t u_\epsilon + \lambda\partial_n u_\epsilon \text{ on } \Gamma \times R_t^+$

(42) $v_\epsilon\Big|_{\partial\Omega\backslash\Gamma} = 0$

The   counter example is given by $v_\epsilon - u_\epsilon$. This proof is   a   copy   of   the   proof   given   by   M.   Taylor   (never published) of a classical result of J. Ralston [14].

The   proof   of the theorem 4 starts from   the   energy estimate:

(43) $\dfrac{1}{2} \displaystyle\int (|\partial_t u(x,t)|^2 + |\nabla_x u(x,t)|^2 dx$

$$+ \int_0^t \int_\Gamma \lambda(\sigma)\, |\partial_n u(\sigma,s)|^2 d\sigma ds =$$

$$\dfrac{1}{2}\int (|\partial_t u(x,\theta)|^2 + |\nabla_x u(x,0)|^2) dx$$

with the fact that $\partial_t u$ and $\partial_n u$ give control on the tangential derivative of u (∂M is non characteristic for the operator $\square$). Then it uses the corollary 1 and it is obtained by adapting the proof of the theorem 1.

VI. Further comments

The above program has reduced the choice of control or stabilization region to a geometrical problem which indeed is a standard, dynamical systems, problem and which in some case may not be easy. For instance one would like to choice Γ as small as possible. It is clear that Γ should contain at least one vertices for any closed ray reflecting on the boundary; and in two space variables one can show (use the Kolmogorov Arnold Moser Theorem cf. Arnold [1] p. 405) that for any non degenerate closed ray Γ must contain a, non arbitrarly small, neighourhood of these vertices.

Keeping this ideas in mind one may conjecture that the smallest possible Γ would be some Γ_ϵ containing union of one vertices for any closed ray plus a non arbitrary small neighbourhood of this vertices if the ray is non degenerate or an arbitrary small (let say of order ϵ) of these vertices when the ray is degenerate. Then of course the time T_ϵ such that the C G(M) property holds for Γ_ϵ x]0, T_ϵ[will depend on ϵ and in the spirit of Nehoroshev [13] it may be possible to show that T_ϵ must be of the order of $e^{C/\epsilon}$. To the best of my knowledge no result in the direction do exist.

In practice the control region may not satify, for any positive T the C G(M) property and may still be used and give in some case convenient results. This may be due to one of the following facts:

(i) Other phenomena like dissipation of energy are present and the problem is not exactly hyperbolic.

(ii) The initial vibration is smooth enough (does not contains to high frequencies modes) so that the asymptotic analysis (based on high frequency expansion) described in the present work do not apply. In fact using a notion of analytical propagation of singularity (cf. Sjostrand [17] one may show that a necessary condition for exact contrability is that the solution is microlocally analytic near any y such that $\Pi(y) \cap \{ \Gamma x]0, T[\} = \{\emptyset\}$. In some sense this condition may also be sufficient.

References

[1] V.I. Arnold: Mathematical Methods of Classical Mechanics. Springer Verlag 1978.

[2] J. Chazarain and Pirion: Introduction a la Theorie des Equations aux derives partielles lineaires. Gauthier Villars 1981.

[3] G. Chen: Energy Decay Estimates and Exact Boundary Value Controlability for the Wave Equation in a Bounded Domain. J.M.P.A. 58 (9) (1964) 249-274.

[4] L. Hormander: The Analysis of Linear Partial Differential Operators. Springer Verlag.

[5] N. Iwasaki: Local Decay of Solutions for Symmetric Hyperbolic Systems with Dissipative and Coercive Boundary Conditions in Exterior Domains. Pub. RIMS Kyoto U.S. (1969) 193-218.

[6] J. Lagnese: Decay of Solutions of the Wave Equation in a Bounded Region with Boundary Dissipation. J. Diff. Eq. 50 (2) (1983) 163-182.

[7] P. Lax and R. Phillips: Scattering Theory Academic Press (1967).

[8] J.L. Lions and E. Magenes: Non Homegeneous Boundary Value Problems and Applications. Springer Verlag (1972).

[9] J.L. Lions: Exact Controlability, Stabilization and Perturbation for Distributed Systems. Von Neumann Lecture Boston SIAM Metting July 1986.

[10] R. Melrose: Singularities and Energy Decay in Acoustical Scattering. Duke Math. J. 46 (1979) 43-59.

[11] R. Melrose and J. Sjostrand: Singularities of Boundary Value Problems I. Comm. Pure Appl. Math. 31 (1978) 593-617.

[12] C. Morawetz, W. Strauss and J. Raltson: Decay of Solutions of the Wave Equation Outside non Trapping Obstacles. Comm Pure Appl. Math. 30 (1977) 447-508.

[13] N.N. Nehoroshev: The Behavior of Hamiltonian Systems that Are Close to Integrable Ones. Funct. Analysis and its Applications 5: 4 (1971). Uspeki Mat. Nauk 32: 6 (1977).

[14] J. Ralston: Solution of the Wave Equation with Localised Energy. Comm. Pure and Appl. Math. 31 (1969) 807-823.

[15] J. Rauch and M. Taylor: Penetration into Shadow Region and Unique Construction Properties in Hyperbolic Mixed Problems. Indiana University Math. J. 22 (1972) 277-285.

[16] J. Rauch and M. Taylor: Exponential Decay of Solutions to Hyperbolic Equations in Bounded Domains. Indiana University Math. J. 24 (1974) 79-86.

[17] J. Sjostrand: Propagation of Analytic Singularities for Second Order Dirichlet Problems. Comm. P.D.E. 5 (1980) 41-94.

C. Bardos
Center of Applied Mathematics
E.N.S., 45 Rue d'Ulm, Paris 75009, France

COMPLEX VARIABLES IN INDUSTRIAL MATHEMATICS

S.D. Howison, Oxford University

1   Introduction: why x + iy in a three-dimensional world?

Complex analysis has long been regarded both as an elegant subject in its own right, and as a powerful tool in applied mathematics. The subject has inescapable limitations, however, and it might be said that it has been superseded by more modern methods. I aim to show here that, even taking these limitations into account, there is a niche for it in modern mathematics, in this case as applied to industrial problems. The modernity lies perhaps more in the novelty of the problems studied (here, for example, free boundary problems from industry) and in the inter-action with, say, numerical methods than in the techniques themselves.

With this in mind, we shall pass by all the techniques such as transform methods in which complex variables appear indirectly, and consider only direct applications. This means that we are necessarily restricted to two-dimensional problems. Although we do indeed live in a three-dimensional world, many situations are to a good degree of approximation two-dimensional and this is not a severe constraint. Secondly, we can only easily deal with Laplace's equation or with simple Poisson equations; but many useful physical situations are modelled by these equations, so we are not unduly restricted here either.

This paper is in final form, and no version of it will be submitted for publication elsewhere.

What may we hope to achieve using complex analysis? Firstly, a complete solution can sometimes be found; this is always useful in itself - there is no arguing with an explicit solution - but may be more useful still as a test case for a numerical scheme designed to cope with more general problems outside the scope of special techniques. Secondly, we may find a quick way to get qualitative information about the behaviour of solutions of the mathematical model, and this may be an important check on the validity of the model itself. We illustrate these remarks in sections 2 and 3, via a complete solution in section 2 and a useful result about the model in section 3.

2   Viscous film coating [1]

One process used to coat a long cylindrical object with a viscous fluid is to draw the object parallel to its generators vertically through a bath of the fluid (Fig. 1). The fluid is drawn up with the wire, and a long way (i.e. several diameters) above the bath surface, the flow is undirectional, with gravity balancing the shear induced by the motion of the cylinder.

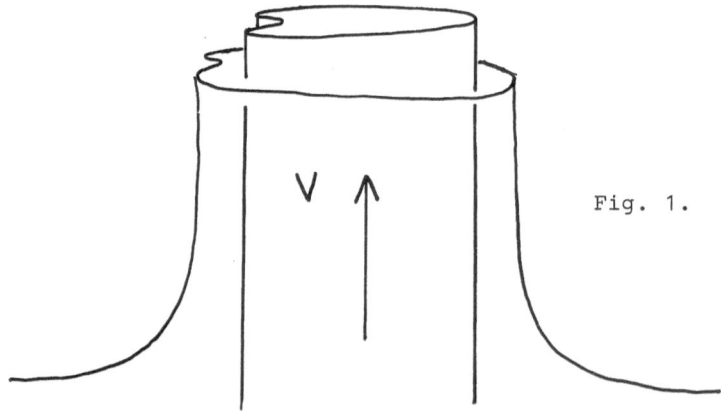

Fig. 1.

An important question is the degree of thinning to be expected in the viscous layer near a sharp corner on the object to be coated. The worst case would be the coating of a plate of zero thickness, and Tuck et al. [1] describe a simple mathematical model of this situation.

If the vertical fluid velocity component is W, far upstream where the flow is undirectional W satisfies

$$\frac{\partial^2 W}{\partial X^2} + \frac{\partial^2 W}{\partial Y^2} = g/\nu$$

in each horizontal cross-section through the plate and fluid; here ν is the kinematic viscosity of the fluid and g the acceleration due to gravity. On the plate (represented by the line $-\infty < X < 0$, $Y = 0$) we set $W = V$, the plate velocity. The fluid-air interface is a free boundary, and the two conditions holding on it are $W = \frac{1}{2}V$, $\partial W/\partial N = 0$ ($\partial/\partial N$ is the outward normal derivative). The second of these conditions is that of zero stress at the boundary, while the first, which is less obvious, corresponds to maximising the flux of fluid, and is justified on stability grounds [1]. We scale the model by writing $(X,Y) = (V\nu/2g)^{\frac{1}{2}}(x,y)$, $W-\frac{1}{2}V = \frac{1}{2}Vu(x,y)$, and then u satisfies

(2.1) $\nabla^2 u = 1$

in the region Ω occupied by fluid (Fig. 2), with

(2.2) $u = 1$

on $-\infty < x < 0$, $y = 0$, and

(2.3) $u = \frac{\partial u}{\partial n} = 0$

on the free boundary Γ.

156

This free boundary problem was solved numerically by Tuck et al. [1], but it can in fact be solved explicitly using techniques from the theory of groundwater movement. We are in particular interested in the minimum film thickness OP and the shape of the free surface.

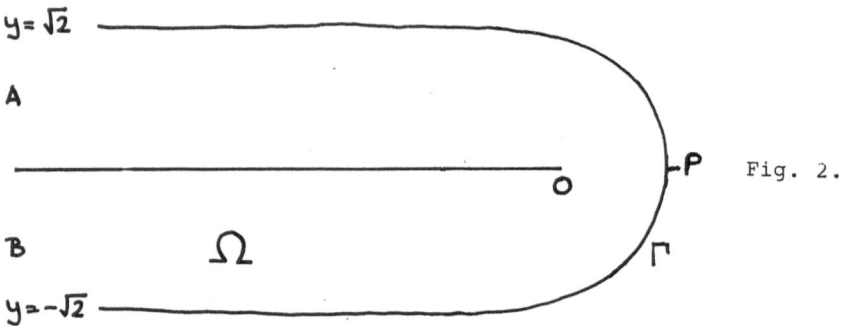

Fig. 2.

We set $\phi = \partial u/\partial x$, and from (2.1)-(2.3) it is easily shown that in Ω

(2.4) $\nabla^2\phi = 0$,

while on the negative x-axis

(2.5) $\phi = 0$,

and on Γ

(2.6) $\phi = 0$, $\partial\phi/\partial n = dy/ds$,

where s is arclength. There is a singularity in ϕ near 0, (otherwise ϕ would be identically zero), and since as $x^2 + y^2 \to 0$, $u \sim 1 + \frac{1}{2}y^2 + O(r^{\frac{1}{2}}\sin\,\theta/2)$, we have

(2.7) $\phi \sim O(r^{-\frac{1}{2}}\sin\,\theta/2)$

as $r \to 0$; here r,θ are polar coordinates centred at 0. Lastly

157

as x → -∞,

(2.8)   φ ~ O(e$^{x\pi/\sqrt{2}}$) .

Now write z = x + iy and introduce w(z) = φ + iψ, where ψ
is the harmonic conjugate of φ. From (2.6) and the Cauchy-
Riemann equations we have that on Γ

(2.9)   φ = 0 ,   ψ = y ,

while from (2.7) w → ∞ as z → 0 in such a way that

(2.10)   z ~ O(1/w^2) ,

and from (2.8) as z → -∞ in Ω,

(2.11)   z ~ ∓ $\frac{\sqrt{2}}{\pi}$ log(w ∓ i√2)

with the + sign in (2.11) corresponding to the part of Ω below
the x-axis.

The problem (2.9)-(2.11) is solved by finding z as a
function of w in the potential plane (Fig. 3).

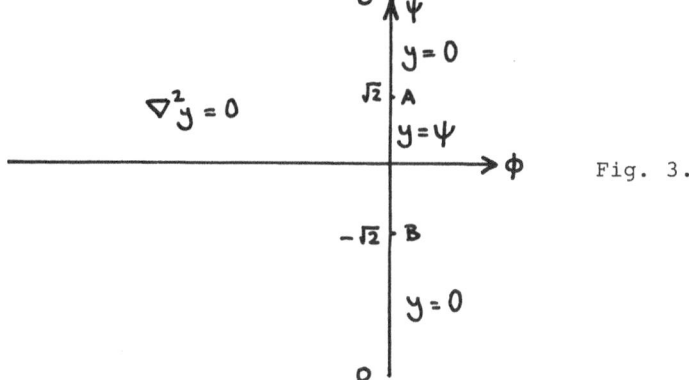

Fig. 3.

In this plane y is a harmonic function of φ and ψ satisfying the conditions shown on the ψ-axis. It is easily verified
that the solution is

(2.12) $z = \frac{2\sqrt{2}}{\pi} - \frac{w}{\pi i} \log(\frac{w + i\sqrt{2}}{w - i\sqrt{2}})$

with the branch cut taken from $-i\sqrt{2}$ to $i\sqrt{2}$. The free surface is given by setting $w = i\psi$ in (2.12), and in particular the distance OP is $2\sqrt{2}/\pi$. The film is thinner by a factor $2/\pi = 0.637...$; Tuck et al. [1] calculated 0.626, but this was apparently an error in transcription and should have read 0.637 [2]. The claim in [1] for 3-figure accuracy for the numerical method is justified in this example.

It is interesting to note that precisely this problem occurs in a completely different physical situation, namely a model of dopant diffusion in semiconductors. The details are beyond the scope of this paper, but one can obtain a free boundary problem for a dopant concentration which, after some transformations, can be reduced to (2.1)-(2.3). This is described in [3], which also contains other examples of the coating problem (the interior and exterior of a right-angled corner) and a different method of solution.

3   Heat and current flow in thermistors

Thermistors are circuit devices made of a ceramic material whose resistivity increases by 5 orders of magnitude as its temperature increases from 100°C to 200°C. They have many applications, and here we focus on their use as circuit breakers. If there is a current surge in a circuit containing a thermistor, the increased Joule heating in the device will cause its resistance to rise with its temperature, thereby cutting the current off. The advantage of using a thermistor for this purpose - compared with, say, a fuse - is that once the surge has gone away the thermistor will cool down and the device can return to normal operation without needing to be replaced or reset.

There are various reasons for making a mathematical model of thermistor operation. One is that the performance characteristics of the device, such as the time taken to switch off for a given current surge, and the maximum leakage current, can be tailored to suit specific requirements by altering the physical characteristics of the device, once the dependence of the performance on the latter is known. Secondly, cracks can occur if the current surges are too high; a mathematical model of heat and current flow is the first stage in a thermoelastic calculation of the stresses generated by a current surge.

A typical thermistor is a cylinder of diameter $R \simeq 5$ mm and thickness $h \simeq 2$ mm. The top and bottom surfaces are covered with a thin conducting layer to which in turn is soldered a wire connecting the thermistor to the external circuit. We shall describe a simple model of the heat and current flow in such a device when voltages $\pm V$ are applied to the upper and lower contacts. We scale distances with h, time with the conduction timescale $\tau = \rho c h^2 / k$ (ρ is density, c specific heat and k thermal conductivity), the electric potential ϕ with V, the temperature T with $\Delta T = T^* - T_a$, where T^* is the 'switching temperature' at which the electrical conductivity starts to fall and T_a the ambient air temperature, and lastly the electrical conductivity $\sigma(T)$ with its value σ_a when the material is at temperature T_a. The temperature scale is chosen so that T_a corresponds to $T = 0$, and $\sigma(T)$ decreases rapidly as T increases beyond 1.

The equations satisfied by T and ϕ inside the thermistor are now

(3.1) $\qquad \nabla \cdot (\sigma(T) \nabla \phi) = 0$

and

(3.2) $\qquad \dfrac{\partial T}{\partial t} = \nabla^2 T + \alpha \sigma(T) |\nabla \phi|^2 \; .$

The final term in (3.2) represents the volumetric heating caused by the current; $\alpha = \sigma_a V^2 / k \Delta T$ is a dimensionless parameter.

In order to make analytical progress we must make some simplifications in these equations and in the boundary conditions (about which we have hitherto said nothing). We shall only consider steady states, reached after a long time (the transient behaviour is in any case complicated by the effects of the external circuit). Also we shall take $\sigma(T)$ to be a step function:

$$\sigma(T) = 1 \qquad \text{for } T < 1 \; ,$$
(3.3)
$$= \delta \ll 1 \qquad \text{for } T > 1 \; .$$

A typical value of δ is about 10^{-5}. This assumption means that there may be a free boundary, on which $T = 1$, separating the 'hot' region $T > 1$ from the 'cold' region $T < 1$.

Turning to boundary conditions, we have

(3.4) $\qquad \phi = \pm 1$

on the top and bottom respectively, and

(3.5) $\qquad \dfrac{\partial \phi}{\partial n} = 0$

on the sides. The temperature boundary condition might in general be modelled by a radiation condition of the form

(3.6) $\partial T/\partial n + \beta T = 0$

on the sides, where β is a dimensionless heat transfer coefficient, but for simplicity we shall let $\beta \to \infty$ and take

(3.7) $T = 0$

on the thermistor boundary. Lastly since from (3.3) and (3.1) ϕ is a harmonic function where $T \neq 1$, we shall use 2-dimensional cartesian coordinates (x,y), $-a < x < a = R/h$, $-1 < y < 1$, in place of cylindrical polar coordinates; we expect that this will only qualitatively change the solution.

If α is small, the whole device will be below $T = 1$. As we increase α a hot spot will appear in the middle, and a reasonable supposition is that as α continues to increase the hot region H will occupy a substantial proportion of the centre of the device as in Fig. 4.

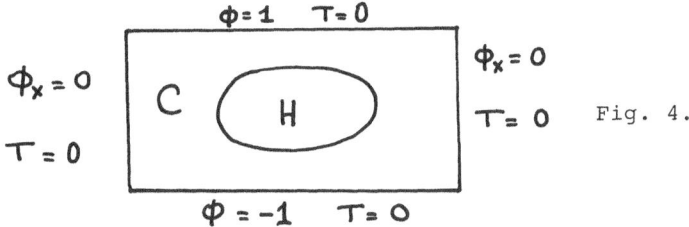

Fig. 4.

We shall now show that this supposition is false.

Since $\delta \ll 1$, we write

(3.3) $T \sim T_0 + \delta T_1 + \cdots$

(3.4) $\phi \sim \phi_0 + \delta \phi_1 + \cdots$

in the hot region H and cold region C. The conditions on Γ, the curve separating H from C, are that

(3.5) $T = 1$, $[\partial T/\partial n]_C^H = 0$

and

(3.6) $[\sigma \, \partial \phi/\partial n]_C^H = 0$.

Substituting (3.3) and (3.4) into (3.1),(3.2),(3.5),(3.6), we find from the lowest order terms that $T_0 \equiv 1$ in H_0, while the problem for T_0 and ϕ_0 in C_0 is

(3.7) $\nabla^2 T_0 = -\alpha |\nabla \phi_0|^2$, $\nabla^2 \phi_0 = 0$

with

(3.8) $T_0 = 1$, $\dfrac{\partial T_0}{\partial n} = 0$, $\dfrac{\partial \phi_0}{\partial n} = 0$

on Γ_0, and

(3.9) $T_0 = 0$, $\phi_0 = \pm 1$ on $y = \pm 1$

(3.10) $T_0 = 0$, $\partial \phi_0/\partial n = 0$ on $x = \pm a$

Let $w_0 = \phi_0 + i\psi_0$ be analytic in C_0, i.e. ψ_0 is the harmonic conjugate of ϕ_0. We notice that equations (3.7) are invariant under conformal mappings (as, indeed, are the full equations (3.1),(3.2)) and this suggests using ϕ_0 and ψ_0 as independent variables. By symmetry we take $\psi_0 = 0$ on Γ and $\psi_0 = \pm Q_0$ (to be found) on $x = \pm a$. Then in the ϕ_0-ψ_0 plane the thermistor boundary corresponds to the boundary of the rectangle $-1 < \phi_0 < 1$, $-Q_0 < \psi_0 < Q_0$, and Γ is mapped onto a slit S_0: $-\hat{\phi}_0 < \phi_0 < \hat{\phi}_0$, $\psi_0 = 0$, for some $\hat{\phi}_0$ also to be found.

Equations (3.7) and (3.8) now become

$$\nabla^2 T_0 = -\alpha ,$$

with

$$T_0 = 1 , \quad \frac{\partial T_0}{\partial n} = 0 \quad \text{on} \quad S_0 ,$$

and the unique solution to these equations is $T_0 \equiv 1 - \frac{\alpha \psi_0^2}{2}$.
However, it is not possible to satisfy (3.9) and (3.10) with
this T_0. We conclude that our original assumption that H_0 was
O(1) was wrong, and we propose instead that H is a thin strip
along part of the x-axis. Specifically let H be the region
$-\delta h(x) < y < \delta h(x)$, $-x_0 < x < x_0$, where h and x_0 are to be
found. Expanding in powers of δ, we find that the lowest order
problem in the cold region is again (3.7), (3.9) and (3.10), but
instead of (3.8) we have

(3.11) $T_0 = 1 , \quad [\partial T_0/\partial y]_{y=0^-}^{y=0^+} = -\alpha [\phi_0]_{y=0^-}^{y=0^+} \frac{\partial \phi_0}{\partial y}$

on $y = 0$, $-x_0 < x < x_0$, with

(3.12) $\partial T_0/\partial y = 0 , \quad \phi_0 = 0$

on $y = 0$, $x_0 < |x| < a$. The second term of (3.11) relates the
heat flux generated in H to the product of current through H and
voltage drop across H; it is obtained by matching with a linear
solution for ϕ in H and using (3.5) and (3.6). Preliminary
numerical work [4] suggests that for α not too large this is a
reasonable problem, but that as α increases, points where $T_0 > 1$
appear in the supposedly cold region. At the moment the
question of how the hot region changes as α increases remains an
interesting unsolved problem. It is at any rate reassuring to

note that in the w_0-plane the boundary of the hot region, which in the physical plane is known apart from its endpoints $x = \pm x_0$, is mapped onto an unknown curve since ψ_0 is no longer known on it (current now passes through the hot region in significant quantities). T_0 now satisfies $\nabla^2 T_0 = -\alpha$ in the w_0-plane, and $T_0 = 1$, $\partial/\partial n\,(T_0 + \alpha\phi_0^2/2) = 0$ on this curve; this is no longer an obviously overdetermined problem.

Acknowledgements

I am indebted to Dr. E.O. Tuck for bringing the coating problem to my attention, likewise to Dr. Miles Drake of STC who brought the thermistor problem to Oxford. I would also like to acknowledge useful discussions on the latter problem with Dr. R. Westbrook, and to thank Shell for financial support.

Postscript

There have been developments in the thermistor problem since I prepared the MS of this paper. Dr. Drake left STC shortly after telling us about the problem, and communication with STC lapsed for some months. This is always dangerous in the early days of a collaborative project with industry: put simply, the mathematicians may, in isolation, solve the wrong problem, through insufficient information and/or the lure of interesting mathematics. To some extent this occurred with the thermistor, and I record it here as a warning appropriate to this volume.

We resumed contact with STC in March 1987, and at the Mathematical Study Group with Industry meeting that year, three facts emerged which, combined, put a completely different complexion on the problem, to the extent that it is not far wrong to say that the ideas in section 3 now only have the status of 'interesting mathematics seeking a physical situation...'. Those facts were:

(a) The dimensionless parameter α is rather large, typically about 600. The discussion in section 3 applies only when α is $O(1)$ so that $T < 1$ outside the thin hot region.

(b) $\sigma(T)$ is not accurately modelled by a step function. A more realistic dependence is

$$\sigma(T) = 1 \qquad 0 < T < 1$$
$$\text{(P1)} \qquad = \exp((1-T)/\varepsilon) \qquad 1 < T < 2$$
$$= \delta \qquad 2 < T$$

where $\varepsilon \sim 10^{-1}$. This (combined with the larger value of α) allows substantially more heating in the hot region $T > 1$ than was previously thought.

(c) The limit $\beta \to \infty$ in equation (3.6) is inappropriate. In fact β is small, about $O(10^{-2})$ on thermistor-air interfaces and $O(10^{-1})$ on thermistor-solder interfaces. This latter models the observed fact that 80% of the heat produced by a thermistor is conducted away down the wires connecting it to the circuit. The small value of β tells us that the thermistor boundary is more nearly insulated than isothermal.

I do not have the space to describe in full the work which has since been carried out on this problem [5]. Briefly, however, numerical work [6] and some simple asymptotics, sketched below, indicate that the temperature profile is quite smooth, and that the whole device operates at a temperature greater than the switching temperature: the 'free boundary' aspect of the problem disappears.

We have to solve (3.1),(3.2), with $\sigma(T)$ now given by (P1) and with (3.4),(3.5) and the full (3.6) as boundary conditions for ϕ and T. We shall take $\alpha \gg 1$, and $\beta \sim \varepsilon \ll 1$ on the solder portion of the boundary $-x_s < x < x_s$, $y = \pm 1$, while $\beta \ll \varepsilon$ on the remaining (air) portions. A consistent asymptotic scheme for the steady problem is then to write

$$T \sim 1 + \varepsilon \, \log(\alpha/\varepsilon) + \varepsilon T_1 + \cdots$$

$$\phi \sim \phi_0 + \cdots$$

where

$$\nabla^2 T_1 + e^{-T_1} |\nabla \phi_0|^2 = 0$$

and

$$\nabla \cdot (e^{-T_1} \nabla \phi_0) = 0$$

with ϕ_0 satisfying (3.4),(3.5), and with

$$\frac{\partial T_1}{\partial n} = \begin{cases} -(\beta/\varepsilon) T_1 & \text{on the solder boundary} \\ \\ 0 & \text{on the air boundary .} \end{cases}$$

Although this problem is only slightly simpler than the full problem, it contains only one dimensionless parameter, $\beta/\varepsilon \sim O(1)$. Assuming that it has a solution, variations in T will be smooth and on a scale $O(\varepsilon)$. This scenario is confirmed by the numerical work mentioned above.

Acknowledgements (bis)

I would like to acknowledge further helpful conversation with Drs E.J. Hinch, A.C. Fowler, R. Casselton (STC components) and H. Macartney (STC).

References

[1]   E.O. Tuck, M. Bentwich and J. van der Hoek, 'The free boundary problem for gravity-driven undirectional viscous flows', IMA J. App. Maths. 30, 191-208, 1983.

[2]   E.O. Tuck, private communication.

[3]   S.D. Howison and J.R. King, 'Explicit solutions to some free boundary problems', submitted to IMA J. App. Maths., 1987.

[4]   R. Westbrook, private communication.

[5]   MSGI report 1987, available from J.R. Ockendon, Mathematical Institute, 24-29 St. Giles, Oxford, OX1 3LB, U.K.

[6]   E.J. Hinch, private communication; also in [5].

S.D. Howison,

Mathematical Institute,

24-29 St. Giles,

Oxford, OX1 3LB,

U.K.

Ellis Cumberbatch, Claremont
MOSFET Modelling

I.   Introduction

The Jet Propulsion Laboratory (JPL) has contracted with
The Claremont Mathematics Clinic* over the past three years
for project work on modelling phenomena in metal-oxide-
semiconductor-field-effect-transistors (MOSFET's).   The work
has provided students and faculty with challenging problems
in a variety of areas (numerical optimization, differential
equations, statistics).   We will review that work here so as
to provide an insight into a burgeoning area of science of
industrial importance whose mathematical problems are
attracting attention from an increasing number of our
colleagues.

II. The Device

Figure 1 indicates the basic physical set up of a
MOSFET device.   There may be 10^5 - 10^6 of these devices on a
single chip, and current technology goals (to reduce L, the
channel length, to sub-micron size) are for higher
densities.   The substrate consists of p-type silicon (doped
with acceptor atoms of density, say, of 10^{14} cm^{-3}) into
which has been diffused/implanted higher concentrations of
donor atoms (of density 10^{19} cm^{-3}) to make two n-type
regions called the source and drain.   On the surface are
then laid insulator (silicon oxide) and metal layers, the
latter acting as contacts.   When a voltage difference, V_D,

*   Using teams of faculty and students, the Clinic has
completed over 80 year-long projects for industry since
1973.   For details of this operation see [2] or [3].

is imposed across the source to drain contacts, a current, I_D, flows. It is required to find the dependence of I_D on V_D, on the gate and substrate voltages V_G, V_B, and on the other physical parameters of the device. The characteristics of I_D are such that the device may be used as a switch or as an amplifier.

The equations governing the flow of current will not be displayed here; a source for these, with a physical emphasis is Sze [21], a mathematical one is Markowich [8]. In the stationary case these equations consist of three Poisson equations for the electrostatic potential $\emptyset$, and the electron and hole quasi-Fermi potentials $\emptyset n$, $\emptyset p$. The source terms in these equations are non-linear functions of the potentials. A simplified version of one of these equations will be given later. For mathematical investigations concerning existence and uniqueness of solutions to these equations, see Markowich [8] and Seidman [18]. The large changes in doping level take place over extremely narrow regions, so much so that the doping may be taken as discontinuous, yielding what is called a p-n junction. When the equations have been appropriately made non-dimensional there occurs a basic small parameter resulting from the high doping level. The mathematical problem is then of singular perturbation character and asymptotic analysis for the p-n junction has been obtained by Please [15] and by Markowich and Ringhofer [9]. Broader questions relating to the singular perturbation character of the problem for the whole device are discussed in [8].

Engineers have obtained somewhat more ad-hoc solutions to these equations; their approach is described in the next section.

III. Current Flow in the Channel

We now describe the various approximations introduced to obtain a formula for the current I_D. Uniform conditions across the width (z direction) are assumed. The current flow under the gate is confined to a narrow layer called the channel. For channel lengths of reasonable size (L of micron size), the channel depth and current density are slowly varying over the channel length compared to their transverse variations, so that the latter may be calculated on a local basis, that is as a one dimensional problem. The equation valid under the gate is therefore equivalent to that related to the MOS diode, (see [21]),

$$(3.1) \qquad \epsilon^2 \, d^2\emptyset/dx^2 = (n_0/p_0) \, (e^\emptyset - 1) - e^{-\emptyset} + 1,$$
$$\text{where } \epsilon^2 = kT\epsilon_s/(p_0 q^2 d^2).$$

Here $\emptyset$ is the electrostatic potential measured in units of $q/(kT)$, and x is scaled in units of d, a representative device dimension. Additionally, k is Boltzmann's constant, T the temperature, ϵ_s the permittivity of the semiconductor, q the electron charge, and n_0, p_0 are the densities of electrons and holes in the semiconductor substrate. The parameter ϵ^2 is small (typically of order 10^{-7}); the singular perturbation character of the equation is evident, and indicates that changes in potential will occur across narrow layers (the channel).

Integration of (3.1) proceeds conventionally, and leads to the inverse relation

$$(3.2) \qquad x = \int^{\emptyset} d\emptyset/F$$

where F is a square-root of terms containing exponentials of $\emptyset$. Of interest here is the relationship of the $x = 0$ values of $\emptyset$ and $d\emptyset/dx$, as the latter field may be viewed as due to a surface space charge, and that together with $\emptyset$ are related to the potential applied to the gate. The extension of (3.1) to the calculation for MOSFET current flow along the channel involves the assumption that $\emptyset$ depends on y only through its dependence on $\emptyset n(y)$, the electron quasi-Fermi potential, where

$$(3.3) \qquad n = n_i e^{\emptyset - \emptyset n}$$

is the electron density replacing $n_o e^{\emptyset}$ on the right hand side of (3.1). The density of current flowing in the y direction is given by

$$(3.4) \qquad j = -q\mu_n n \, d\emptyset n/dy$$

where μ_n is electron carrier mobility (taken to be constant). Integration of (3.4), first in the x direction and then in the y direction, yields

$$(3.5) \qquad I_D = - q\mu_n/L \int_0^{V_D} \int_{\emptyset}^{\emptyset F} e^{\emptyset - \emptyset n}/F \; d\emptyset d\emptyset n$$

where $\emptyset F$ is the bulk Fermi potential, and where F is a nasty non-linear function of $\emptyset$ and $\emptyset n$, and other parameters. This formula for I_D was established over twenty years ago by Pao and Sah [12]. Numerical integration of the double integral

has shown it to be accurate compared with experimental data,
for parameter values chosen appropriately. Various
approximations to (3.5) have also been produced; some are
based on alternative physics, some on mathematical
approximations to F. (See [1], [6], [13], [22]).

IV. Clinic Project #1

In order to use (3.5), or one of its derivative
formulae, for accurate current/voltage characteristics as
design information in integrated circuits employing
arrangements of MOSFET devices, it is necessary to know the
physical parameter values to some accuracy. Measurements of
some of these are difficult due to the small size of the
device; also the variabilities of fabrication mean that
transferability of data is suspect. It has become necessary
to solve an inverse problem: to infer parameter values from
measured data. That is, given a formula such as (3.5) or
one of its equivalents,

(4.1) $I_D = G (V_D, V_G, V_B;$ parameters),

and a set of data (I_D, V_D, V_G, V_B), a "best" set of
parameter values is desired which fits the formula to the
data. There were 7 or 9 parameters depending on the model
chosen.

The project was developed in the summer of 1984
resulting in an excellent survey of the mathematical
background of MOSFET physics, and a comprehensive plan for
the numerical work. (Morris and Everson, [10]). The
numerical schemes were run, tested, and compared with the

data during the 1984-5 academic year. The results are available in the Clinic Report, [17]. A portion of the work has been written for publication [5].

The scheme for parameter extraction was a variation of standard routines. The error norm was the sum of squares of the differences between predicted and measured values of I_D. The non-linear optimization procedure to minimize the error was carried out in two stages. The first stage employed a constrained Gauss-Newton method to search the minimum within a specified volume of parameter space, starting from a distribution of initial guesses to try to ensure a global result. The error tolerance was set large enough to avoid interminable computing. The best result was then subjected to a steepest descent (Levenberg-Marquardt) algorithm which refined it to a higher accuracy. Both stages used standard IMSL subroutines. A measure of the fitness of a particular extracted optimum set of parameter values was taken to be the root-mean-square (RMS) of extracted values against experimental ones. RMS values below 10% were considered reasonable, and to indicate that the model (4.1) under scrutiny was accurate at the given voltages. However, it was found that the system was insensitive to two of the parameters. Moreover, the RMS values increased substantially at larger values of V_B and smaller values of V_D, indicating the non-validity of the model at these voltages.

These results were extremely useful to JPL engineers in their analysis of MOSFET characteristics in VLSI design, and

the parameter extraction routines designed by the Clinic
have been transferred for use at JPL.

V.   Clinic Project #2

The goal of the 1985-6 Clinic for JPL was to obtain a
more complete two-dimensional theory of current flow under
the gate and to improve the formula for I_D, (3.5), from that
analysis. The one-dimensional analysis outlined in Section
III becomes less accurate for short-channel MOSFET's, as the
results of the parameter extraction work confirmed.   The
Clinic team embarked on a careful study of the various 1-D
models as preparation, and in doing this observed a number
of omissions and problem areas. With this information the
team was able to construct a more satisfactory 1-D model,
both physically and mathematically, able to provide the
drain current formula in regimes not previously accessible.
When incorporated into the parameter extraction routines
this new model provided acceptable results over larger
ranges of gate and base voltages. The goal of generating a
two-dimensional model was not achieved, and this remains an
open problem.

Additionally, current flow in regions under the source
and drain contacts were examined. Current encounters
resistance as it crosses the metal/semiconductor interface
(contact resistance), and as it is bent around and
compressed into the narrow channel (crowding resistance).
These parasitic resistances become relatively more important
as the channel length is reduced, and their inclusion in
design rules is mandated. Also the earlier formulae for

drain current assumed that the source and drain voltages
apply at the relevant ends of the channel, clearly an error
if the source and drain regions contribute significant
resistance. The first modelling for these regions was again
one-dimensional, substituting series and parallel resistors
(the latter under the contact) for the distributed region.
This is known as the transmission line model (TLM), [11].
The Clinic considered two problems. The first examined
current flow under the contact extending the TLM to include
depth effect. The approriate boundary value problem is
shown in Figure 2. Current, uniformly distributed, flows
along the strip from the right and exits across the contact,
$0 < y < r$. The boundary condition on the contact is a
statement of Ohm's law, given that the top of the contact is
at zero potential. The constant $\lambda = (\rho_c \, o)^{-1}$, where ρ_c is
the contact resistivity and o is the strip conductivity. It
can be shown that the function minimizing the integral

$$(5.1) \qquad J(u) = \int_{-\infty}^{\infty} \int_{0}^{d} |\nabla u|^2 \, dxdy \; + \lambda \int_{0}^{r} u^2(0,y) dy$$

provides a solution to the boundary value problem. The TLM
solution is obtained by assuming that the trial function u
is x- independent. A better approximation is obtained by
taking

$$(5.2) \qquad u = m(y) + n(y) \, (1 - x/d)^2$$

as the equipotential surfaces may have a parabolic shape in
the contact region. The differential equations for m and n
resulting from the Euler-Lagrange conditions for the minimum
of J are straightforward. However, the algebra became

lengthy and solutions were obtained only in the limit of d small. These indicate that the dominant part of the current flow exists across the contact in a region of width $\sqrt{d}$ close to the contact at $y = r$. (Note: an alternative solution, based on a singular perturbation approach for small d, has been obtained separately, [14]). The second problem for source/drain resistance taken up by this Clinic was the crowding of current lines into the channel. This was modelled as current flow from a narrow to a wide strip. A solution using the Schwarz-Christoffel transformation is available in Smythe [20]. Graphs of resistance for various geometries were supplied. The results of this section are available in [16].

VI. Clinic Project #3

The work on contact resistance was continued into 1986-7. Source and drain contacts do not span the width of the transistor as pictured in Figure 1; instead they are separated from the edges by small distances (flanges) as shown in Figure 3 (plan view). In order to reduce the extra flange resistance (caused by current lines bending to surround the contact) of the standard design, Figure 3(a), the "dog-bone" design of Figure 3(b) was introduced. Here the contact is as wide as the bulk width, with extra flanges added to allow for misalignments. Figure 3(c) indicates a design where a tab is added to attach measurement contacts. The charge from JPL was to find a formula for source/drain resistance for such geometries and to provide formulae for the "flange effect." This effect becomes more important when the flanges are not equal; that is when the contact is misaligned. This occurs due to fabrication variations. The Clinic has constructed a model, extending the two-dimensional work in [14]. This is available in [4].

The main emphasis in 1986-7 was a statistical analysis of the variations of source/drain resistance, [19]. In addition to misalignments, contact resistance variations may be due to imperfect removal of the insulting material at contact cut-outs and to metallurgical inhomogeneities which seem to cause spatial variation over the wafer. Lieneweg and Hannaman of JPL have developed a test structure to measure the resistance of individual contacts, [7]. This data (on the variation of conductance) was analyzed by means of regression surfaces, and it was found that quadratic surfaces fairly well represented the systematic part of the variation over the wafer. The random part was approximated by a normal distribution and used to estimate the contact yield, that is the percentage of working chips.

Acknowledgement

The author would like to thank Cesar Pina, Martin Buehler and Udo Lieneweg of the JPL VLSI Technology group for their support and encouragement of the Clinic teams during the years of these projects. These were funded by the Defense Advanced Research Projects Agency through the National Aeronautics and Space Administration. This article was written whilst the author had support from the Air Force Office of Scientific Research, Grant 87-0222.

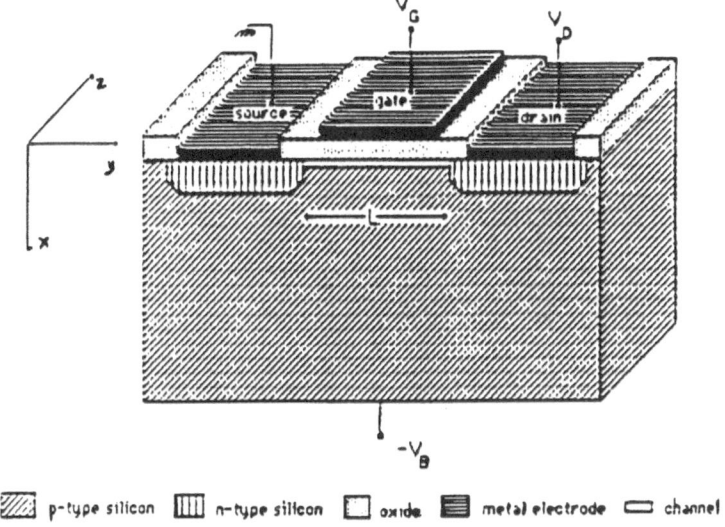

Figure 1: MOSFET schematic

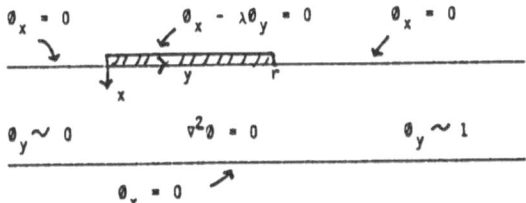

Figure 2: Boundary-value problem

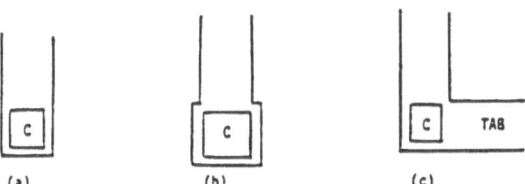

Figure 3: Contact geometries (plan view)

References

[1] Brews, J.: A Charge-Sheet Model of the MOSFET. Solid-State Electronics 21 (1978) 345 - 355

[2] Borrelli, R. L.; Spanier, J.: The Mathematics Clinic: A Review of Its First Decade. The UMAP Journal 4 (1983) No. 1 29 - 48

[3] Cumberbatch, E.: The Claremont Colleges' Mathematics Clinic. Article in Teaching and Applying Mathematical Modelling (ed. J. S. Berry et al) E. Horwood Ltd. (1984)

[4] Cumberbatch, E.; Fang, W.: Three Dimensional Modelling for Contact Resistance of Current Flow into a Source/Drain Region. Claremont Graduate School Math Clinic Report (1987)

[5] Gribben, R.; Martelli, M.: Optimal Parameter Extraction for the Brews Charge-Sheet MOSFET Model. To be published in Mathematical Engineering in Industry

[6] Ihantola, H. K.; Moll, J. L.: Design Theory of a Surface Field-Effect Transistor. Solid-State Electronics 7 (1964) 423 - 430

[7] Lieneweg, U.; Hannaman, D.: Yield Analysis of Interfacial Contact Resistance Measurements on Cross-Contact-Chains. VLSI Workshop on Test Structures. (1986)

[8] Markowich, P. A.: The Stationary Semiconductor Device Equations. Springer-Verlag (1986)

[9] Markowich, P. A.; Ringhofer, C. A.: A Singularly Perturbed Boundary Value Problem Modelling a Semiconductor Device. SIAM J. Appl. Math. 44 No. 2 (1984) 231 - 256

[10] Morris, H.; Everson, R.: The Modelling of Short Channel MOS Devices for Use in VLSI. Claremont Graduate School Clinic Report (1984)

[11] Murrmann, H; Widmann. D.: Messung Des Übergangswiderstandes Zwischen Metall Und Diffusionsschicht in Si-Planarelementen. Solid-State Electronics 12 (1969) 879 - 886

[12] Pao, H. C.; Sah, C. T.: Effects of Diffusion Current on Characteristics of Metal-Oxide (insulator)-Semiconductor Transistors. Solid-State Electronics 9 (1966) 927 - 937

[13] Pierret, R. F.; Shields, J. A.: Simplified Long-Channel MOSFET Theory. Solid-State Electronics <u>26</u> (1983) 143-147

[14] Pimbley, J. M.; Cumberbatch, E.; Hagan P. S.: Analytical Treatment of MOSFET Source-Drain Resistance. IEEE Transactions on Electron Devices. <u>ED-34</u>, No. 4 (1987) 834 - 838

[15] Please, C. P.: An Analysis of Semiconductor P-N Junctions. IMA J of Appl. Math. <u>28</u> (1982) 301 - 318

[16] Ruddock, G., et al: Modelling Short Channel MOSFETS for Use in VLSI. Claremont Graduate School Clinic Report (1986)

[17] Rykken, C., et al: Parameter Extraction and Transistor Models. Claremont Graduate School Clinic Report (1985)

[18] Seidman, T. I.: Steady State Solutions of Diffusion-Reaction Systems with Electrostatic Convection. Nonlinear Analysis <u>4</u>, No. 3 (1980) 623 - 637

[19] Smith, R., et al: Analysis of Resistances in the Cross Contact Chain. Claremont Graduate School Clinic Report (1987)

[20] Smythe, W. R.: <u>Static and Dynamic Electricity</u>, 3rd ed. McGraw-Hill (1968)

[21] Sze, S. M.: <u>Physics of Semiconductor Devices</u>, 2nd ed. J. Wiley (1981)

[22] Van De Wiele, F.: A Long-Channel MOSFET Model. Solid State Electronics <u>22</u> (1979) 991 - 997

Ellis Cumberbatch
Department of Mathematics
The Claremont Graduate School
Claremont, CA 91711

REACTION-DIFFUSION MODELS FOR THE SPREAD OF A CLASS OF INFECTIOUS DISEASES

V. Capasso

Summary: An outline of the results obtained in the various phases of a
research project to define optimal control policies for the eradication
of a class of man environment man diseases is given. The main mathematical
model is based on a reaction diffusion system involving a linear parabolic
equation and a non linear ordinary differential equation coupled via an
integral boundary feedback mechanism.

A threshold theorem is discussed which gives necessary and sufficient
conditions for the extinction of the epidemic. Problems of identification
of parameters are presented and an optimal control problem has been faced
in which the cost function opposes the cost of the epidemic to the cost
of the sanitation program.

1.   Introduction

In the present paper an outline of the work performed by a CNR

(National Research Council of Italy) research unit is presented, concerning

the problem of the control of infectious diseases whose mechanism of spread

is mainly due to an interaction of the human population with a polluted

environment. Fecal-oral transmitted diseases such as typhoid fever,

infectious hepatitis, cholera, etc. belong to this class that we shall

call in general man-environment-man diseases (MEM). For an update class-

ification of infectious diseases refer to [1].

Based on the statistical analysis of data on typhoid fever and

infectious hepatitis in the city of Bari (located on the Adriatic coast

of Southern Italy) [6, 14] a first elementary model had been proposed

in [9] described by the following ODE's:

"This paper is in final form and no version of it will be submitted for
publication".

$$\frac{dz_1}{dt}(t) = -a_{11} z_1(t) + a_{12} z_2(t)$$

(1.1)

$$\frac{dz_2}{dt}(t) = -a_{22} z_2(t) + g(z_1(t))$$

for $t > 0$, subject to suitable initial conditions.

Here $z_1(t)$ denotes the (average) concentration of the infectious agent in the environment at time $t \geq 0$; $z_2(t)$ denotes the infective human population at time t; $\frac{1}{a_{11}}$ is the mean lifetime of the agent in the environment; $\frac{1}{a_{22}}$ is the mean infectious period of the human infectives; a_{12} is the multiplication parameter of the infectious agent due to the human population, and finally $g(z_1)$ is the "force of infection" of the human population due to the agent.

It will be assumed in the sequel that $g: \mathbb{R}_+ \to \mathbb{R}_+$ satisfies the following assumptions

(H) (i) $g(0) = 0$

(ii) g is increasing

(iii) g is twice continuously differentiable on $\mathbb{R}_+$ with $0 < g'(0)$
      and $g''(z) < 0$ on $\mathbb{R}_+ - \{0\}$.

Other relevant epidemic phenomena can be modelled as in (1.1), such as schistosomiasis, helminthic infections, etc. (see e.g. [12]) with suitable modifications. Actually the whole class of MEM diseases, when ignoring spatial and/or temporal structure at the system, may be modelled mathematically by ODE systems of the form

$$\frac{dz_1}{dt}(t) = f_1(z_1(t), z_2(t))$$

(1.2)

$$\frac{dz_2}{dt}(t) = f_2(z_1(t), z_2(t))$$

where $\underline{f}(\underline{z}) := (f_1(z_1,z_2), f_2(z_1,z_2))'$ is a quasimonotone nondecreasing function of $\underline{z} := (z_1,z_2)'$; this means that $f_1(z_1,\cdot)$ is a nondecreasing function of z_2 and $f_2(\cdot,z_2)$ is a nondecreasing function of z_1. This expresses in mathematical terms the positive feedback of the environment on the epidemic.

Classically in the literature [3] epidemiologists have been interested in defining a "threshold" parameter which discriminates situations in which the epidemic, however started, tends to extinction and situations in which it tends to some endemic, possibly non trivial, state. Along these lines in [9] it was shown that system (1.1) actually admits such a parameter, and it is given by

$$(1.3) \qquad \theta : = \frac{g'_+(0) \, a_{12}}{a_{11} \, a_{22}} .$$

It is such that if we denote by $\mathbb{K} := \mathbb{R}_+ \times \mathbb{R}_+$ the positive cone in $\mathbb{R}^2$ the following "threshold theorem" holds.

Theorem 1.1. - Under the assumptions (H) on g,

(a) If $0 < \theta < 1$ then system (1.1) admits only the trivial equilibrium solution $0 := (0,0)$; in the positive quadrant $\mathbb{K}$. It is there globally asymptotically stable.

(b) If $\theta > 1$ then two equilibrium solutions exist in $\mathbb{K}$ for system (1.1). They are given by the origin 0 and by the only nontrivial solution $z^* := (z^*_1, z^*_2)'$ $\mathbb{K}^* := \mathbb{K} - \{0\}$ of the following system

$$(1.4) \qquad \begin{aligned} - a_{11} \, z_1 + a_{12} \, z_2 &= 0 \\[2mm] - a_{22} \, z_2 + g(z_1) &= 0 \end{aligned}$$

In this case, 0 is unstable, while z^* is g.a.s. in $\mathbb{K}^*$.

We may like to stress here that Theorem 1.1 strongly depends on the strict concavity of g assumed in (H)(iii). In fact if we consider an S-shaped g, defined for example as

$$(1.5) \qquad g(z_1) = a_{12} \, \frac{z_1^2}{\alpha^2 + z_1^2}$$

then the parameter θ given in (1.3) does not clearly discriminate between a situation of extinction of the epidemic and the existence of a nontrivial endemic state. Fig. 1.1 illustrates the different behaviours of the solutions in this case.

Anyway θ gives us important informations, more in the case of a strictly concave function g.

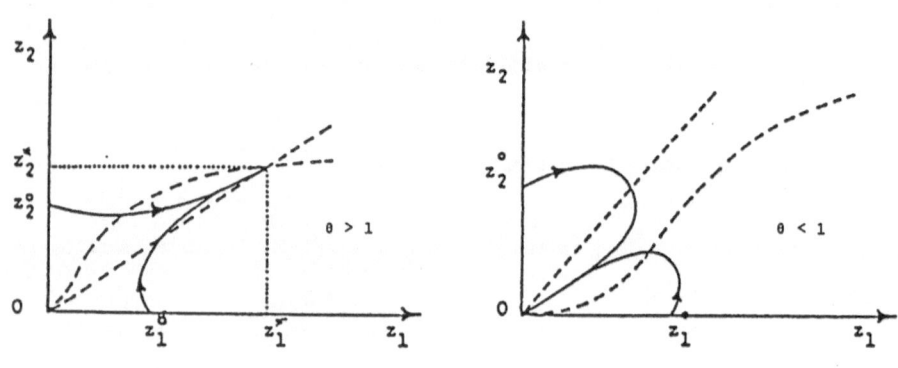

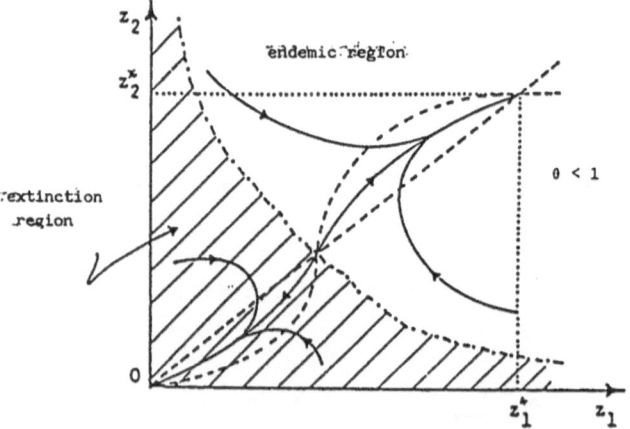

Fig. 1.1.- Phase plane portraits of the solutions for model (1.1)+(1.5). Dotted lines show the isoclines.

2. Reaction-diffusion models

Given the importance of a threshold theorem such as Theorem 1.1, our analysis has gone further on system (1.1) trying to show how the threshold parameter is modified for more complex models which take into account possible space and/or time heterogeneities of the system as shown by the statistical analysis.

Two kinds of models have been proposed for the spatial spread of the epidemic.

Model 1. - This model, proposed in [5|, takes into account the possibility that reservoirs of the infectious agent are present in the habitat, to which all the multiplied agent is sent and from which it diffuses to the whole habitat. This model is more suitable for schistosomiasis as discussed in [4], for which the reservoirs are ponds, lakes or artificial water reservoirs.

Mathematically the model is described by

$$\frac{\partial}{\partial t} u_1(x;t) = A_1(u_1) - a_{11}u_1(x;t) + a_{12}\int_\Omega k(x,x')u_2(x',t)dx'$$

(2.1)

$$\frac{\partial}{\partial t} u_2(x;t) = A_2(u_2) - a_{22} u_2(x;t) + g(u_1(x;t))$$

with standard boundary conditions.

Model 2. - This model, proposed in [7], applies to an habitat along the sea shore and takes into account the fact that the infectious agent when multiplied by the infective human population is sent to the sea through an untreated sewage. A feedback diffusion of the infectious agent to the habitat occurs when the human population uses to eat raw sea food which restarts the epidemic process. This model clearly applies to typhoid fever and infectious hepatitis for towns along the European Mediterranean coasts.

Mathematically the model is described by

$$\frac{\partial}{\partial t} u_1(x;t) = A_1(u_1) - a_{11} u_1(x;t)$$

(2.2)

$$\frac{\partial}{\partial t} u_2(x;t) = A_2(u_2) - a_{22} u_2(x;t) + g(u_1(x;t))$$

with boundary conditions

(2.3a) $$B_1(u_1) = \int_\Omega k(x,x') \ u_2(x';t) dx'$$

(2.3b) $$B_2(u_2) = 0$$

For both models Ω denotes the habitat, A_1 and A_2 denote mathematical operators which describe transport and/or diffusion.

The kernel $k(x,x')$ describes the mechanism of transfer of the infectious agent generated by the human population at $x' \in \Omega$, to the point $x' \in \Omega$; in Model 2 the point $x \in \partial\Omega$, the boundary of Ω. Clearly the boundary $\partial\Omega$ may be split into two disjoint parts: Γ_1, denoting the sea shore at which $k(x;\cdot) \geq 0$ and its complement $\Gamma_2 = \partial\Omega - \Gamma_1$ at which $K(x,\cdot)$ may be taken identically zero.

The quantities u_1 and u_2 now denote spatial densities of the interacting populations; i.e.

(2.4) $$z_i(t) = \int_\Omega u_i(x;t) dx \quad , \quad i = 1,2.$$

Operators B_i denote standard boundary operators such as

(2.5) $$B_i(u_i)(x) = \frac{\partial}{\partial\nu} u_i(x) + \alpha_i(x) u_i(x), \quad x \in \partial\Omega.$$

We shall refer to Model 2 with greater details since it looks more realistic with respect to the kind of epidemic that we wish to control; secondly because it shows interesting mathematical features.

This model had been proposed for the first time in [7] with $A_i \equiv \Delta$ the classical Laplace operator which describes random dispersal; and $B_2 = \dfrac{\partial}{\partial\nu}$ to describe complete isolation of the human population. On the other hand, since the random dispersal of the infectious agent is more likely than that of the human population, the same model has been analyzed in [8] with $A_1 = \Delta$ and $A_2 \equiv 0$.

In that paper existence and uniqueness of the solution of the evolution problem related to system (2.2)-(2.3) are shown and conditions are given so that the solution operator associated with the system generates a strongly continuous semigroup of nonlinear operators on the space of $L^1(\Omega) \times L^1(\Omega)$ as well as of $C(\bar{\Omega}) \times C(\bar{\Omega})$ functions.

In fact it can be shown that the solution $\{u(t) = (u_2(t))'$, $t \in \mathbb{R}_+$ of system (2.2), (2.3) subject to the initial conditions $u_1(0;x) = u_1^o(x)$, $u_2(0;x) = u_2^o(x)$, $x \in \Omega$, is classical in $(0, +\infty)$ under suitable regularity assumptions on the data; this means that

$$u \quad (C^{1,2}((0,+\infty) \times \Omega, \mathbb{R}) \quad C^{0,1}((0,+\infty) \times \bar{\Omega}, \mathbb{R})) \times C^{1,0}((0,+\infty) \times \bar{\Omega}, \mathbb{R}).$$

In order to study the asymptotic behaviour of system (2.2),(2.3) we need to introduce the ordered Bonach space $X := C(\bar{\Omega}) \times C(\bar{\Omega})$ with supnorm and the partial order induced by the positive cone $X_+ := \{u \in X, u = (u_1, u_2)' \mid u_i(x) \geq 0, x \in \Omega, i=1,2\}$. Under natural assumptions on the parameters of the system and conditions (H) on the function g, it has been shown in [8] that the nonlinear evaluation semigroup $\{U(t), t \in \mathbb{R}_+\}$ is strictly monotone on X_+. By this we mean that (for classical solutions)

(2.6)   If u^o, $v^o \in X_+$, $u^o \leq v^o$, $u^o \neq v^o$

then $U(t)u^o \ll U(t)v^o$, for $t > 0$.

In particular we have the strict positivity

(2.7)   If $0 \neq u^o \quad X_+$, then $0 \ll U(t)u^o$, for $t > 0$.

Another important result is related to the concavity of the solution operator U(t) on X_+.

(2.8)   For any $v° \in X_+$, $v° \neq 0$, $\sigma \in (0,1)$, and $t > 0$ there exists a constant $\alpha = \alpha(v°,t,\sigma) > 0$ such that

U(t) $\sigma v° \geq (1+\alpha)$ $\sigma U(t)v°$.

Consider now the linear operator obtained by linearizing the right hand side of system (2.2), (2.3) at the origin.

(2.9a) $Au := (A_1 u , A_2 u)'$, $u=(u_1,u_2)'$

with

$A_1 u := \Delta u_1 - a_{11}u_1$,   in Ω

$B_1 u_1 = Hu_2$,   in $\partial\Omega$

(2.9b) $A_2 u := a_{21}u_1 - a_{22}u_2$,   in $\bar{\Omega}$.

It can be shown [10] that the operator A admits a dominant simple real eigenvalue $\lambda_1 > -a_{22}$ (for any $\lambda \in \sigma(A)$: Re $\lambda \leq \lambda_1$) in X; the associated eigenvector ϕ X can be chosen to be strongly positive, i.e. $\phi \gg 0$ in $\bar{\Omega}$, and with norm $\| \phi \|_X = 1$.

These results allow us to prove the extension of the "threshold" Theorem 1.1 to the reaction diffusion model (2.2), (2.3).

Theorem 2.1   - [10].Under the "basic" assumptions on the parameters, if the dominant eigenvalue λ_1 of A is negative, then the trivial solution is g.a.s. in X_+. If $\lambda_1 > 0$ it is unstable.

An explicit estimate of the sign of λ_1 can be given for the particular case in which $\alpha_i(x) \equiv 0$, i=1,2 in the boundary operators.

Corollary 2.1. - [10]. Under the above assumptions, if furthermore   in (2.3), $\alpha_i(x) = 0$, i=1,2, then

(a) $\lambda_1 < 0$ if $g'_+(0)$ $\sup_{y\in\Omega} \int_{\partial\Omega} k(x,y)d\sigma(x) < a_{11}a_{22}$

while

(b) $\lambda_1 > 0$ if $g'_+(0)$ $\inf_{y\in\Omega} \int_{\partial\Omega} k(x,y)d\sigma(x) > a_{11}a_{22}$.

Conditions for the existence and uniqueness of a nontrivial endemic state in case (b) are still under investigation.

We may compose the new "threshold parameter" suggested in (b) for the instability of the extinction state

(2.10) $\quad \tilde{\theta} := \dfrac{g'_+(0) \inf\limits_{y\in\Omega} \int_{\partial\Omega} k(x,y)d\sigma(x)}{a_{11}a_{22}}$

with the one defined in (1.3) for the space homogeneous case.

3. An optimal control problem

An optimal control problem arises if one wishes to reduce the epidemic phenomena described by model (2.2), (2.3) by reducing the boundary feedback along the sea shore, i.e. by reducing the "strength" of the kernel $k(x,y)$, so that the threshold parameter $\tilde{\theta}$ reduces below its critical level.

This corresponds to a sanitation program by means of the treatment of the sewage before sending it to the sea [11]. The sanitation program implies a cost that has to be compared with the cost of the epidemic itself. Thus the optimal control problem appears as follows. If we assume that $\{w(t), t \in \mathbb{R}_+\}$ is the control parameter in the kernel k

(3.1) $\quad \tilde{k}(x,y;t) = w(t)k(x,y), \quad t \in \mathbb{R}_+, \quad x \in \Gamma_1, \quad y \in \Omega$

then

Problem (P): For any fixed $T > 0$, minimize

(3.2) $\quad \int_0^T \int_\Omega f(u_2(t,x))dxdt + \int_0^T h(w(t))dt + \int_\Omega l(u_2(T,x))dx$

190

for all (u_1, u_2, w) subject to the state system (2.2),(2.3) where k has been substituted by $\tilde{k}$.

In problem (P), f and l are related to the cost of the epidemic while h describes the cost function of the sanitation program tending to reduce w(t).

Problem (P) has been faced in [2] where necessary optimality conditions are given and a numerical algorithm to implement them is suggested.

4. Numerical simulations and identification problems

Numerical simulations have been carried out by means of a software developed in collaboration with the Mathematical Institute of the University of Iasi (Romania) to show the asymptotic behaviour of system (2.2),(2.3) below and above threshold [2] .

Computer experiments are reported in Fig. 4.1 which shows that above threshold, i.e. for large values of the parameter $\tilde{\theta}$ as defined in (2.10), actually the epidemic tends to a nontrivial endemic state.

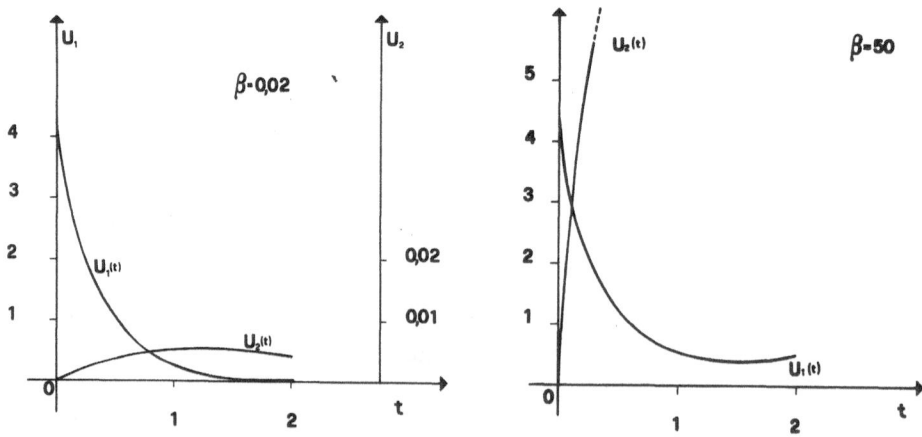

Fig.4.1.- Computer experiments for system (2.2),(2.3), below and above threshold. Average values are shown.

The numerical simulations in Fig. 4.1 refer to a square domain with one of its sides corresponding to the sea shore Γ_1 through which the boundary feedback occurs.

The kernel has been chosen of the form

$$k(x,y) = \sum_{i=1}^{m} \alpha_i \, \chi_i(y)\tilde{\chi}_i(x)$$

where χ_i is the characteristic function of the line interval $[x_i, x_{i+1}]$, $x_i \in \Gamma_1$, $i=0,1,\ldots,m-1$; $x_o=0$, $x_i = i\frac{L}{m}$, $i=1,\ldots,m$. While $\tilde{\chi}_i$ is the characteristic function of the rectangular interval $T_i=[0,L]\times[x_i,x_{i+1}] \subset \Omega$, $i=0,\ldots,m-1$.

In the second equation of system (2.2) we have chosen for the force of infection $g(u_1)$ the following expression

$$g(u_1) = \beta \, \frac{u_1}{\gamma u_1 + 1}$$

where β has been used as the parameter that we increase in order to let the threshold parameter $\tilde{\theta}$ go above its critical value. The two pictures in Fig. 4.1 respectively show the behaviour of $z_i(t) = \int_\Omega u_i(x,t)dx$, $i=1,2$, for values of $\tilde{\theta}$ below and above the critical value.

A relevant problem related to model (2.2),(2.3) is the identification of the transfer kernel $k(x,x')$ which sustains the positive feedback via the boundary condition (2.3a).

This problem has been carried out in collaboration with a research group in Graz [13]. It has been assumed the knowledge in real time of the concentration $u_1(x;t)$ of the infectious agent in the sea along the sea shore, $x \in \Gamma_1$ and the evolution of the epidemic $u_2(x;t)$ in the habitat $x \in \Omega$, for any $t \geq 0$.

It is also assumed that the kernel shows space homogeneity in Ω:

$$k(x,x') = k(x), \qquad x \in \Gamma_1$$

Fig. 4.2 shows the results of the identification procedure performed
for the case in which Ω is a unit square with one side only corresponding
to the sea shore Γ₁.

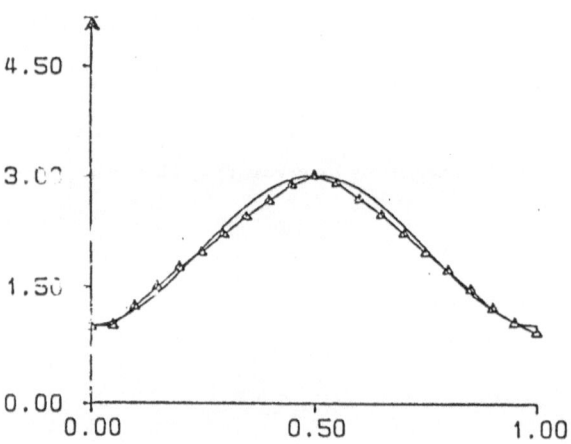

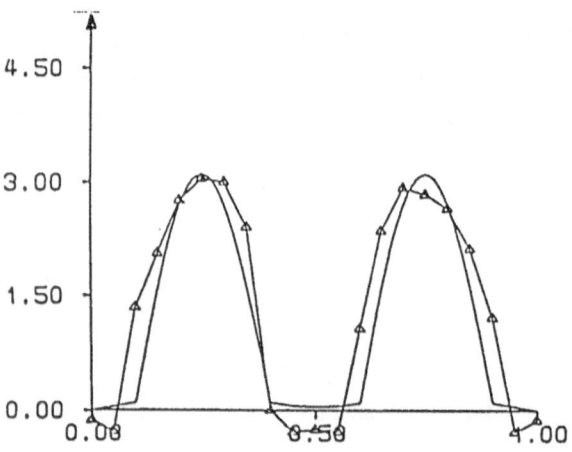

Fig.4.2.- Identification of the kernel k(x,x').
_____ theoretical   Δ Δ Δ   estimated

193

Acknowledgements — Work performed in the context of the Special Program
"Control of Infectious Diseases" of the National Research Council of
Italy (CNR).

References

[1]   Anderson, R.M. and May, R.M., eds. Population Dynamics of Infectious
      Diseases Agents Dahlem Konferenzen (Springer-Verlag, Heidelberg,
      1982).

[2]   Arnautu, V., Barbu V. and Capasso, V., : Controlling the spread of
      a class of epidemics. Preprint.

[3]   Bailey, N.T.J., The Mathematical Theory of Infectious Diseases,
      Griffin, London 1975.

[4]   Barbour, A.D., Macdonald's model and the trasmission of bihlarzia.
      Trans. Roy. Soc. Trop. Hyg. 72 (1978) 6-15.

[5]   Capasso, V., Asymptotic stability for an integrodifferential reaction-
      -diffusion system. J. Math. Anal. Appl. 103 (1984), 575-588

[6]   Capasso, V., Grosso, E., Serio, G., I modelli matematici nell'indagine
      epidemiologica. Il tifo addominale: studio delle serie temporali.
      Annali Sclavo, 22 (1980)189-206.

[7]   Capasso, V., Kunisch, K., A Reaction-diffusion system modelling
      man-environment epidemics Ann. of Diff. Eqs. 1, 1985 pp. 1-12 (China).

[8]   Capasso, V., and Kunisch, K., A reaction-diffusion system arising
      in modelling man-environment diseases.Preprint.

[9]   Capasso, V., and Paveri-Fontana, S.L., A mathematical model for the
      1973 cholera epidemic in the European Mediterranean region. Rev.
      Epidem. et Santé Publ. 27 (1979), 121-32. errata, Ibidem 28 (1980), 330.

[10] Capasso, V., and Thieme, H., A threshold theorem for an epidemic
      system with a boundary feedback. To appear.

[11] Cvjetanovic, B., Grab, B., and Uemura, K., Dynamics of Acute
      Bacterial Diseases. Epidemiological Models and their Application
      in Public Health. Suppl. N°1 to Vol. 56 of the Bulletin of the
      World Health Organization. Geneve: WHO, 1978.

[12] Nasell, I. Hybrid Models of Tropical Infections. Lect. Notes in
      Biomathematics, 59. Heidelberg: Springer-Verlag, 1985.

[13]   Schelch, H., Parameter estimation in a special reaction-diffusion
       system modelling man-environment diseases specific for the
       Mediterranean region, Thesis, Technical University of Graz. (In German).

[14]   Serio, G., Considerations on the behaviour of the Hepatitis A in
       Bari by the time series analysis
       Rivista di Statistica Applicata, 19, 1, 1986, pp. 27-39.

Vincenzo CAPASSO
Dipartimento di Matematica
Università di Bari
Campus
70125 BARI, Italy

NUMERICAL SOLUTION OF PIPELINE SYSTEM PROBLEMS
BY MONOTONE DIFFERENCE APPROXIMATIONS

Gisbert Stoyan, Budapest

Summary: We consider equations of parabolic type which may
degenerate into first order hyperbolic (or ordinary differential)
equations. These equations are given on a one - dimensional
graph (representing, for example, a gas, oil or water pipeline
system or blood vessels or a system of rivers, channels and
lakes) and model the convective and diffusive transport of
concentrations and heat.
For such problems, a monotone difference scheme is described
which continuously connects well-approved approximations of the
parabolic, hyperbolic and further cases, including the method of
characteristics.

1   Introduction

Quite a number of long- and short-term processes can be
modelled by (possibly degenerating) parabolic equations of the
type

$$(1.1) \qquad A \frac{\partial u}{\partial t} = \frac{\partial}{\partial x} \left(AD \frac{\partial u}{\partial x} \right) + Q \frac{\partial u}{\partial x} + Ar$$

on graphs: Radioactive corrosion in the water cycles of an
atomic reactor, adsorption processes in gas pipelines, spread of
medicaments with the blood circulation or of plumes of poisonous
industrial wastes in river systems (this latter problem being
the original motivation for the author [2]). In (1), A is the
cross section, D an (eddy) diffusion coefficient, Q flow rate,
r reaction rate; all these coefficients are functions of x and
t; u and r may be n-vectors and then A, D and Q are diagonal
matrices.

To be acceptable, approximations to (1) not only must be stable
and effectively solvable but also should satisfy conditions of
discrete physical modelling like conservation of positivity and
monotonicity, the maximum principle, and avoid oscillations and
numerical diffusion at steep fronts. All these properties must
hold for great and small or zero coefficients (e.g., a), independently on grid spacing.

This paper is in final form and no version of it will be
submitted for publication elsewhere.

A difference scheme fulfilling these conditions will be described below in Sect. 3 (after shortly surveying in Sect. 2 the computation of flow rates and pressures in river and pipeline systems, resp.); boundary conditions will be considered in Sect. 4 and the solution of the special linear systems in Sect. 5.

2   Computation of flow rates

The transient motion of fluids in pipes and open channels can be described by systems of one-dimensional first order hyperbolic equations [1, 6, 20, 28, 30]. Being nonlinear, discretization methods must be used for their solution, e.g. variants of the Godunov scheme [1], of the method of characteristics [19] or the Preissmann scheme [6].

In case of stationary flow (on which we are concentrating), hydraulic formulas are in common use which give a relation between flow rate Q (or cross-sectional mean velocity v), hydraulic diameter R (or inner pipe diameter d) and pressure p (or height h):

For water flow in open channels and rivers

$$(2) \qquad v = k \ R^{2/3} (h/\ell)^{1/2}$$

is used (where k is the Manning-Gauckler-Strickler roughness coefficient and ℓ length of the river reach considered), and, similarly, for incompressible fluids in circular pipes [3]

$$(3) \qquad \frac{1}{\rho}(p_1 - p_2) = \alpha Q^2 \ , \quad \alpha := 8\lambda\ell/(\pi^2 d^5),$$

where ρ is density, p_1 and p_2 the pressure at the beginning and end of the pipe; λ is the friction coefficient. For k and λ see tables in [15], for λ most often the following implicit formula [4] is used:

$$(4) \qquad \frac{1}{\sqrt{\lambda}} = - \ 2\lg \left\{ \frac{2.51}{Re\sqrt{\lambda}} + 0.27 \ \frac{k}{d} \right\}, \ Re := \frac{vd}{\nu}$$

where ν is kinematic viscosity and Re the Reynolds number. This equation is solved by iteration starting from the limit case Re $= \infty$. Remark that here λ depends on Q through Re and v.

For compressible fluids (natural gas, say) in the isothermal case, for small Mach numbers, neglecting the pressure head and

elevation of the pipe with respect to the horizontal plane, the
following formula is accepted [21, 14, 29, 20]:

$$(5) \quad \frac{1}{2p_1\rho_1} (p_1^2 - p_2^2) = \alpha Q^2$$

(with α from (3)). Using once more an empirical friction formula,
e.g. the Weymouth formula [14]

$$\lambda = const \; d^{-1/3} \; ,$$

we arrive at

$$(6) \quad Q = const \; d^{8/3}((p_1^2 - p_2^2)/\ell)^{1/2} \; .$$

Instead of 8/3 and 1/2, other exponents are also in use as shows
the standard „Panhandle" equation

$$(7) \quad Q = const \; d^{2.6182}((p_1^2 - p_2^2)/\ell)^{0.5394}$$

and similar formulas, see [14], p. 305.
Equations (3) or (5) - (7) are given along any section of a pipe.
The system of pipes forms, mathematically, a one-dimensional
graph which is usually not of tree structure: Pipeline systems
contain cycles to prevent break-down of public utilities like
gas and water supply, for economy of energy or to reach high
purification levels (recycling in the power industry and the
chemical processing industry).
At any junction of the pipes (i.e., any node of the graph) we
have by mass conservation

$$(8) \quad \Sigma Q_i = 0$$

(Kirchhoffs first law). Considering that the pressure drop between
points 1 and 2 of a cycle must be independent of the path between
1 and 2, we get from (3) for every cycle one equation of the
type

$$(9) \quad \Sigma \beta_i Q_i^2 = 0$$

(Kirchhoffs second law) and similar equations from (5) and (6).
At some nodes pressures may be prescribed (water-towers, piston
pumps) whereas prescribed velocities or consumptions are in-

cluded into (9).

All these equations form a nonlinear system which is usually solved by some form of Newton iteration [5, 8, 32] or by a direct linearization [7] of Q_i^2 in (9) into $Q_i Q_i^*$ where Q_i^* is iterated in an outer iteration - as are the β_i which may depend on Q, see (4). Such a direct linearization is also useful in connection with noninteger exponents like $1/0.5394 \approx 1.854$ arising from (7).

3 Monotone difference approximation

For the diffusion-convection equation (1) we describe now a difference approximation which is monotone in the sense that the corresponding matrix is an M-matrix (see, e.g., [10]), independently on discretization parameters, for $a := AD \geq 0$. Writing also $c := A$, $b := Q$, $f := Ar$, we consider on one pipe section of the graph the equation

$$(10) \quad c\, \frac{\partial u}{\partial t} = \frac{\partial}{\partial x} \left(a\, \frac{\partial u}{\partial x} \right) + b\, \frac{\partial u}{\partial x} + f$$

and start from the stationary constant coefficient case $c = 0$, $a = \text{const} \geq 0$, $b = \text{const}$:

$$(11) \quad 0 = au" + bu' + f, \quad 0 \leq x \leq 1, \quad u' := \frac{du}{dx}\ .$$

The length of the section is here normed to 1. For (11) we write the so-called exact difference scheme [27, 17], i.e. a difference approximation the solution $y = (y_0, \ldots, y_N)^T$ of which has the property

$$y_i = u(x_i), \quad x_i := ih, \quad i = 0, 1, \ldots, N, \quad h := 1/N,$$

where $u(x)$ is the solution of (11), with first kind boundary conditions added for definiteness,

$$(12) \quad u(0) = u_0, \quad u(1) = u_N\ .$$

The x_i, $i = 1, \ldots, N-1$ will be called inner grid points whereas x_0 and x_N are simultaneously grid points and junction nodes of our graph.

For (11), (12) and f linear in x, the exact difference scheme has the form

(13) $0 = (h \wedge y)_i + (hMf)_i$,

$(h \wedge y)_i := ((a + bh(\frac{1}{2} + \alpha))y_x)_+ - ((a - bh(\frac{1}{2} - \alpha))y_x)_-$,

$(hMf)_i := (hf(\frac{1}{2} + \alpha))_+ + (hf(\frac{1}{2} - \alpha))_-$,

$i = 1, \ldots, N-1;\ y_0 = u_0,\ y_N = u_N$,

$\alpha = \alpha(q) := (\rho - 1)/2q,\quad \rho := q \coth q$,

$q := 0$ if $a = b = 0$ and $q := bh/2a$ else.

Here

$$|\alpha| \le \frac{1}{2},\ \alpha(0) = 0,\ \operatorname{sgn} \alpha = \operatorname{sgn} b,\ \lim_{|q| \to \infty} \alpha = \frac{1}{2} \operatorname{sgn} b.$$

In (13) we have used the following notations:

$$g_\pm := g(x_i \pm \frac{h}{2}),\ (y_x)_\pm := \pm(y_{i+1} - y_i)/h ,$$

the definition of q is formulated as it is to have a useful approximation in the instationary case (c>0 in (10)) if there a = b = 0. The hyperbolic function $\rho(q)$ may be replaced by $(3 + 3|q| + 3q^2 + 2|q|^3)/(3 + 3|q| + 2q^2)$ without practical loss of accuracy. For the scheme (13) the following estimate has been shown to hold for two times differentiable f [24]:

(14) $\displaystyle\max_{0 \le i \le N} |y_i - u(x_i)| \le \min(\frac{1}{4a}, \frac{1}{|b|})h^2 ||f''||_{L_1}$.

The entries $m_{i,i+1}$ of the matrix (m_{ij}) corresponding to (13) have the values $-(a \pm bh(\frac{1}{2} \pm \alpha)) \le 0$, the other off-diagonal entries and for $1 \le i \le N-1$ the row sums are zero, the first and last row are given by $m_{ij} = \delta_{ij}$, $i = 0$, N. Since by (14) the matrix is nonsingular for $a + |b| > 0$, from [26] it follows that it is indeed an M-matrix.

The transition from (13) to the variable coefficient case is obvious by extending the definition of the „+" index from f and y_x to a, b, α and also to h: $h_\pm := |x_{i+1} - x_i|$ in the nonequidistant case.

The difference operator $h\wedge$ of (13) has its best approximation not at the point x_i but at $x_{i+\alpha} := x_i + \alpha h$ (correspondingly we have the identity $(hMf)_i = hf(x_{i+\alpha})$ for linear f), hence to proceed to the instationary case (10) we need an approximation of $c\,\frac{\partial u}{\partial t}$ at $x_{i+\alpha}$. This can be achieved by starting from

$$(h\bar{M}g)_i := h\,\max(0,\alpha)g_{i+1} + h[\tfrac{1}{2} + \alpha - \max(0,\alpha) +$$

$$+ \tfrac{1}{2} - \alpha - \max(0,-\alpha)]g_i + h\,\max(0,-\alpha)g_{i-1} \approx$$

$$\approx hg_{i+\alpha}$$

and substituting

$$cy_t := c(y^{j+1} - y^j)/\tau$$

for g where c will be evaluated at the „$\pm$" points, $\tau > 0$ is the time step and

$$y^j := y(x,\ t_j),\ t_j := j\tau\ .$$

To guarantee stability and monotonicity, a weight $\sigma \in [0,1]$ is now introduced:

$$hM_\sigma\ cy_t := (2\sigma hmax(0,\alpha)c)_+\ y_{t,i+1} + [(h(\tfrac{1}{2} + \alpha - 2\sigma max(0,\alpha))c)_+ +$$

$$+ (h(\tfrac{1}{2} -\alpha-2\sigma max(0,-\alpha))c)_-]y_{t,i} + (2\sigma hmax(0,-\alpha)c)_-y_{t,i-1}$$

The same weight σ is introduced into the spatial approximation $h\wedge y$ in the usual way as $h\wedge y^{(\sigma)}$ where

$$y^{(\sigma)} := \sigma y^{j+1} + (1 - \sigma)y^j\ .$$

The monotone approximation of (10) is then

(15) $hM_\sigma\ cy_t = h\wedge y^{(\sigma)} + hMf$.

For constant coefficients, this is a second order approximation independently of σ if τ is determined from [22]

(16) $\frac{b\tau}{ch} =: p = 2\alpha + c_0 \cdot bh,\ c_0 = const$,

with the Courant number p. In (15), f and the coefficients c, a, b are evaluated at $t_{j+1/2}$. The matrix corresponding to (15) is an

M-matrix if $c > 0$ or $c = 0$, $a + |b| > 0$ everywhere and if σ is selected in every grid cell (with center $(x_{i+\frac{1}{2}}, t_{j+\frac{1}{2}})$) from [23]

(17) $\sigma = \begin{cases} 0 & \Delta > 0 \\ \frac{1}{2} & \Delta = 0 \\ 1 - \min(\frac{1}{2}, (\frac{1}{2} - |\alpha|)\frac{c}{\tau}/|\Delta|), & \Delta < 0, \end{cases}$

$$\Delta := 2\frac{c}{\tau}|\alpha| - (a + |b| h(\frac{1}{2} + |\alpha|))/h^2 .$$

If the M-matrix property is fulfilled, the scheme preserves positiveness; then also the maximum principle holds [25], since the necessary and sufficient condition, the nonnegativity of the row sums, is satisfied. Further if $f = 0$ and $|p| \le 2|\alpha|$, then the property holds that monotone solution profiles at time t_j are transformed into monotone profiles at t_{j+1}; here the Courant number restriction is not necessary if modifying the σ selection rule [23].

The scheme (15), (17) is of three-point, two-level type and thus can readily be solved by the usual shortened Gauss elimination for tridiagonal systems if considered for only one section with given values of the solution at the endpoints; see also Sect. 5.

We mention now three important special cases:

a) If c, $a > 0$ and $b = 0$, the scheme coincides with the familar weighted difference scheme [17] for selfadjoint parabolic equations but σ is calculated in every grid cell according to (17) where $\Delta < 0$ (due to $b = q = \alpha = 0$). Hence $\sigma \ge 1 - ch^2/2a\tau$ or $\tau \le ch^2/2a(1 - \sigma)$ which is the well-known monotonicity condition [17], p. 179.

b) If $c > 0$, $b > 0$ (say), the equation is first order hyperbolic and, since $q = \infty$, $\alpha = \frac{1}{2}$, now (15), (17) reduce to the monotone scheme

(18) $(1 - \sigma)cy_{t,i} + \sigma cy_{t,i+1} = by_x^{(\sigma)} + f^{j+\frac{1}{2}}$,

$$\Delta = \frac{c}{\tau} - \frac{b}{h} = \frac{c}{\tau}(1 - p), \quad \sigma = \begin{cases} 0 & 0 < p < 1 \\ \frac{1}{2} & p = 1 \\ 1 & p > 1 \end{cases}$$

(Compared to (15), we have deleted the common index „+" and the factor h_+.) For $p = 1$ this is just the method of characteristics,

$$y_{i+1}^{j+1} = y_i^j + hf^{j+\frac{1}{2}}$$

which enables us to compute steep fronts without numerical diffusion.

Selecting in case $a > 0$ the time step according to (16), $c_0 = 0$ (this time step is usually acceptable for $|b| \gg a > 0$), we get automatically $|p| = 1$ if $a \to 0$.

If computing the movement of a steep (concentration or heat) front through the pipeline system with this time step, the front must be traced. Reaching a junction node, the time step is re-evaluated according to the data of the pipe section with the greatest flow rate downstream. This results into one steep front proceeding since the fronts in the branching sections with smaller flow rates will disappear after a few time steps due to numerical diffusion.

This approach has been applied to the simulation of pollution problems in the Berlin river system [2] modelled by a graph with 27 nodes and 28 edges.

c) If $c > 0$ and $a = b = 0$, our approximation takes the (second order) form

(19) $$\frac{1}{2}[(ch)_+ + (ch)_-]^{j+\frac{1}{2}} y_t = \frac{1}{2}[(hf)_+ + [hf]_-]^{j+\frac{1}{2}} .$$

Finally, if $c = 0$ and $a + |b| > 0$, (15) is (13) with $y = y^{j+1}$ since $\sigma = 1$ by (17).

We remark that in the considered special cases we do not refer to constant coefficients; selecting the weighting parameters from (17), the scheme adapts locally to the varying types of equation (1) in dependence on the coefficients and step lengthes.

4 Boundary conditions

We have three types of boundary conditions at the different nodes of the graph:

1) inflow nodes: first kind conditions (prescribed concentrations and temperature);

2) outflow nodes: in the original problem formulation as obtained from the engineer there is usually no prescription of conditions for outflow nodes, even the certainty is expressed that the values there should already be determined from what happens earlier and upstream. Mathematically this means: take zero diffusion (a = 0) at an outflow node, i.e.

$$(20) \qquad c \frac{\partial u}{\partial t} = b \frac{\partial u}{\partial x} + f \ .$$

This turns out to be a first order (in small a) outflow boundary condition, see [13], where outflow conditions of any order are derived.

For us not only the case $a \ll |b|$ is of interest: think of heat conduction in a pipe section of a cooling system in case of pump failure. We obtain the boundary condition at an outflow node by deleting the nondefined expressions from (15), say at a left outflow node (i = 0) the expressions indexed „-" which would refer to $x_{-1/2}$. This approach seems formal but is used also in finite elements at isolation boundaries and leads for (15) to the most natural closure of the conservation law derived from the inner grid points [22, 23].

Therefore, at a left outflow point there remain the „+" expressions

$$(21) \qquad (2\sigma h max(0,\alpha)c)_+ y_{t,i+1} + (h(\tfrac{1}{2} + \alpha - 2\sigma max(0,\alpha))c)_+ y_{t,i} =$$

$$= ((a + bh(\tfrac{1}{2} + \alpha))y_x^{(\sigma)})_+ + (hf(\tfrac{1}{2} + \alpha))_+^{j+1/2} \ .$$

Considering once more the special cases of Sect. 3, we have from (21):

a) if c > 0, a > 0, b = 0 then

$$\tfrac{1}{2}(ch)_+ \, y_{t,i} = (ay_x^{(\sigma)})_+ + \tfrac{1}{2}(hf)_+^{j+1/2} \qquad\qquad (i = 0)$$

which is the usual second order approximation of $a\partial u/\partial x = 0$;

b) if c, $b > 0$, $a = 0$ then (due to $q = \infty$, $\alpha = \frac{1}{2}$) we get just (18) - i.e., we have the same approximation (now of (20)) at inner grid points and at the outflow node $i = 0$, thus excluding artificial boundary layers. (We observe that the attempt to pose the outflow condition at an inflow point ($c > 0$, $b < 0$ at $i = 0$) results into (comp. (21))

$$(h(\frac{1}{2} - |\alpha|)c)_+ y_{t,i} = ((a - |b|h(\frac{1}{2} - |\alpha|))y_x^{(\sigma)})_+ +$$
$$+ (hf(\frac{1}{2} - |\alpha|))_+$$

since $\alpha < 0$. If now $a \to 0$ (i.e. $q \to -\infty$, $\alpha \to -\frac{1}{2}$) then there remains only $0 = 0$ and we get a zero row in the matrix.)

c) if $c > 0$, $a = b = 0$ we obtain

$$\frac{1}{2}(ch)_+ y_{t,i} = \frac{1}{2}(hf)_+^{j+\frac{1}{2}}$$

in close accordance to (19).

3) At junction nodes of the graph we have, in the continuous formulation, continuity of function values and of the heat and concentration fluxes. These latter conditions are approximated by adding the above discrete outflow boundary conditions of all pipe sections meeting at the junction.
Between the several contributions added there are those from the sections for which the junction node is an inflow node. As mentioned, such inflow sections in the limit $a \to 0$ will give zero contributions. This is just what is needed:
The concentration at the junction node will then only be influenced by the concentrations at neighboring upstream points. For the downstream points the concentration at the junction node serves as an inflow boundary condition. This is achieved without making any distinction between the different sections meeting at such a node, only by solving the linear system corresponding to the described difference scheme and its boundary conditions.
We illustrate this considering the special case

$$c = a = 0, \quad b \neq 0$$

for a junction node with s pipe sections meeting there, those sections for which the node is an outflow point being numbered $k = 1, \ldots, r < s$.

For a moment, we denote the concentration at the junction node by y_0, the concentrations at the neighboring upstream points by y_k and correspondingly the grid spacing, flowrates and source rates in the neighboring interval of the outflow sections by h_k, b_k and f_k, $k = 1, \ldots, r$.

Due to $a = 0$, $b \neq 0$, we get zero contributions from those sections for which the junction node is an inflow point.

The k-th outflow section gives (comp (21)) $b_k h_k (y_k - y_0)^{(\sigma)} / h_k + h_k f_k$ - if the length coordinate is measured from the junction node (we write $k \in s^+$), and here $b_k > 0$; or we have $- (-b_k h_k (y_0 - y_k)^{(\sigma)} / h_k) + h_k f_k$ - if the junction node is the last point of that section ($k \in S^-$); here $b_k < 0$. By (17) there holds $\sigma = 1$ due to $c = 0$, $\Delta < 0$, and hence $y^{(\sigma)} = y^{j+1}$. We will omit this upper index.

Adding the contributions we obtain

$$0 = \sum_{S^+} b_k (y_k - y_0) + \sum_{S^-} b_k (y_0 - y_k) + \sum_1^r h_k f_k$$

and from here, taking into account the signs of b_k,

(22) $$y_0 = \sum_1^r (|b_k| y_k + h_k f_k) / \sum_1^r |b_k| \ .$$

This is just the formula for complete mixing at the junction node.

Now let be y_ℓ the concentration at an neighboring downstream point (with corresponding notations for grid spacing etc. in the adjacent interval of the inflow pipe section). Then for the considered downstream point the difference scheme reduces to

$$0 = -(-h_\ell b_\ell (y_\ell - y_0)/h_\ell) + h_\ell f_\ell$$

for $\ell \in s^+$ (and here $b_\ell < 0$) and

$$0 = h_\ell b_\ell (y_0 - y_\ell)/h_\ell + h_\ell f_\ell$$

if $\ell \in s^-$ (and then $b_\ell > 0$). In both cases the value (22) of y_0 will serve as a first kind boundary condition.

5   <u>Solution of the linear equations</u>

The equations of the difference scheme give rise to a
large linear system with a sparse M-matrix of special structure:
As long as $a > 0$, the matrix has 3 nonzeros per row for inner
grid points and $s + 1$ nonzeros for a row corresponding to a
junction node with s branching pipe sections. In general, the
matrix is not symmetric.

Applying the Fryazinov algorithm [11] we compress the large
system to a small one which involves the nodes of the graph only.
The new matrix is once more a nonsymmetric sparse M-matrix (the
graph of which is the undirected simplified graph of the pipeline system: loops with only one node and multiple connections
of possibly different orientation between the same nodes - all
this being admitted in the Fryazinov algorithm - are missing
now). The matrix also is structurally symmetric and hence is
suited to application of decomposition methods considered in
[12], see especially exercise 1 in § 4.5.

The Fryazinov algorithm seems to be known in western countries
only in water research [6], pp. 117-121; it might also be useful
in solving a number of problems of the chemical processing
industry where graphs containing cycles are usually decomposed
into parts of tree structure [18, 31, 9] by prescribing the
values of unknowns corresponding to selected points of the cycles
and then iterating these values in an outer iteration.

This approach is mostly efficient but known [16] to be not as
reliable as direct solution of the full equations without taking
into account the sparse structure.

The Fryazinov algorithm, as described in [11], is confined to
three point equations (at the inner grid points) but leads to
a direct solution of the equations and, in combination with [12],
fully exploits the special sparse structure.

In case of concentrations transported by a stationary flow field
it is obvious to compute the large and the compressed system
along with its decomposition only once. Then at all time levels
only back substitution has to be performed which (for the inner
grid points) needs four operations per point in the Fryazinov
algorithm.

References

[1] Alalykin, G. B.; Godunov, S. K.; Kireyeva, I. L.,;
     Pliner, L. A.: Solution of one-dimensional problems of gas
     dynamics on moving grids. Moscow: Nauka 1970 (in Russ.)

[2] Baumert, H.; Braun, P.; Glos, E.; Müller, W.; Stoyan, G.:
     Modelling and computation of water quality problems in
     river networks. In: Lecture Notes in Control and Informa-
     tion Sciences 23. Berlin: Springer 1981, 482-491

[3] Becker, E.: Technische Strömmgslehre. Stuttgart: Teubner
     1976

[4] Colebrook, C. F.: Turbulent flow in pipes, with particular
     reference to the transition region between the smooth and
     rough pipe laws. J. Inst. Civ. Eng. 11 (1938/39) 133-156

[5] Cross, H.: Analysis of flow in networks of conduits or
     conductors. Bulletin Univ. of Illinois. Engng. Experimental
     Station, No 286 (1936)

[6] Cunge, J. A.; Holly, F. M.; Verwey, A.: Practical Aspects
     of Computational River Hydraulics. London: Pitman 1978

[7] Demuren, A. O.; Ideriah, F. J. K.: Pipe network analysis
     by partial pivoting method. J. Hydraul. Eng. 112, 5 (1986)
     327-334

[8] Epp, R.; Fowler, A. G.: Efficient code for steady-state
     flows in networks. J. Hydraul. Div. ASCE 96 (1970) 43-56

[9] Evans, L. B. et al.: ASPEN: An advanced system
     for process engineering. Computer in Chem. Eng. 3 (1979)
     319-327

[10] Fiedler, M.: Special Matrices and Their Applications in
     Numerical Mathematics. Dortrecht: Martinus Nijhoff Publ.,
     and Prague: SNTL 1986

[11] Fryazinov, I. V.: Algorithm for the solution of difference
     problems on graphs. Zhurnal vyc. matem. i matem. fiziki 10,
     2 (1970) 474-477 (in Russ.)

[12] George, A.; Liu, J. W. H.: Computer Solution of Large
     Sparse Positive Definite Systems. Englewood Cliffs:
     Prentice-Hall, 1981

[13] Halpern, L.: Artificial boundary conditions for the linear
     advection diffusion equation. Math. Comput. 46, (1986)
     425-438

[14] Katz, D. L. et al.: Handbook of Natural Gas Engineering.
     N.Y.: McGraw-Hill, 1959

[15] Kittner, H.; Starke, W.; Wissel, D.: Wasserversorgung.
5. Aufl. Berlin: VEB Verlag für Bauwesen 1985

[16] Perkins, J. D.; Sargent, R. W. H.: SPEEDUP: A computer
program for steady-state and dynamic simulation and design
of chemical processes. In: Selected Topics on Computer-
-Aided Process Design and Analysis (Mah, R. S. H.;
Reklaitis, G. V., eds.) AIChE Symposium Series 78, 214
(1982) 1-11

[17] Samarskij, A. A.: Theorie der Differenzenverfahren (Transl.
from Russ.). Leipzig: Teubner 1984

[18] Sargent, R. W. H.: The decomposition of systems of
procedures and algebraic equations. In: Lecture Notes in
Mathematics 630. Berlin: Springer 1978, 158-178

[19] Schmitz, G.; Edenhofer, J.: Flood routing in the Danube
river by the new implicit method of characteristics (IMOC).
In: Hydrothermodynamic Modelling of Natural Waters.
Hamburg: Institut für Meereskunde der Universität Hamburg
1983, 231-243

[20] Schröder, R.: Grundgleichungen für die Berechnung von Rohr-
und Gerinneströmungen. In: [32], 1-28

[21] Shapiro, A. H.: The Dynamics and Thermodynamics of
Compressible Fluid Flow, v.I. N.Y.: Ronald Press Co. 1953

[22] Stoyan, G.: On a maximum norm stable, monotone and
conservative difference approximation of the one-dimensional
diffusion-convection equation. In: Proceedings of the conference on "Simulation der Migrationsprozesse im Boden-
und Grundwasser", TU Dresden 1979, 139-160

[23] Stoyan, G.: Beiträge zur Theorie und Anwendung von
Differenzenverfahren. Diss. B, Akademie d. Wiss. d. DDR,
Berlin 1982

[24] Stoyan, G.: Explicit error estimates for difference schemes
solving the stationary constant coefficient diffusion-
-convection-reaction equation. ZAMM 64 (1984) 173-191

[25] Stoyan, G.: On maximum principles for monotone matrices.
Linear Algebra Appls. 78 (1986) 147-161

[26] Taussky, O.: A recurring theorem on determinants. Amer.
Math. Monthly 56 (1949) 672-676

[27] Tichonov, A. N.; Samarskij, A.A.: On homogeneous difference
schemes of high order of accuracy. Doklady AN USSR 131, 3
(1960) 514-517 (in Russ.)

[28] Tsharnyj, A.: Transient Motions of Real Fluids in Pipes.
Moscow: Gos. Isd. Techn. Lit. 1951 (in Russ.)

[29] Tscharnyj, A.: Fundamentals of Gas Dynamics. Moscow: Gos. Top. Tech. Isd. 1961 (in Russ.)

[30] Vasil'jev, O. F. et al.: Nonisothermic Gas Flow in Pipes. Novosibirsk: Nauka 1978 (in Russ.)

[31] Westerberg, A. W. et al.: Process Flowsheeting. Cambridge: Cambridge Univ. Pr. 1979

[32] Zielke, W. (ed.): Elektronische Berechnung von Rohr- und Gerinneströmungen. Berlin: E.-Schmidt - Verlag 1974. (see herein the papers by H. Vielhaber, W. Endres, R. G. Cembrowicz)

G. Stoyan
ELTE Computing Center
H-1117. Budapest
Bogdánfy u. 10/b.

DYNAMIC MODEL OF A METALLURGICAL SHAFT REACTOR WITH IRREVERSIBLE
CHEMICAL KINETICS AND MOVING LOWER BOUNDARY *)

Dr. Svenn Anton Halvorsen, Elkem a/s

Summary: A (preliminary) unidimensional dynamic model for
simulating a metallurgical reactor has been implemented. A
system of nonlinear hyperbolic differential equations describes a
shaft reactor with gas reacting in counter current with
solid/liquids. Due to irreversible chemical kinetics the first
derivative of the reaction rates will be discontinuous. The
lower shaft boundary can be moving.
    Application of control volume spatial discretization and a
suitable (standard) time integration routine for (stiff) ordinary
differential equations, has proven successful for the gas
flow/heat transfer problem. The approach does not work "as is"
when including irreversible, fast gas reactions. Some method for
stabilizing the gas variables is necessary.
    If the model incorporates control volumes that can approach
zero volume, Newton type methods are unsuitable for solving the
nonlinear equations involved.

1   Introduction

    Within metallurgy, as for other established industries,

there is a general trend that standard production grows less

profitable as the technology becomes more widespread. A company

can meat this challenge in various ways: through marketing

strategy, cost reductions, production improvements (new

processes, running a process closer to its theoretical limit)

and/or concentrate on specialized products (e.g. higher purity).

For the two last cases, where improved process understanding will

be vital, mathematical process models can be powerful tools.

    In many metallurgical processes a very large amount of

energy is deposited in the hearth of a furnace, where metal, slag

and gas is formed through chemical reactions. The gas, moving

upwards, reacts in counter current with the raw materials.

Processes of this type are used for producing pig iron, FeMn,

FeCr, FeSi, Si, etc.

    A total process model will constitute an extremely complex

coupled 3-dimensional problem. One will need to describe

interaction between heat flow, electric current, chemical

---

*) This paper is in final form and no further version of it will be
submitted for publication

reactions and material flows (gas, liquids, particulate solids). One should not attempt to develop such a model before mastering simpler models describing only a part of the total problem in detail.

In our present model development we intend to make a coarse description of the furnace hearth (point model) and a fairly detailed unidimensional description of a shaft above the hearth, including gas and solid/liquid flows, chemical reactions, heat and material balance.

2   Model equations

In the following description we will for simplicity only consider the shaft equations and neglect the presence of liquid(s).

For each gas and solid component we will need material balance equations **)

(1)
$$\frac{\partial c_A}{\partial t} + \frac{\partial}{\partial z}(v_s c_A) = \sum_i a_i R_i$$

(2)
$$\frac{\partial c_Q}{\partial t} + \frac{\partial}{\partial z}(v_g c_Q) = \sum_i q_i R_i$$

Heat balance is described by the enthalpy transport equations

(3)
$$\frac{\partial H_s}{\partial t} + \frac{\partial}{\partial z}(v_s H_s) = Q_{gs}$$

(4)
$$\frac{\partial H_g}{\partial t} + \frac{\partial}{\partial z}(v_g H_g) = -Q_{gs}$$

where the source term Q_{gs} describes enthalpy transport due to heat transfer and mass fluxes between the gas and the solids. Enthalpies are chosen as state variables instead of temperatures as we can always compute the temperature knowing the enthalpy (and concentrations), while the reverse computation will not be defined at transition temperatures (e.g. melting).

The gas flow is given by the Ergun equation (see for instance [7]),

---

**) A symbol list is given at the end of the paper

$$(5) \qquad \frac{\partial p}{\partial z} = - \frac{150}{d_p^2} \frac{(1-\varepsilon)^2}{\varepsilon^3} \mu v_{gs} - \frac{1.75}{d_p} \frac{(1-\varepsilon)}{\varepsilon^3} \frac{pM_g}{RT_g} v_{gs}|v_{gs}|$$

where the pressure, p, is computed by applying the ideal gas law.

In our first implementation we only consider discontinuous solid flow (i. e. $v_s = 0$). Materials from the shaft will enter the furnace hearth through melting. Thus the shaft/hearth interface will be moving upwards at a velocity given as a pure temperature function (simplest assumption). After a certain time or when the shaft/hearth interface reaches a certain level, the shaft materials drop down and cold raw materials are filled on the top.

The reaction rate for a typical heterogeneous reaction

(i) $\qquad n_A\,A(s) + n_P\,P(g) \;=\; n_B\,B(s) + n_Q\,Q(g)$

will be given as

$$(6) \qquad R_i = F_i(K_i\,p_P^{n_P} - p_Q^{n_Q}) \,/\, \left(\frac{1}{k_{ti}} + \frac{1}{k_{si}}\right)$$

The equilibrium constants K_i and the chemical reaction coefficients k_{si} can vary rapidly with temperature, making the reaction rates highly nonlinear. In addition reaction rates will generally have discontinuous derivatives at equilibrium (where $R_i = 0$). Due to different mechanisms the transport and/or chemical reaction coefficients can depend on the direction of the reaction. Discontinuous derivatives will also be observed when one of the involved solids is not present. In this case the reaction can only run in one direction.

The set of equations constitutes a nonlinear hyperbolic system with a moving boundary. At the shaft/hearth interface the boundary conditions are given as a set of ordinary differential equations describing the furnace hearth and interactions with the shaft.

3 Numerical solution procedure

The numerical discretization is done in two steps. First the partial differential equations are approximated by a system of ordinary differential equations. This system is then solved applying an appropriate code for ordinary differential equations.

We demand that the discretizations shall be such that material and heat balances are *exact*. For the spatial discretization this is ensured by following the procedure of control volume integration [5]. The shaft is divided into a finite number of control volumes (see figure 1). Let $M_{A,i}$ be the amount of the solid component A in control volume i. Then integrate equation (1) over control volume no. i. Approximating interface concentrations with their upstream values and the integrated reaction rates by the midpoint rule, gives

$$(7) \qquad \frac{dM_{A;i}}{dt} = (\frac{dz_i}{dt} - v_{s;i}) c_{A;i-1} - (\frac{dz_{i+1}}{dt} - v_{s;i+1}) c_{A;i}$$

$$+ \Delta z_i \sum_j a_j R_j (c_{A;i}, \ldots)$$

Observe that the interface fluxes are formulated so that the flux leaving control volume no. i enters control volume no. i+1, thus maintaining exact material balance. If the concentration $c_{A;i}$ is chosen as the dynamic variable (instead of the amount $M_{A;i}$) this property will be violated whenever the control volume interfaces z_i and z_{i+1} are not moving with the same velocity.

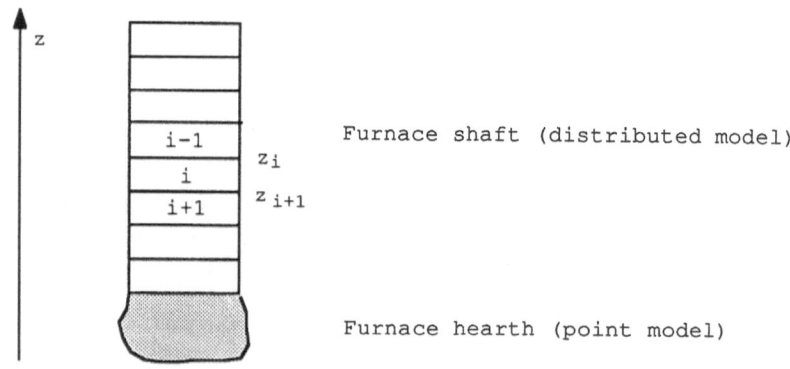

Furnace shaft (distributed model)

Furnace hearth (point model)

Figure 1 – Furnace hearth and shaft partitioned into control volumes

The material balance equations for the gas components and the heat balance equations are discretized similarly. The partial differential system is thus reduced to a system of nonlinear ordinary differential equations.

The time derivatives for gas concentrations and enthalpy can safely be neglected, as the gas dynamics is extremely rapid compared to remaining process dynamics, and the model is not sophisticated enough for a proper description of the rapid gas dynamics. However, we have got the advice to keep the time derivatives, as a stiff system of ordinary differential equations is more easily handled than a differential algebraic system [2].

For solving the ordinary differential system, we have chosen SIMPLE [3,4], a recently developed code, based on a diagonally implicit Runge–Kutta scheme. This routine has the following properties, which seem appropriate for the problem:
- variable time steps (adjusting to problem dynamics)
- A-stability (suitable for stiff equations)
- One step method (more flexible than multistep methods when handling discontinuities)
- Comparatively efficient (on standard test problems)

For each time step the SIMPLE code solves 3 nonlinear equations applying quasi Newton iteration. The same Jacobian matrix can be used for all 3 systems. The Jacobian is kept for as many steps as possible. New evaluation is performed when the iterations converge slowly or diverge.

It can be shown that any Runge–Kutta method preserves the material and heat balances provided the internal nonlinear equations are solved exactly.

4   Implementation/testing

4.1   Gas flow and heat balance

It seems to be a common experience that computational software should be implemented in stages, with appropriate testing at each stage. We chose to start extremely simple with a pure heat transfer problem with constant heat transfer and constant gas velocities. For this problem the analytical solution can easily be found. The program performed excellently after a minor debugging session.

Our problems started when implementing a realistic formulation for heat transfer (correlation proposed by Rowe et. al. [6]) and the Ergun equation for gas velocities. The test problem was letting a constant flow of warm gas enter the "furnace hearth", in order to heat up a shaft of carbon particles.

The time integration did not work properly, often leading to abnormal program stop. Uninformative FORTRAN error messages like "Square root of negative value" occurred frequently. (Caused by unphysical/inconsistent input to the function evaluation routine(s) during diverging iterations.) Some redesign was necessary.

It was observed that a small error in the total gas concentration would be magnified by a factor of 10^4 when computing the gas velocities. Our first redesign was therefore a change of dynamic variables. Instead of using n_g gas concentrations, we used Δp (deviation from atmospheric pressure) and n_g-1 gas concentrations. The implementation worked, but not efficiently. The time steps were too small. After some redesign of the step changing strategy, the step lengths could be increased to a reasonable size. The step length was now governed by the rate of convergence for solving the internal nonlinear equations, and not by accuracy (problem dynamics). Frequent evaluation of the Jacobian matrix was necessary (approximately every time step after the transient).

Our next redesign was based on the observation that the (discretized) Ergun equation can be viewed as an algebraic equation derived by dropping the time derivative in a differential equation. We introduced v_{gs} as additional dynamic variables and added the term $\rho_g \frac{\partial v_{gs}}{\partial t}$ to the left hand side of equation (5). The implementation proved successful. The program chose a very short time step length for the first steps ($\approx 10^{-4}$ s). This first transient consists of sound waves, stabilizing gas concentrations and pressure. (Initial values did not correspond to a smooth startup.) Thereafter the step length increased steadily as the solution approached the final steady state (whole shaft at the same temperature as the temperature of the furnace hearth). The final time step length was approximately 40000 s for a 100000 s simulation. (Shaft height 1 m, steady state reached after approximately 20000 s.)

The computational time was further reduced by ignoring SIMPLE's accuracy test for the gas velocities (performed with appropriate scaling). This trick increased program speed considerably during the initial stabilization. After 1 s simulation time there was no influence on the program performance (or accuracy).

4.2   Moving boundary

The feature of a moving lower shaft boundary was first implemented with a fixed spatial grid, except for the lowest shaft control volume where the lower boundary would be moving. Some modifications in the computer code were necessary in order to compute the time when the lowest control volume would reach zero height (discontinuity search).

The procedure did not work. The routine reduced time step length steadily as the time for zero height was approached. Close to the discontinuity each time step was approximately half of the time left before zero height. If a longer time step was tried, the iterations on the nonlinear equations diverged. A mathematical analysis revealed what was the problem: The Jacobian grows singular as the discontinuity is approached. Some elements will be proportional to $1/\Delta z_n$, and a quasi Newton solution procedure is doomed to fail.

An alternative is a "rubber band" grid, where each control volume interface is moving with a velocity proportional to the distance from the shaft top. The implementation was straightforward. Program tests have shown that the feature works properly.

4.3   Irreversible kinetics

Discontinuity handling has been implemented. For each reaction and control volume we have defined a switch determining whether the reaction rate formulation for forward or backward reaction should be applied. After completion of calculations for a time step, the equilibrium expressions are evaluated to check whether the direction of each reaction is in accordance with the switch settings. If not, the time, t_d, for zero reaction rate is found (for the appropriate control volume(s) and reaction(s)). The appropriate switches are then changed and integration can continue from the time t_d. During the zero search the dynamic variables are evaluated by an interpolation polynomial. (See for instance Enright et. al. [1] for a further discussion of the principles for discontinuity handling.)

The implementation has been tested for a fast reaction where gas is consumed. A typical reaction has the form:

(ii) n_A A(s) $+ n_P$ P(g) = n_B B(s)

Fast reaction means that the partial pressure of the gas P will be very close to equilibrium,

(8) $p_P^{n_P} \approx \dfrac{1}{K_{ii}}$

when the reaction is running forward (achieved by choosing appropriate reaction rate parameters). In the test example the backward rate was set to zero. The initial temperature of the solids in the shaft was chosen such that the reaction would run forward. As the shaft is heated, the equilibrium constant decreases, and the pressure of P will increase until the temperature reaches the level where K_{ii} corresponds to maximum partial pressure of P (here: the boundary value at the furnace hearth). At this point the reaction will stop.

Figure 2A shows the computed concentration (in the shaft) of gas P in a test run with one shaft control volume. In this case the fast reaction stops at 1200 s.

The straightforward implementation did not work properly. Two computational problems were encountered:

- p_P crossing the equilibrium pressure prematurely
- transient behaviour (with small time steps) after proper crossing of the equilibrium

For the former case, the routine reduces time step length drastically when crossing the equilibrium. After crossing, the computed pressure almost immediately comes back to the right side of the equilibrium, while the routine is maintaining very short time steps. Using more than one shaft control volume, oscillatory behavior was observed, with the gas pressure frequently crossing the equilibrium.

It can be noticed that for a fast reaction, a small error in the computed pressure of P, will correspond to a comparatively large error in the deviation from equilibrium (and correspondingly in the reaction rate). The premature crossing of the equilibrium is probably caused by this property of the system. Hopefully the problem can be solved through some stabilizing procedure. Applying the backward Euler algorithm on the gas equations at appropriate intervals could do the job, as this algorithm is extremely stable (L-stable) and should therefore effectively damp out deviations from the quasi stationary solution of the gas variables.

The latter problem will be considered in more detail. First notice that turning off a fast reaction implies a significant change in some of the largest elements of the Jacobian matrix.

Figure 2B shows what happens when the Jacobian is not evaluated immediately after turning off the fast reaction. The time for passing the discontinuity is properly evaluated, but the gas concentration shows no sign of discontinuity (in the derivative(s)) nor any sharp transient.

When passing the discontinuity, the previous computed Jacobian becomes obsolete and we would expect SIMPLE to do a reevaluation. The routine finds, however, no signs of bad convergence, and "converge" to an erroneous solution (after 1-3 iterations). A possible explanation can be as follows: The gas variables follow a quasi stationary solution with some internal relationship between the variables. This relationship must further be preserved during the first few Newton iterations when applying the associated Jacobian matrix. When passing the discontinuity the true solution gets a new relationship, but the routine preserves the old one until the Jacobian is reevaluated.

This idea of a relationship which is preserved by the associated Jacobian, can also explain the premature crossing of equilibrium: Using a slightly obsolete Jacobian, the computed gas variables may drift away from the true solution. After reevaluating the Jacobian, the deviation can be such that the equilibrium is crossed as the routine tries to get back to the quasi stationary solution (for the gas variables). This is

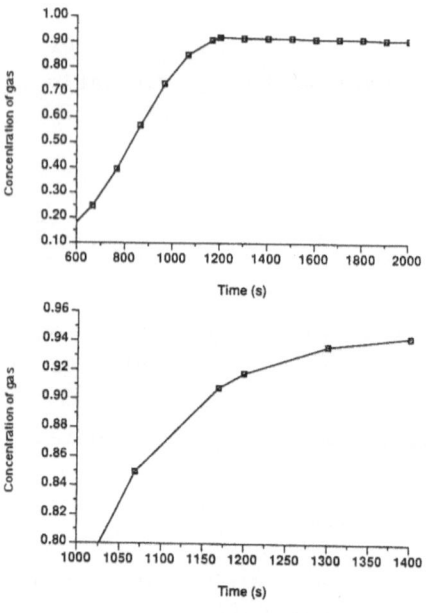

A – Concentration of the gas P
in the shaft control volume.
Jacobian computed at
discontinuity.

B – Concentration of the gas P
in the shaft control volume,
when passing the discontinuity
without Jacobian reevaluation.

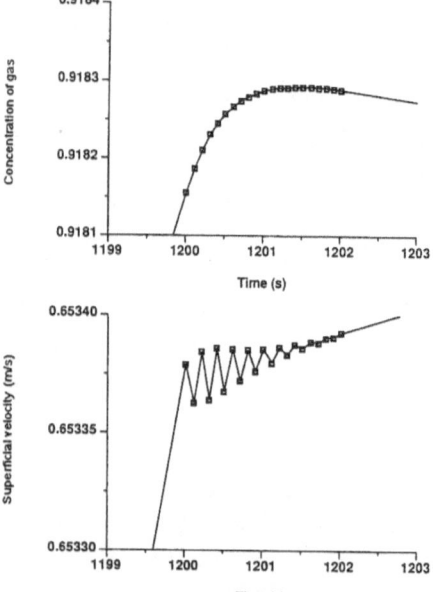

C – Concentration of the gas P
in the shaft control volume.
Details at discontinuity.
Jacobian evaluated at the
discontinuity.

D – Velocity of gas leaving
the the furnace.
Details at discontinuity.
Jacobian evaluated at
discontinuity.

Figure 2 – Test example with one shaft control volume.
Computed solution.

consistent with observations indicating that premature crossing of equilibrium only occurs immediately after a Jacobian evaluation.

Further investigations are needed to verify these concepts, and, provided verified, find the nature of the relationship between the variables. It may turn out that the observations are implementation and/or problem dependent.

After passing a discontinuity the true solution is (approximately) computed if and only if the Jacobian is reevaluated. The computed gas concentration in the test example shows a very sharp transient at 1200 s (figures 2A and 2C). Mathematical analysis reveals that such transients follow when making significant, discontinuous changes in some of the largest elements of the Jacobian.

Some stabilizing procedure is necessary in order to find the new quasi stationary solution for the gas variables. In the test example we have chosen 20 short time steps (0.1 s). Oscillatory behaviour of the computed gas velocity indicates that significant larger time steps can not be applied (figure 2D).

More efficient stabilizing procedures should be tried. A procedure utilizing an L-stable method (e.g. backward Euler) could be the optimal choice.

5   Conclusions

A (preliminary) unidimensional dynamic model for simulating a metallurgical reactor has been implemented. Further coding and program verification is needed before having a model ready for adjustments against experiments/process observations.

Application of control volume spatial discretization and a suitable (standard) time integration routine for (stiff) ordinary differential equations, has proven successful for the heat transfer problem.

The approach does not work "as is" for problems involving irreversible, fast gas reactions. Some method for stabilizing the gas variables seems necessary. An L-stable method may prove necessary.

If the model incorporates control volumes that can approach zero volume, Newton type methods are unsuitable for solving the nonlinear equations involved.

6   Final remarks

A project for developing a model for an industrial process is a typical multidiscipline project.   The ability of communicating across boundaries between traditional professions is a *must*. The development of our model has involved specialists on process understanding (metallurgy/chemistry), model formulation and analysis (metallurgy/physics/chemistry/research experience), and numerical/mathematical analysis.

A mathematician on his own will be of minute benefit for an industrial project.   In cooperation with specialists on appropriate fields the mathematician can be of great value for ensuring success.

7   Symbol list

ε	void fraction
μ	gas viscosity
ρ_g	gas density
a_i	stoichiometric coefficient
A	solid component
B	solid component
c_A	concentration of solid component A (moles/m^3)
$c_{A;i}$	concentration of solid component A in control volume i = $M_{A;i}$ / Δz_i
c_Q	concentration of gas component Q (moles/m^3 shaft volume)
d_p	particle diameter
F_i	effective surface area for reaction i
H_s	enthalpy for the solids (J/m^3)
H_g	gas enthalpy (J/m^3)
k_{si}	chemical reaction coefficient for reaction i
k_{ti}	mass transfer coefficient for reaction i
K_i	equilibrium constant for reaction i
$M_{A;i}$	amount of solid component A in control volume i (moles/m^2)
M_g	average molecular gas weight
n_A	stoichiometric coefficient
n_B	stoichiometric coefficient
n_P	stoichiometric coefficient
n_Q	stoichiometric coefficient
p	gas pressure

p_P partial pressure of the gas P
p_Q partial pressure of the gas Q
P   gas component
q_i stoichiometric coefficient
Q   gas component
Q_{gs} enthalpy flux from gas to solids
R   molar gas constant
R_i reaction rate for reaction i
t   time
T_g gas temperature
v_g vertical gas velocity
v_{gs} superficial gas velocity (= $\varepsilon\ v_g$)
v_s vertical velocity of solid components
$v_{s;i}$ vertical velocity of solid components at the location z_i
z   vertical coordinate
z_i location of the interface between shaft control volumes i-1
        and i
Δz_i height of control volume i

8 References

[1]   Enright W.H.; Jackson K.R.; Nørsett S.P., Thomsen P.G.:
      Effective Solution of Discontinuous IVPs Using a Runge-
      Kutta Formula Pair with Interpolants, Numerical Analysis
      Report No. 113, Department of Mathematics, University of
      Manchester, England, 1986

[2]   Hindmarsh Alan and Petzold Linda (Lawrence Livermore
      National Laboratory): Private communications

[3]   Nørsett S.P.; Thomsen P.G.: Embedded SDIRK-methods of Basic
      Order Three, BIT 24 (1984) 634-646

[4]   Nørsett S.P.; Thomsen P.G.: User's Guide for SIMPLE – a
      stiff system solver, to appear as institute report,
      Department of Numerical Mathematics, The Norwegian
      Institute of Technology, Trondheim, 1987

[5]   Patankar S.V.: Numerical Heat Transfer and Fluid Flow,
      Series in computational methods in mechanics and thermal
      sciences, McGraw-Hill, New York, 1980

[6]   Rowe P.N.; Claxton K.T.; Lewis J.B.: Heat and Mass Transfer
      from a Single Sphere to Fluid Flowing through an Array,
      Trans. Inst. Chem. Eng. 43 (1965) T321-T331,

[7]   Szekely J.; Evans J.W.; Sohn H.Y.: Gas-Solid Reactions,
      Academic Press, New York, 1976

Address of the author: Elkem a/s, R & D Center
                       P.O.Box 40, Vågsbygd
                       N-4602 KRISTIANSAND S
                       NORWAY

Wave Induced Washout of Submerged Vegetation

in Shallow Irish Lakes.

by

P.F. Hodnett, Department of Mathematics, National Institute for
Higher Education, Limerick, Ireland, and J.J. King, Central
Fisheries Board, Glasnevin, Dublin, Ireland.

ABSTRACT

The Irish Central Fisheries Board which is responsible for
monitoring and maintaining fish stocks in Irish lakes, has noted
that at times vegetation growing on lake bottoms is washed away
by the action of the wind on the lake surface. Since the
vegetation is necessary for the health and survival of fish
stocks it is desirable to prevent vegetation washout but to do
so, it is necessary to understand the mechanism through which it
occurs. It is, therefore, necessary to establish a model which
can predict washout for given wind speed, wind direction and lake
geometry. To achieve this, it is necessary to understand how the
action of wind driven waves on the surface of the lake is
transmitted down to the lake bottom to create bottom stresses
which cause washout of vegetation from the lake bottom. This
paper details initial attempts to create such a predictive model
for submerged vegetation washout.

1. INTRODUCTION

The submerged vegetation can play an important role in the lake ecosystem. The macrophytes may, depending on biomass and cover density, play a major role in primary production. In addition, their three-dimensional form creates an extensive physical framework which provides shelter and attachment for a range of invertebrated animals. The diversity of the invertebrate life renders such vegetated areas important as fish feeding grounds (as well as for several species of wild fowl). Vegetation also has a role in sediment stabilization.

Many of the larger Irish trout fishery lakes lie on limestone. They have hard water (high values of alkalinity and conductivity). Such systems favour the development of large crops of the macroscopic algae known as Charophytes (Stoneworts).

In recent years the Charophyte vegetation has been washed up on the shoreline of certain lakes after periods of strong winds. Washout occurred on shorelines facing the prevailing wind of that time. Such washout has a deliterious effect on the ecology of the lake. Loss of vegetation leads to loss of invertebrate life, destablizing of sediment and potential release of nutrients (in particular phosphate) from decay of washed-out plants and from disturbance of the exposed sediment.

The untimate aim of the present equiry is to develop a system
which would enable prediction of plant washout and the depths to
which it would occur for winds of know direction and force.
In this initial study the well established mathematical tools of
linear wave theory are used to develop such a predictive system.

2. OBSERVATIONAL INFORMATION

In Fig. 1 a number of the Irish lakes (L. Sheelin; L. Conn; L.
Owel; L. Ennel; L. Arrow; L. Cullin; L. Corrib Lwr.) are drawn to
scale. For reference the direction north is marked as is a line
3 km. in length. The areas marked ⬛️ in the lakes are areas
of Charophyte vegetation growth. The areas marked with xx on the
lake shores are areas of observed vegetation washout after
storms. Vegetation washout has been observed in L. Sheelin,
L. Ennel and L. Corrib Lower. A distinguishing physical feature
of all these lakes is that they are shallow in relation to their
surface area. In fact both L. Cullin and L. Corrib Lower are
extremely shallow with maximum depths of only 3m.

Fig. 2 gives detailed information in relation to observed washout
of vegetation in L. Ennel. Fig. 2(a) shows the Charophyte
vegetation growth before a storm. Fig. 2(b) shows washout of
vegetation on the north eastern shore after storms (winds
approximately from the west) in September 1983, October 1984,
August - October 1985. From the records of a local
Meteorological Station the storm in September 1983 occurred on the
3rd and 4th days of that month with the wind direction varying

between 20° south of west to 20° north of west with maximum winds
speeds of 20 knots (storm force 5) or approximately 10 m.sec^{-1}.
Fig. 2(c) shows washout of vegetation on the south western shore
of the lake after storms (winds from the north and north west) in
April 1983, March 1984, March 1986.

These observations suggest that for washout of vegetation to
occur at a point in the lake the wind speed must exceed some
minimum value and that in addition the distance, along the wind
direction, from the shore line to that point on the lake (called
the _fetch_) exceeds a minimum value. No doubt the storm must also
persist for a minimum time period.

3. WIND,WATER WAVE EMPIRICAL DATA

As reported by Kang et.al. (1982) widely used empirical data
relating to wind generated water waves date from measurements by
Bretschneider (1958) in Lake Okeechobee (U.S.A.) and the Gulf of
Mexico. Bretschneider obtained graphical relations for the
significant wave height, period, wind speed, wind fetch and mean
depth. Ijima and Tang (1966) transformed these graphical
relations to the following more convenient numerical forms (c.f.
also U.S. Army CERC, 1973).

$$\frac{gH_s}{U^2} = 0.283 \tanh\left[0.53\left(\frac{gD}{U^2}\right)^{0.75}\right] \tanh\left[\frac{0.0125\left(\frac{gF}{U^2}\right)^{0.42}}{\tanh\left(0.53\left(\frac{gD}{U^2}\right)^{0.75}\right)}\right] , \quad (1)$$

$$\frac{gT_s}{2\pi U} = 1.2 \tanh \left[0.833\left(\frac{gD}{U^2}\right)^{0.375}\right] \tanh \left[\frac{0.077\left(\frac{gF}{U^2}\right)^{0.25}}{\tanh\left(0.833\left(\frac{gD}{U^2}\right)^{0.375}\right)}\right] , \quad (2)$$

where H_s is the significant wave height, T_s is the significant
wave period, U is the wind speed, F is the fetch length, D is the
mean depth along the fetch length and g is gravitational
acceleration. The significant wave height is the average height
of the one-third highest waves in an observed wave train (say,
containing approximately 100 waves), and is approximately equal
to the average height observed by trained personnel. The
significant wave period is the average period corresponding to
the highest one-third waves. As the mean depth, D, increases
(i.e. D large), equations (1) and (2) reduce to the simpler deep
water relations (not used in this paper).

A wave train with variable wave heights is therefore modelled
with a wave train with uniform wave height H_s and corresponding
wave period T_s determined through expressions (1), (2) when
U,F,D, are given.

4. WATER WAVES - LINEAR THEORY

In two space coordinates (x,y) where x denotes horizontal
distance and y vertical distance measured positive upwards with
velocity vector $\underline{u} = (u,v)$ given in terms of a velocity potential
ϕ by $\underline{u} = \nabla\phi$, linear water wave theory (c.f. Whitham, 1974,
Chap. 13) gives the following. Let $y = \eta(x,t)$ represent the

elevation of the water surface (i.e. y=0 represents the undisturbed water surface) and $y = -d(x)$ represent the lake bottom. The surface displacement η is given by

$$\eta = a \sin (kx-wt), \tag{3}$$

representing a progressive wave of small amplitude, a, wave number, k, frequency, w, propagating in the x direction. The wave height is 2a, the wavelength, $L = 2\pi/k$ and the period $T = 2\pi/w$. The velocity potential ϕ is given by

$$\phi = -\frac{aw \cosh k (y+d)}{k \sinh (kd)} \cos (kx-wt), \tag{4}$$

where the dispersion relation between frequency and wave number is

$$w^2 = gk \tanh (kd). \tag{5}$$

The phase speed, $c(k)$ is given by

$$c(k) \equiv w/k = \left(\frac{g}{k} \tanh kd\right)^{1/2}.$$

Linear wave theory is valid provided that the waves are of small amplitude so that squares of η and ϕ may be neglected. The particular solution (4) for ϕ is valid provided that the lake bottom is a slowly varying function of x i.e. provided that the derivative of $d(x)$ is small in comparison to one.

The horizontal component, u of the velocity $\underline{u}$ is given from (4) by

$$u \equiv \phi_x = aw \cosh k(y+d)\sin(kx-wt)/\sinh(kd), \tag{6}$$

whose value at the lake bottom, denoted by u_b, is given by

$$u_b = aw \sin(kx-wt)/\sinh(kd). \tag{7}$$

The maximum value of u at the lake bottom, denoted by u_{bm}, given by (7), is

$$u_{bm} = aw/\sinh(kd). \tag{8}$$

5. INITIAL PREDICTIVE MODEL

For a given lake, from known storm washout conditions (Meteorological records) of wind in speed and direction, use expression (1) and (2) to calculate H_s (the significant wave height) and T_s (the significant wave period) for each point in the lake surface. In these calculations the fetch length, F, is defined as the distance in the direction of the wind between the shoreline and a particular point on the lake surface while the mean depth, D is obtained by averaging the water depth along the fetch. Then values of H_s, T_s (computed at each point of the lake surface) give for the linear wave equations, the amplitude $a = \frac{1}{2} H_s$, period $w = 2\pi/T_s$ and wave number k given by expression (5) at the local depth, d. Using these values of a,w,k (and the local depth, d) the maximum value of u at the lake bottom, u_{bm}, is computed at each point on the lake bottom from expression (8). This yields a map of u_{bm} for a given lake for known storm conditions. By inspection establish if a critical value of u on the lake bottom u_{cr}, can be identified so that areas of the lake bottom where $u_{bm} \geq u_{cr}$ coincide with known areas of vegetation washout during the storm. If this is possible then a predictive model for washout of vegetation has been established where for a given lake when u_{bm} computed from expression (8) exceeds a critical value u_{cr} then washout of vegetation occurs at that point of the lake bottom.

A refinement of the above process is to use the bottom shear
stress rather than the bottom velocity (as above) as a predictor
of occurrence of washout. For an oscillatory boundary layer, the
maximum shear stress exerted at a point in the lake bottom, τ_b,
is given by

$$\tau_b = \rho\, C_f\, u^2_{bm},\tag{9}$$

where ρ is the water density and C_f is a bottom friction
coefficient which depends on both the bottom surface roughness
and the flow characteristics in the bottom boundary layer. As
reported by Kang et.al. (1982) an expression for C_f (based on the
eddy viscosity assumption) in terms of the Reynolds number, R, was
proposed by Kajiura (1968) and is

$$\tfrac{1}{2}\alpha\, C_f^{-1/2} - \tfrac{1}{2}\ln C_f = \tfrac{1}{2}\alpha N + \tfrac{1}{2}\pi C_1 - \gamma + \tfrac{1}{2}\ln(\alpha/N) + \ln R,\tag{10}$$

where $\alpha = 0.4$, $N = 12$, $C_1 = 0.0593$ and $\gamma = 0.57222$.
Expression (10) assumes that the lake bottom is smooth. The
Reynolds number, R is defined by $R = \delta u_{bm}/\nu$ where ν is the
kinematic viscosity of water and $\delta = (\nu/w)^{1/2}$. When $R<200$
expression (10) is adequately approximated by

$$C_f = R^{-1}\tag{11}$$

In the refined process compute τ_b at each point of the lake
bottom from expressions (9,) (8), (10) (or (11) as appropriate).
This yields a map of τ_b for a given lake for known storm
conditions. Complete the process as before by establishing if a
critical value of τ_b, say τ_{cr}, can be established so that areas
of the lake bottom where $\tau_b \geq \tau_{cr}$ coincide with known areas of
vegetation washout during the storm. If this is possible a
predictive model for vegetation washout (based on the bottom
stress τ_b) has been established.

6. DISCUSSION

This initial attempt at developing a predictive model for vegation washout in shallow lakes is based in the assumption that linear water wave theory provides an adequate description of the effect of wind generated surface waves on the lake bottom. Only after the completion of the numerical calculations contained in the predictive model can the adequacy or otherwise of linear wave theory for this problem be decided. It may transpire that linear wave theory is not adequate and that the more complex theory of non linear shallow water waves may be necessary. Indeed another basic assumption in the model described here i.e. that the most important factor in determining vegetation washout is the value of the velocity (or shear stress) at the lake bottom is also questionable. It may be that the integrated effect of the wave forces over the total length of the vegetation (and not only at the lake bottom as is assumed here) may be important and may need to be taken into account.

REFERENCES

1. Bretschneider, C.L. 1958. Revision in wave forecastings; deep and shallow water. In Proc. 6th Conf. Coastal Engineering, pp 30-67, ASCE.

2. Ijima, T. and F.L.W. Tang. 1966. Numerical calculations of wind waves in shallow water. In Proc. 10th Coastal Eng. Conf., pp. 38-45, Tokyo.

3. Kang, S.W., Y.P. Sheng and W. Lick. 1982. Wave action and bottom shear stresses in Lake Erie, J. Great Lakes Res., $8(3)$, pp. 482-494.

4. Kajiura, K. 1968. A model of the bottom boundary layer in waves. Bulletin Earthquake Res. Inst. Tokyo Univ. 46, pp. 75-123.

5. U.S. Army Coastal Engineering Research Centre. 1973. Shore Protection Manual. Vol. 1. Fort Belvoir, Virginia.

6. Whitham, G.B. 1974. Linear and Nonlinear Waves. J. Wiley, New York.

235

CAPTIONS FOR ILLUSTRATIONS

Fig. 1. A set of Irish lakes where areas marked ⬛ in the lakes are areas of Charophyte vegetation growth. Areas marked xx on the lake shores are areas of observed vegetation washout after storms.

Fig. 2. (a) L. Ennel with Charophtye vegetation growth before a storm. (b) L. Ennel with vegetation washout on the north eastern shore after a storm from the west. (c) L.Ennel with vegetation washout on the south western shore after a storm from the north/north-west.

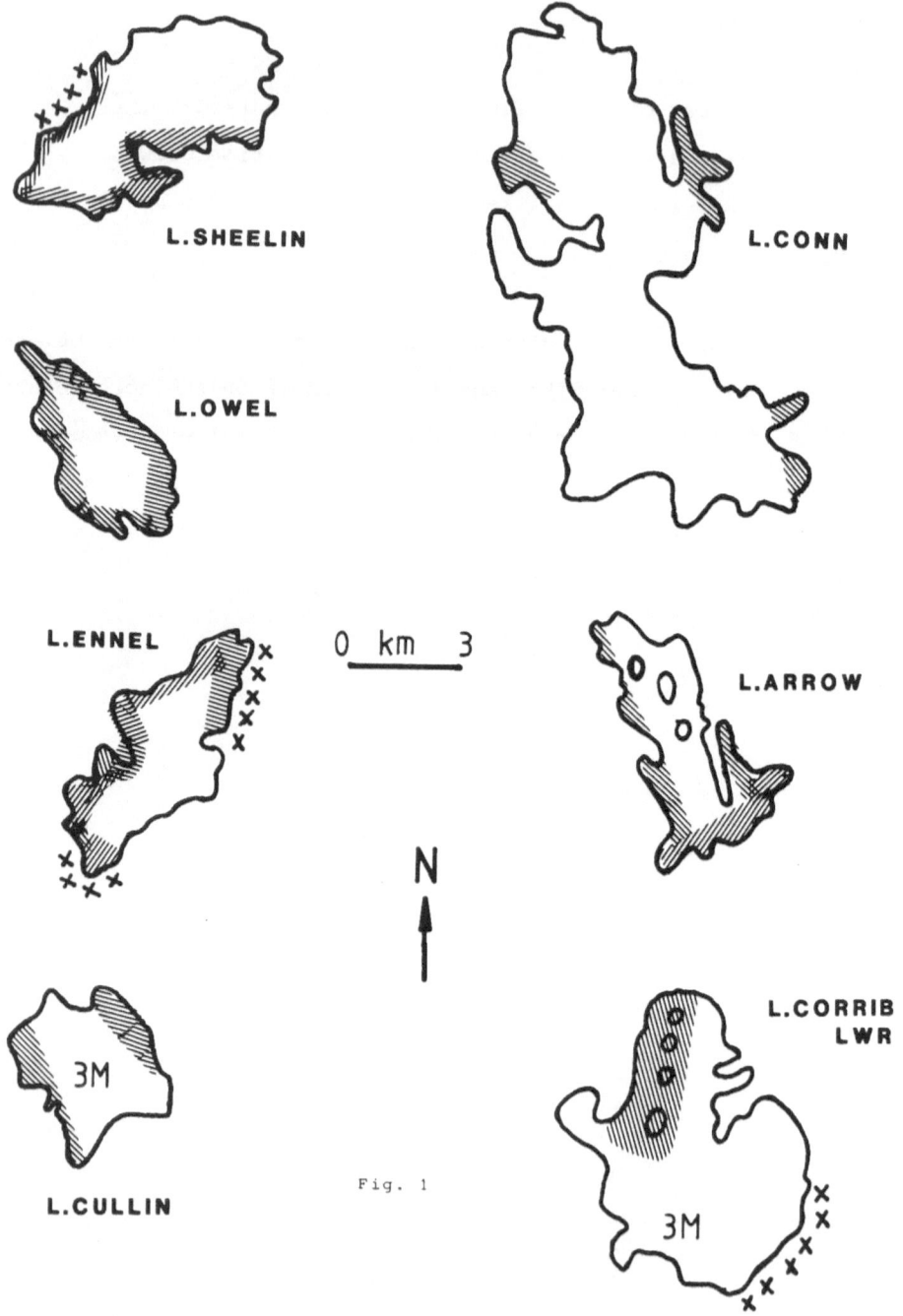

Fig. 1

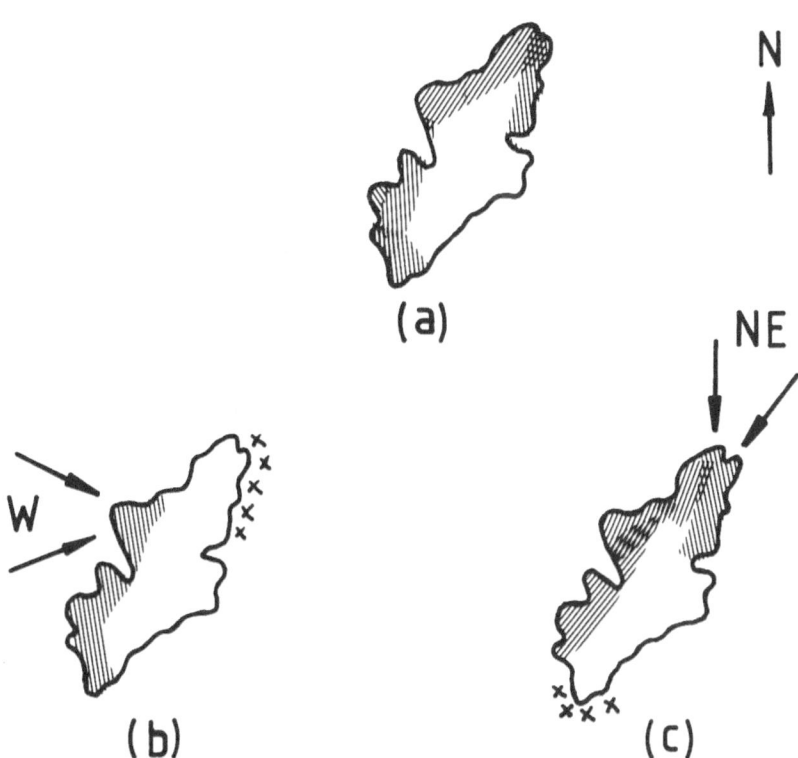

Fig. 2

APPROXIMATE CONVERSION OF SPLINE CURVES

Josef Hoschek

Fachbereich Mathematik, Technische Hochschule, D6100 Darmstadt

In German car body industries (VDA) different manufacturers and their subcontractors have different geometric modeling systems for curve and surface representations. For exchanging data between the different geometric modeling systems conversions of curve and surface representation are required in order to compensate differences in the types of polynomial bases, maximum polynomial degrees and mesh sizes. Conversion means reducing the degree of a spline curve (and splitting more than one segment) or elevating the degree of more than one spline segment (and merging to one segment). For this purpose a set of methods was developed by DANNENBERG and NOWACKI [2]. They have extended a conversion method introduced by HÖLZLE [8] for plane curves to surfaces by interpreting a surface as a net of curves. The extension of the algorithm to surfaces is implemented in the VDA-Software, but often a great number of new patches is obtained. So the question arises how to develop a method which yields to a more economical patch number. In the present paper a new conversion method for spline curves is introduced, which works very effectively for plane spline curves. The method can be extended to approximate conversion of spline surfaces and to approximation of offset curves and offset surfaces by spline curves and spline surfaces (s. [6,7]).

1. Bernstein-Bézier-techniques

The Bernstein-Bézier-techniques [1,6] use the Bernstein polynomials $B_j^n(t)$ as basis functions for the curve- or the spline-curve-representation. A *Bézier-curve* $\mathbf{X}(t)$ of degree n has the parametric representation

$$\mathbf{X}(t) = \sum_{i=0}^{n} \mathbf{b}_i \, B_i^n(t) \qquad\qquad (t \in [0,1]), \qquad (1)$$

with the Bernstein polynomials of degree n

$$B_i^n(t) = \binom{n}{i} (1 - t)^{n-i} \, t^i$$

and the (vector valued) *Bézier-points* $\mathbf{b}_i$. The linear hull of the Bézier-points is called the *Bézier-polygon*. The Bézier-curves have some very important properties:

- the *convex-hull-property*: the whole Bézier-curve lies within the convex hull of the Bézier-polygon;
- the *variation-diminishing-property*: if the Bézier-polygon has k as (maximal) number of the intersection points with an arbitrary line g , the Bézier-curve has k as the upper bound of the intersection points of g.

Fig. 1 contains some examples of Bézier-curves.

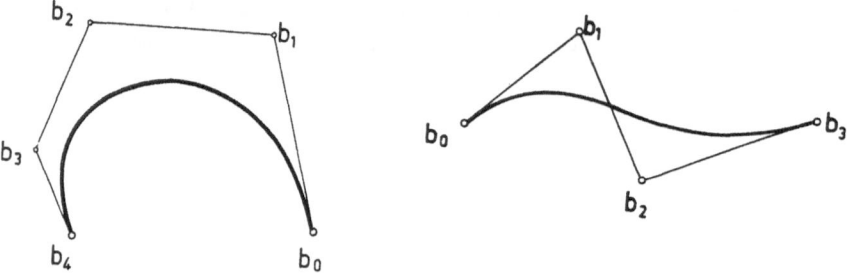

Fig. 1: Bézier-curves of degree 4, 3.

A generalization of (1) are the *rational* Bézier-curves of degree m with the parametric representation

$$\mathbf{X}(t) = \frac{\displaystyle\sum_{i=0}^{m} \beta_i \, \mathbf{b}_i \, B_i^m(t)}{\displaystyle\sum_{i=0}^{m} \beta_i \, B_i^m(t)} \,. \qquad (2)$$

The weights β_i are additional design parameters for changing the slope of the curve. For $\beta_i > 0$ the convex-hull-property and the variation-diminishing-property hold too.

2. Conversion and continuity conditions

The goals of the conversion algorithms are
- to subdivide a given spline segment of degree n to one or more spline segments of degree m (n > m, *degree reduction*) with respect to a given error tolerance ε,
- to merge as many as possible spline segment of degree M to one spline segment of degree N(M < N, *degree elevation*) with respect to a given error tolerance ε.

The key idea of the algorithm is to use the parametrization as a design parameter: The shape of an approximation curve of a set of points will be changed, if we change the parameter values of the points during the approximation process. If we use the parameter values as design parameters we need spline conditions which are invariant to the parametrization: such an invariant osculating condition is wellknown in differential geometry [3, 5] as the contact of order s of two curves. Meanwhile in Computer Aided Geometric Design the name *geometric continuity* GC^s is used instead of contact of order s.

Two curves **X** and Y have the *geometric continuity* GC^s [4], if the following conditions hold (at a common point of **X** and Y):

$$s = 1: \quad \mathbf{X}' = \lambda_1 \mathbf{Y}' \quad,$$

$$s = 2: \quad \mathbf{X}'' = \lambda_1^2 \mathbf{Y}'' + \mu_1 \mathbf{Y}' \qquad \text{(and the condition for s=1)},$$

$$(3)$$

with arbitrarily chosen parameters λ_1, μ_1. These conditions can be developed by using an osculating line or an osculating conic or out of the first terms of the Taylor expansion.

We now assume that the given curve **X** is a Bézier-curve of degree n and has the parametric representation

$$\mathbf{X} = \sum_{i=0}^{n} \mathbf{V}_i \, B_i^n(t) \qquad (4)$$

with V_i as given Bézier-points. The required approximation curves Y may have the parametric representation

$$Y = \sum_{i=0}^{m} W_i \, B_i^m(t) \tag{4b}$$

with the unknown Bézier-points W_i and $m < n$. The geometric continuity (3) will be transformed by (4a,b) into the boundary condition

$$s = 1: \quad W_0 = V_0, \quad W_1 = V_0 + \lambda_1(V_1 - V_0) , \tag{5a}$$

$$W_m = V_n, \quad W_{m-1} = V_n + \lambda_2(V_n - V_{n-1}) ,$$

$$s = 2: \quad W_2 = V_1 + \lambda_1^2 \, \omega_1(V_2 - V_1) + \mu_1(V_1 - V_0) ,$$

$$W_{m-2} = V_{n-1} + \lambda_2^2 \, \omega_1(V_{n-2} - V_{n-1}) + \mu_2(V_n - V_{n-1}) , \tag{5b}$$

$$\text{with} \quad \omega_1 = \frac{m(n-1)}{n(m-1)} .$$

3. Reduction of degree

In this section we will develop algorithms for solving the reducing problem: Given one spline segment of degree n and a given error toler-ance ε_0, find the minimal number k of knots which subdivide the given spline segment X in k new segments ($k \geq 1$) of spline curves Y of degree m ($n > m$) and a required order of contact ($s = 1,2$).

3.1 Contact of order 1

First we choose $s = 1$ and $m = 3$ and obtain from (5a) four conditions for the unknown Bézier-points W_0, W_1, W_2, W_3 of Bézier segment Y. These Bézier-points are determined by
- the parameters λ_1, λ_2,
- the parametrization of the Bézier segment Y.
To evaluate an optimal pair λ_1, λ_2 we choose on the given curve X (n+1) points P_i with (aequidistant) parameter values $t_i = i/n$ ($i = 0(1)n$) and postulate for $Y(t)$ as approximation curve of these points

$$P_i = \sum_{j=0}^{3} W_j \, B_j^3(t_i) + \delta_i \tag{6}$$

with δ_i as vectors of error. If we insert (5a) equation (6) is transformed into

$$D_i = \lambda_1(V_1 - V_0) \, B_1^3(t_i) + \lambda_2(V_n - V_{n-1}) \, B_2^3(t_i) + \delta_i \tag{7}$$

with the known vector

$$D_i: = P_i - V_0 \, B_0^3(t_i) - V_0 \, B_1^3(t_i) - V_n \, B_2^3(t_i) - V_n \, B_3^3(t_i) \; .$$

As absolute value of the error vectors in (7) follows

$$\sum_{i=0}^{n} |\delta_i|^2 =: \delta = \sum_{i=0}^{n} (D_i - \lambda_1(V_1 - V_0) \, B_1^3(t_i) + \lambda_2(V_{n-1} - V_n) \, B_2^3(t_i))^2 .$$

The minimum of δ is determined by the conditions

$$\frac{\partial \delta}{\partial \lambda_1} = 0 \; , \qquad \frac{\partial \delta}{\partial \lambda_2} = 0 \; . \tag{8}$$

These conditions lead to a linear system for λ_1, λ_2, which must be solved - but the solution depends on the given parametrization.

In general the process (8) doesn't minimize the shortest distances between the approximation curve and the given points P_i, while the error vectors in (6) are not orthogonal to the approximation curve $Y(t)$.

3.2. Parameter optimization

To transfer the (oblique) error vectors referred to equation (6) in (approximately) orthogonal error vectors, we have to change parametrization of the points P_i by the help of a suitable method: a direct path would be to drop the perpendiculars to the approximation curve $Y(t)$ from each point P_i, evaluate the parameter values t_i^* of the intersection point of each perpendicular, insert t_i^* in (6) and start the process (8) again. To avoid the computation of these intersection points, we replace the approximation curve $Y(t)$ at each point $Y(t_i)$ by the tangent T_i, drop the perpendicular from the P_i to the tangent T_i and use the distance Δc_i

between the point $Y(t_i)$ and the foot F_i of the perpendicular as an approximation of the parameter correction (s. Fig. 2). Δc_i has to be related to
the parameter interval [0,1] of the curve $Y(t)$, therefore it must be normed
by a factor ω, which can be an estimate of the arc length of the approximation curve $Y(t)$. By the help of this consideration a first approximation
of an optimal parameter value can be evaluated by

$$t_i^* = t_i + \frac{\Delta c_i}{\omega} \quad . \qquad (9)$$

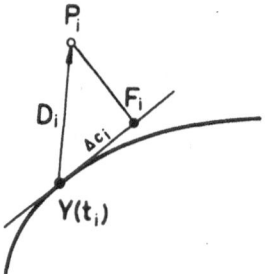

Fig. 2: Correction of the parametrization
of the approximation curve $Y(t)$.

Now we insert these parameter values in (6), repeat the minimization procedure (8) and obtain a corrected approximation curve Y^*. If we repeat the
whole procedure several times we will get error vectors δ_i which converge
to the normals of the approximation curve. Those a parametrization is obtained which leads to the minimization of the shortest distances between
the given points and the approximation curve. The obtained approximation
curve is the best with respect to the (euclidean) distance-norm.

3.3 Algorithm for GC^1-conversion of spline curves

With these results we can formulate the following algorithm for GC^1-con-
version of a spline curve $X(t)$ of degree n to a spline curve $Y(t)$ of
degree m (m < n):

①   Choose: Parameter values $t_i = \frac{i}{n}$ (i = 0(1)n) ,

           Error ε_o, $P_i = X(t_i)$,

           Index J = 1,   maximal value L.

②   Least square method $|P_i - Y(t_i)| \Longrightarrow$ min.

③   Parameter optimization   leads to t_i^* (with respect to (9)),

           J = J + 1 ,

           $t_i = t_i^*$.

245

④ Evaluate the maximal deviation

$$d_j = \max_{j=0(1)n} |P_j - Y(t_j)| \text{ , corresponding parameter value } t_j$$

If $(d_j < \varepsilon_0)$ stop .

⑤ If $(J < L)$ go to 2 .

⑥ Split the given spline curve $X(t)$ into two segments at $t = t_j$.

Fig. 3 shows the reduction of a Bézier-curve of degree 19 to 7 segments of degree 3 with a maximum error δ < 0.015 mm. The Hölzle algorithm as implemented in the VDA-Software leads to 50 segments with δ_H < 0.29 mm. For this comparison the Hölzle algorithm was used as implemented in the VDA-Software and δ_H is a output error of this algorithm. Both curves (the given and the approximating curve) are plotted one above the other, the reduction of the figures can be recognized by the given scale. The boundary points of the segments are denoted by squares with crosses, the Bézier-points of the given curve are denoted by crosses, the Bézier-points of the approximation curves are denoted by squares.

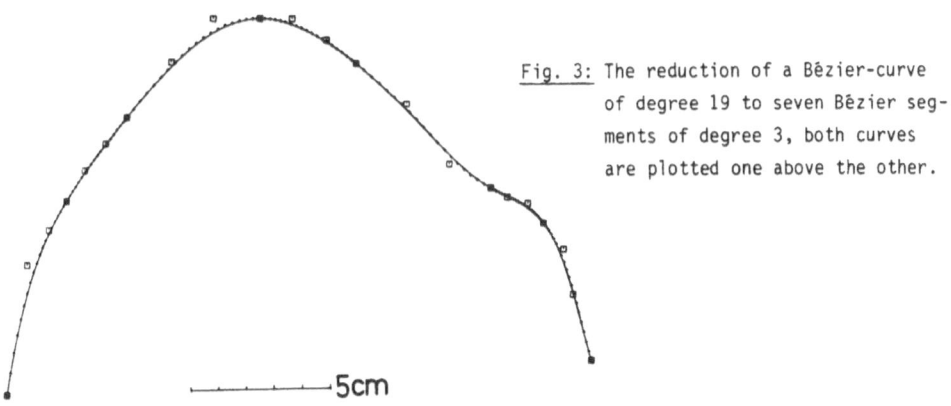

Fig. 3: The reduction of a Bézier-curve of degree 19 to seven Bézier segments of degree 3, both curves are plotted one above the other.

5cm

3.4 Contact of order 2

If we turn to contact of order 2 continuity conditions, equations (5a) and (5b) must be satisfied. Again we choose $(n + 1)$ points P_i of the given curve $X(t)$ with the parameter value t_i = i/n (i=0(1)n) and insert these points into the parametric representation of the required curve $Y(t)$. We introduce the known vector

$$D_i := P_i - V_o\, B_0^m(t_i) - V_o\, B_1^m(t_i) - V_1\, B_2^m(t_i) - V_{n-1}\, B_{m-2}^m(t_i)$$
$$- V_n\, B_{m-1}^m(t_i) - V_n\, B_m^m(t_i) \tag{10}$$

and obtain as the representation of the error vectors δ_i

$$\delta_i = D_i - \lambda_1(V_1 - V_o)\, B_1^m(t_i) - \lambda_1^2\, \omega_1(V_2 - V_1)B_2^m(t_i)$$
$$- \lambda_2^2\, \omega_1(V_{n-2} - V_{n-1})\, B_{m-2}^m(t_i) - \lambda_2(V_n - V_{n-1})B_{m-1}^m(t_i)$$
$$- \mu_1(V_1 - V_o)\, B_2^m(t_i) - \mu_2(V_n - V_{n-1})\, B_{m-2}^m(t_i) \ . \tag{11}$$

If we choose for the approximation curve $Y(t)$ the degree $m = 5$, then the deviation

$$\delta = \sum_{i=0}^{n} \delta_i^2 \ ,$$

is a nonlinear function of λ_1 and λ_2, but linear in μ_1 and μ_2. To avoid nonlinear least square methods, we subdivide the minimization algorithm into two steps:

First we choose λ_1, λ_2 suitably and minimize the error δ (for λ_i = const.) with least square methods by the help of the conditions

$$\frac{\partial\delta}{\partial\mu_1} = 0, \quad \frac{\partial\delta}{\partial\mu_2} = 0 \ . \tag{12}$$

Equation (12) leads to a 2×2 linear system for the parameters μ_1 and μ_2. The solution of (12) will be minimized by parameter optimization according to (9). If the given error tolerance ε_o is greater than δ the procedure stops, if $\varepsilon_o < \delta$ the λ_i will be suitably changed by a nonlinear optimization algorithm in direction of decreasing error value δ. We have used for our examples the *cyclic coordinate ascent* [9], which minimizes the goal function δ by cyclic changing one of the variables λ_i while the other is constant.

3.5 Algorithm for GC^2-conversion of spline curves

For GC^2-conversion of a spline curve $X(t)$ of degree n to a spline curve $Y(t)$ of degree m ($m < n$) we can use the following algorithm:

①   Choose: Parameter values $t_i = \frac{i}{n}$ $(i = 0(1)n)$,
               Error ε_0, $P_i = X(t_i)$,

               Index $J = 1$, max. value L; Index $K = 1$, max. value M,
               λ_{10}, λ_{20}.

②   Least square method $|P_i - Y(t_i)| \Longrightarrow \min$, determines μ_1, μ_2 .

③   Parameter optimization leads to t_i^* (with respect to (9)),
               $J = J + 1$, $K = K + 1$,
               $t_i = t_i^*$,
               If $J < L$ go to 2 .

④   Evaluate the maximum deviation
                    $d_j = \max_{j=0(1)n} |P_j - Y(t_j)|$, corresponding parameter value t_j,

⑤   If $d_j < \varepsilon_0$ Stop .

⑥   Change λ_1, λ_2 by the help of an optimization algorithm,
               If $K < M$, $J = 1$ go to 2.

⑦   Split the given spline curve $X(t)$ into two segments at t_j ,
               Go to 1 .

If $m > 5$ the parameters μ_j and the components of W_3, W_4,...,W_{m-3} are linear unknown variables. We differentiate δ with respect to μ_j and with respect to the components of the unknown Bézier-points W_j and receive for a plane curve a linear system of 2m-8 equations. This system has to be solved, while the optimization algorithm is unchanged.

Fig. 4 contains the reduction of the curve X of degree 19 out of Fig. 3 to a spline curve with three segments of degree 7 and an error value $\delta < 0.014$ mm.

We also can extend our reduction algorithm to contact of order 3 and (at least) $m = 7$. According to the extension of (3) to $s = 3$ in (1) we receive a linear system in μ_1, μ_2, μ_3, μ_4 and nonlinear conditions for λ_i.

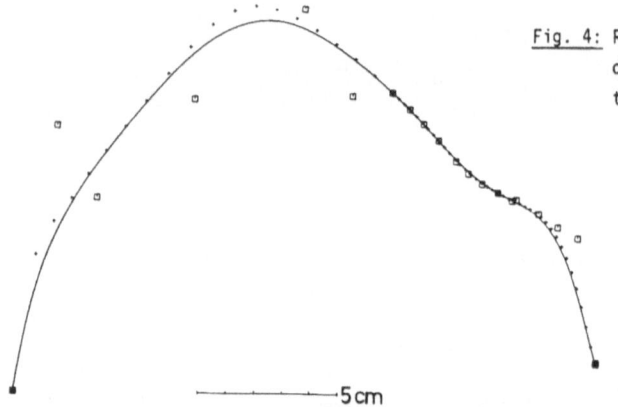

Fig. 4: Reduction of the Bézier-curve
of degree 19 out of Fig. 3 to
three spline segments of degree

—————5cm

4. Merging of spline curves

We can also use the described method in Section 3.4 for merging of spline
curves combined with degree elevation. For example we start with at least
k = 2 segments of degree M = 3 and would like to merge these segments to
one spline curve of degree N = 3 or N = 5. At the boundary points of the
new segment we will demand contact of order 2. We choose on each segment
(M + 1) points with the parameter values t_i = i/M (i = O(1)M) and repeat
for these k(M + 1) points the procedure of Section 3.5. If the error δ is
less than the given error tolerance ε_o we raise k to (k + 1) segments and
repeat the procedure until δ ~ ε_o.

5. Concluding remarks

The method introduced in this paper can be extented to spline approximation of offset curves [7]: We suppose that the given curve has the parametric representation $\mathbf{X}$ = $\mathbf{X}(t)$ (according to (2)), then the corresponding
offset curve $\mathbf{X_d}$ at (oriented) distance d along the unit normal vector $\mathbf{n}(t)$
is given by

$$\mathbf{X_d}(t) = \mathbf{X}(t) + \mathbf{n}(t)\ d \qquad\qquad t \in [0,1]\ . \qquad\qquad (13)$$

The approximation curve $\mathbf{Y}(t)$ of the offset curve has the parametric representation

$$\mathbf{Y}(t) = \sum_{i=0}^{m} \mathbf{W_i}\ B_i^m(t)\ ,$$

where the unknown Bézier-points $\mathbf{W}_i$ can be determined analogously to chapter 2, 3: The boundary Bézier-points of $\mathbf{Y}(t)$ are given by

$$\mathbf{W}_0 = \mathbf{V}_0 + \mathbf{n}(0)\, d$$
$$\mathbf{W}_3 = \mathbf{V}_n + \mathbf{n}(1)\, d\,. \tag{14}$$

If we put (14) in the corresponding position of (5a) the algorithm for GC^1-conversion leads to cubic spline approximation of the given offset curve $\mathbf{X}_d(t)$. For higher degree approximation we have to exchange in (5b) the factor ω_1 for

$$k_i = \frac{m(n-1)}{n(m-1)}\, (1 + \kappa_i\, d)^{-1}$$

with κ_i as curvature of the given curve $\mathbf{X}(t)$ at $\mathbf{X}(0)$ respectively $\mathbf{X}(1)$. The GC^2-conversion-algorithm leads to (at least quintic) spline approximation of the given offset curve $\mathbf{X}_d(t)$.

References

[1] Böhm, W., Farin, G., Kahmann, J.: A survey of curve and surface methods in CAGD. Computer Aided Geometric Design 1, 1-60 (1984)

[2] Dannenberg, L., Nowacki, H.: Approximate conversion of surface representations with polynomial bases, Computer Aided Geometric Design 2, 123-132 (1985)

[3] do Carmo, M.: Differential Geometry of Curves and Surfaces, Prentice Hall, Englewood Cliffs, NJ. 1976

[4] Geise, G.: Über berührende Kegelschnitte einer ebenen Kurve. Zeitschr. Angew. Math. Mech. 42, 297-304 (1962)

[5] Goetz, A.: Introduction to Differential Geometry, Addison Wesley, Reading, 1968

[6] Hoschek, J.: Approximate conversion of spline curves. Computer Aided Geometric Design 4, 59-66 (1987)

[7] Hoschek, J.: Spline approximation of offset curves. Computer Aided Geometric Design 4 (soon published)

[8] Hölzle, G.E.: Knot placement for piecewise polynomial approximation of curves, Computer-aided Design 15, 295-296 (1983)

[9] Zangwill, W.I.: Nonlinear Programming, An Unified Appoach, Prentice Hall, Englewood Cliffs, NJ. 1969

Interpolation Algorithms for the Control of a Sewing Machine

P. Rentrop, U. Wever

Technomathematik, Universität Kaiserslautern

Abstract: Properties of a 'visually pleasing' interpolant for a given data set (x_i, y_i), $i = 0, \ldots, n$ are discussed. There are two different ap= proaches to construct a visually pleasing interpolant. One can use local $C^1[x_0, x_n]$ – or global $C^2[x_0, x_n]$ methods. Both approaches have their own merits and cannot be played off against another. At typical examples the performance of several algorithms for both approaches is studied. Finally the tested in= terpolation schemes were used for curve interpolation and are prepared for the control of a sewing machine.

1 Introduction

Future sewing machines are coupled with a small programmable micro= processor and a screen. The designer can fix points at the screen and the mi= croprocessor creates an interpolating curve and computes the coordinates for the needle position automatically. The typical points of a sewing pattern are divided into normal points ✳ and marked points ○ , see sketch 1.

Sketch 1:   ✳ normal points,   ○ marked points

This paper is in final form and no version of it will be submitted for publication

The marked points o determine the single parts of the sewing pattern, while the normal points # describe the details. Mathematically spoken, an interpo= lating curve should be continuous at the marked points and more smoothly bet= ween two marked points. There are further requirements for an interpolant. The curve should appear immediately at the screen and should preserve the sha= pe of the normal points.

To study these conditions more carefully, we discuss desirable properties of interpolation schemes at part 2. Unfortunately it is not possible to combine all desired properties. In part 3 local C^1-procedures are considered. Typical C^1-interpolants are quadratic splines and cubic Hermite polynomials. C^2-inter= polants like the cubic spline and the exponential spline are discussed in part 4. In part 5 the algorithms are generalized to curve interpolation and tested for several examples. An efficient method to determine the needle position is derived in part 6. Finally, in part 7, some conclusions for the presented in= terpolation schemes are drawn.

2   Properties of Interpolation Schemes

        In literature, several properties for a 'good' interpolant have been introduced. Fritsch, Carlson [6] create a 'visually pleasing' interpolant and Knuth [8] considers a 'most pleasing' curve. Thus 'visually pleasing' or 'most pleassing' is not a well defined mathematical term. Therefore some properties are discussed, that a 'visually pleasing' interpolant should satisfy.

Property I: The interpolation scheme should be linear in the data.
Let $F(y)$ denote the set of all interpolating functions $f(x)$ with $f(x_i) = y_i$.
For given data sets (x_i,y_i), (x_i,u_i), (x_i,v_i), $i=0,\ldots,n$ it should hold:
(Ia) additive linearity:

$$y_i = u_i + v_i, \quad i=0,\ldots,n \implies Y = U + V, \quad \text{where} \quad U \in F(u), V \in F(v), Y \in F(y)$$

(Ib) multiplicative linearity:

$y_i = c\ u_i$, i=0,...,n for some real constant c $\implies$ Y = c U, where
$$Y \in F(y),\ U \in F(u).$$

Essentially property (Ib) guarantees the invariance of the interpolation to different scaling of the data.

Property II: The interpolant should be free of sharp corners, bumps or oscil= lations. The tangents and possibly the second derivatives at the nodes x_i should be continuous. The interpolant is expected to be smooth. Linear seg= ments of data should be recognized and reproduced.

Property III: A segment of the interpolant should depend only on the immedi= ately preceeding and following points. This local behaviour allows a simple updating for additional points.

Property IV: The interpolant should be shape preserving. Monotonicity and convexity / concaity of the data should be transferred to the interpolant. If only data segments behave monotone increasing or convex, we expect our inter= polant to do the same. Schumaker [14] denotes this property with co-monotone or co-convex.

Property V: The complexity of the interpolation scheme should be of order O(n), where n denotes the number of data points.

For different applications one can enlarge this list of properties i. e. to roundness: the deviation between an interpolant and a circle defined by three points should be small, see Knuth [8].

Although the five properties are rather easy to justify on intuitive grounds, there is no good way to satisfy them all. Property I suggests the use of po= lynomials. In order to control the oscillations of a polynomial and to guaran= tee the shape preserving property IV, the polynomial degree should not exceed three. If continuous second derivatives are required, the local property III is violated. If the local property III is valid, only continuous first deri= vatives are possible. This contradicts property II. Therefore two different approaches are studied. The first approach results in local schemes, which produces C^1-interpolants and is treated in part 3. The second approach leads to global C^2-interpolants, see part 4. Typical algorithms resulting from the= se approaches are tested in part 5.

Although property IV looks quite evident, in practical examples it is diffi= cult to switch back and forth between convex and concave data segments or to decide whether monotonicity is a desirable requirement. To get some insight, look at sketch 2 and sketch 3.

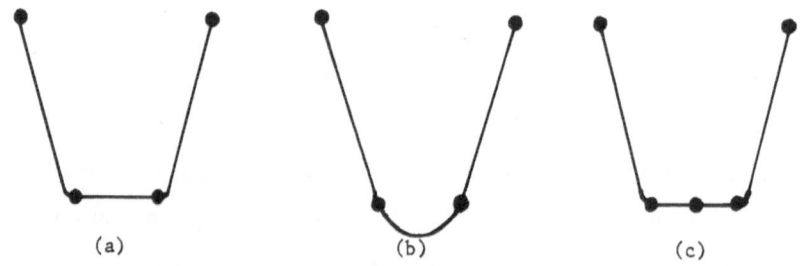

Sketch 2: Different data segments, ● given data

Obviously monotonicity and convexity arguments meet in sketch (2c), whereas it is not clear, whether monotonicity requirements are too stringent in sketch (2a). In our feeling (2b) is more visually pleasing than (2a).

The change between convex and concave data segments in sketch 3 must create an inflection point in the interpolating curve and therefore two extremal va= lues. Monotonicity requirements can damp these extremals.

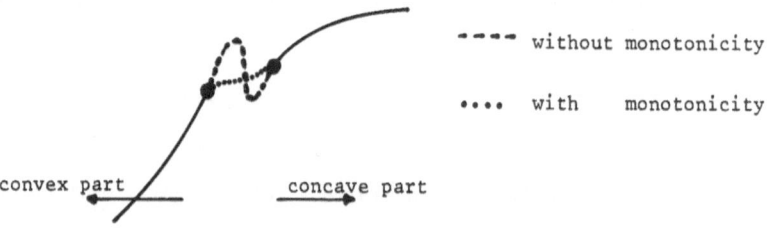

Sketch 3: Convex-concave data segments

3   Local C^1 - Interpolation

In literature essentially two typesof local C^1-interpolation pro=
cedures are distinguished: cubic Hermite polynomial and quadratic spline in=
terpolation.

In the case of the cubic Hermite interpolation we follow Akima [1]. Let
(x_i,y_i), $i=0,\ldots,n$ the given data points with ordered nodes $x_o \leq \ldots \leq x_n$.
In each subinterval a cubic Hermite polynomial is defined by:

$$p(x) = y_i(3(1-t)^2-2(1-t)^3) + y_{i+1}(3t^2-2t^3) + c_i h_i((1-t)^2-(1-t)^3) +$$
$$+ c_{i+1} h_i(t^3-t^2) \tag{2.1}$$

$$x_i \leq x \leq x_{i+1}, \quad h_i = x_{i+1}-x_i, \quad t = (x-x_i)/h_i, \quad i=0,\ldots,n-1$$

(2.1) defines a C^1-function in $x_o \leq x \leq x_n$ by construction. There are $(n+1)$
degrees of freedom for an appropriate choice of the tangents c_i, in order to
satisfy the properties I, III, V and partially II and IV. It is not possible
to enlarge the global smoothness beyond one, unless we violate the local pro=
perty or enlarge the polynomial degree. An interesting scheme to preserve the
monotonicity of the data has been proposed by Fritsch, Carlson [6]. The flow
chart 1 describes the determination of the tangents c_i. The c_i - choice re=
sults from the study of the quadratic equation of the first derivative of $p(x)$.

Flow Chart 1: Algorithm of Fritsch, Carlson [6]

data: (x_i,y_i), $i=0,\ldots,n$

initialize:

$$c_i = \frac{y_{i+1}-y_{i-1}}{x_{i+1}-x_{i-1}}, \quad d_i = \frac{y_{i+1}-y_i}{x_{i+1}-x_i}, \quad i=1,\ldots,n-1$$

$$c_o = \frac{y_1-y_o}{x_1-x_o}, \quad c_n = \frac{y_n-y_{n-1}}{x_n-x_{n-1}}$$

monotonicity region: $A_i=c_i/d_i$, $B_i=c_{i+1}/d_i$, $i=0,\ldots,n-1$

$$R = \{ A^2 + B^2 \leq 9 \}$$

$$\text{if } (A_i,B_i) \notin R \text{ then } A_i = \frac{3A_i}{\sqrt{A_i^2+B_i^2}}, \quad B_i = \frac{3B_i}{\sqrt{A_i^2+B_i^2}}$$

corrected tangent: $c_i = A_i d_i$ (or $c_{i+1} = B_i d_i$), $i=0,\ldots,n-1$

Before we finish the discussion of the cubic Hermite interpolation, we men=
tion that there are infinitely many ways to choose the tangents c_i. For other
examples see Akima [1], Ellis,McLain [5], Knuth [8]. In Heß, Schmidt [7] glo=
bal arguments are introduced - the minimization of an energy functional - in
order to achieve uniqueness of the interpolant. But, unfortunately, this de=
stroys the local property III.

Another type for local C^1-interpolation is the quadratic spline interpolation
as described in McAllister, Roulier [9], [10] and Schumaker [14]. The solution
of the problem (3.2) is performed by quadratic polynomials.

$$p(x_i) = y_i, \quad p(x_{i+1}) = y_{i+1}$$
$$p'(x_i) = c_i, \quad p'(x_{i+1}) = c_{i+1} \quad i=0,\ldots,n-1 \tag{3.2}$$

In order to compute the coefficients of the quadratic polynomial

$$p_i(x) = A_i + B_i(x-x_i) + C_i(x-x_i)^2, \; i=0,\ldots,n-1 \tag{3.3}$$
$$x_i \leqslant x \leqslant x_{i+1}$$

one has to insert additional knots. The location of the at most n knots can
be used to preserve the shape of the data. Additionally one has the freedom
to choose the n+1 tangents c_i. Theoretically the properties I, III, IV and V
can be satisfied. A complete flow chart for a quadratic spline algorithm
can be found in Schumaker [14]. Schumakers method satisfies the properties
I and IV only partially. Because of the use of the length

$$L_i = SQRT(\;(x_{i+1}-x_i)^2 + (y_{i+1}-y_i)^2)$$

the linearity of the interpolant is violated. The superposition principle is
not valid. The interpolant is co-monoton and co-convex in those intervals,
where $p_i(x)$ has no inflection point. Because of these restrictions Schumaker
embeds his algorithm into an interactive system, where the user can adjust
the slopes and the knots. This interactive algorithm produces really visual=

ly pleasing curves.

As in the case of cubic Hermite interpolation, there are infinitely many ways

to determine the tangents c_i. One can achieve uniqueness by introducing glo=

bal aspects.

4   Global C^2 - Interpolation

If one drops the local property III, it is possible to construct

smooth, shape preserving, global C^2-interpolants using spline techniques,

which i.e. are described in Bulirsch, Rutishauser [2], DeBoor [4], Stoer, Bu=

lirsch [17]. In order to avoid possible 'overshooting' of a cubic spline, we

concentrate our interest to 'splines under tension' or so-called 'exponential

splines'. An exponential spline through the given data can be interpreted as

a tie rod under tensile forces with minimal bending energy.

$$\text{min.} \int_{x_0}^{x_n} (y''(x)^2 + P(x)\, y'(x)^2)\, dx \,, \tag{4.1}$$

where $P(x)$ is proportional to the tensile forces in each interval: $P(x) = p_i$,

for $x_i \leq x \leq x_{i+1}$, $i=0,\ldots,n-1$. The exponential spline was introduced by

Schweikert [15], properties of the exponential spline have been studied by

Cline [3], Pruess [11], Späth [16], and in [12]. As a representation of the

exponential spline we choose:

$$E(x) = y_{i+1}\, t + y_i\, (1-t) + \frac{d_{i+1}}{p_i^2} \left(\frac{\sinh(u_i t)}{\sinh(u_i)} - t \right) + \frac{d_i}{p_i^2} \left(\frac{\sinh(u_i(1-t))}{\sinh(u_i)} - (1-t) \right) \tag{4.2}$$

where: $x_i \leq x \leq x_{i+1}$, $h_i = x_{i+1}-x_i$, $t = (x-x_i)/h_i$,

$u_i = p_i h_i$, $i=0,\ldots,n-1$

$y_i = E(x_i)$ given data, $d_i = E''(x_i)$ unknown second derivatives

For given boundary conditions, i.e. $d_o = d_n = 0$ the second derivatives d_i are

uniquely determined by the solution of the symmetric, positive definite tri=

diagonal system:

$$T \cdot d = b \qquad\qquad (4.3)$$

where:
$$d = (d_1,\ldots,d_{n-1})^T, \quad b = (b_1,\ldots,b_{n-1})^T, \quad b_i = \frac{y_{i+1}-y_i}{h_i} - \frac{y_i-y_{i-1}}{h_{i-1}}$$

$$q_i = \frac{u_i\cosh(u_i)-\sinh(u_i)}{u_i^2\sinh(u_i)}\,h_i, \quad r_{i+1} = \frac{\sinh(u_i)-u_i}{u_i^2\sinh(u_i)}\,h_i, \quad u_i = p_i h_i$$

$$T = \begin{bmatrix} q_1+q_0 & r_2 & & & \\ r_2 & & \ddots & & \\ & \ddots & \ddots & & \\ & & & \ddots & r_{n-1} \\ & & & r_{n-1} & q_{n-1}+q_{n-2} \end{bmatrix}$$

The representation (4.2), (4.3) is numerically unstable in the limit cases $p_i \longrightarrow 0$ and $p_i \longrightarrow \infty$. A stable algorithm, which allows different types of boundary conditions is presented in [12]. Formula (4.2) possesses (n-1) degrees of freedom - the tension parameters p_i. The p_i are chosen in such a way, that we get a shape preserving interpolant. This is possible because of the limit cases: $p_i = 0$ leads to the usual cubic spline, $p_i \longrightarrow \infty$ give polygons. In the latter case the C^2-property is lost. In order to avoid sharp edges of the interpolant, the tension parameters should be as small as pos= sible, but large enough to create a monotone, convexity preserving interpo= lant. Different strategies for the determination of the tension parameters have been studied in [13]. The main idea for an 'a priori' estimation of the p_i uses the structure of the tridiagonal system (4.3). The right hand sides b_i of (4.3) are essentially the second order difference quotients of the da= ta. We want to choose the p_i in such a way, that the solutions of the tridi= agonal system d_i have the same sign as the b_i. This would preserve convexity/ concavity. For the derivation we use an equivalent representation of the co= efficients q_i and r_{i+1} from (4.3).

$$r_{i+1} = \frac{(1-\exp(-2u_i))/u_i-2\exp(-u_i)}{u_i(1-\exp(-2u_i))}\,h_i, \qquad i=0,\ldots,n-1 \qquad (4.4)$$

$$q_i = \frac{1-(1-\exp(-2u_i))/u_i+\exp(-2u_i)}{u_i(1-\exp(-2u_i))} h_i$$

For $u_i > 3$ holds $0 \leq \exp(-2u_i) \leq 2.5E\text{-}3$. Therefore we replace r_{i+1} and q_i by more simple expressions

$$\hat{q}_i = \frac{u_i-1}{u_i^2} h_i, \qquad \hat{r}_{i+1} = \frac{1}{u_i^2} h_i$$

and study the simplified tridiagonal system

$$\hat{T} \cdot \hat{d} = b. \tag{4.5}$$

To satisfy $\text{sign}(b_i) = \text{sign}(\hat{d}_i)$, it is sufficient in each row of the system that

$$(u_i-1)|\hat{d}_i| \geq |\hat{d}_{i+1}| \quad \text{and} \quad (u_i-1)|\hat{d}_i| \geq |\hat{d}_{i-1}|.$$

Combining the inequalities of two preceeding rows, gives the inclusion

$$\frac{|\hat{d}_{i+1}|}{u_i-1} \leq |\hat{d}_i| \leq (u_i-1)|\hat{d}_{i+1}|.$$

If we replace the unknown second derivatives by the second order difference quotients – which is justified for large u_i – we have

$$\hat{d}_i \sim p_i b_i.$$

So we achieve a rough and cheap estimate for the tension parameters:

$$u_i = \max \left(\frac{|b_i|}{|b_{i+1}|}, \frac{|b_{i+1}|}{|b_i|} \right) + 1 \tag{4.6}$$

In order to preserve the monotonicity of the interpolant in the inner inter= val $]x_i+\delta, x_{i+1}-\delta[$ we simplify under the assumption $u_i \geq 3$:

$$E'(x) \sim \frac{y_{i+1}-y_i}{h_i} + \frac{d_i-d_{i+1}}{h_i^2 p_i}, \qquad x_i+\delta \leq x \leq x_{i+1}-\delta$$

As a constant sign condition for $E'(x)$ we get:

$$p_i \geq \left| \frac{b_{i+1}-b_i}{y_{i+1}-y_i} \right| \tag{4.7}$$

The whole 'a priori' estimation procedure is listed in the flow chart 2. The estimation scheme can be easily implemented together with the algorithm from [12].

Flow Chart 2: Estimation of the tension parameters $u_i = p_i h_i$

for i=2 to (n-2)

'abnormal cases': if$(b_i b_{i+1} = 0$ or $|y_{i+1} - y_i| \leq$ eps max $(|y_i|, |y_{i+1}|)$)

then u_i = umax , goto end i,

endif

'monotonicity condition':

$$u_i = h_i \left| (b_{i+1} - b_i) / (y_{i+1} - y_i) \right|$$

'convexity/concavity condition':

if $b_i b_{i+1} > 0$ then $q = |b_i / b_{i+1}|$

$$u_i = \max (u_i, q, 1/q)$$

endif

'bounds': $u_i = \min(u_i$, umax)

if $u_i \leq 3$ then $u_i = 0$

end i;

'Proposal': umax = 30 , $u_1 = u_2$, $u_{n-1} = u_{n-2}$

eps = 5.E-7 for 24 bit mantissa

5   Test Results and Curve Interpolation

To study the performance of the three algorithms two critical examp= les from literature have been tested. The first example, whose data are given in Table 1, was constructed by Akima [1] in order to trap the cubic spline. The interpolants are presented in Figure 1.

261

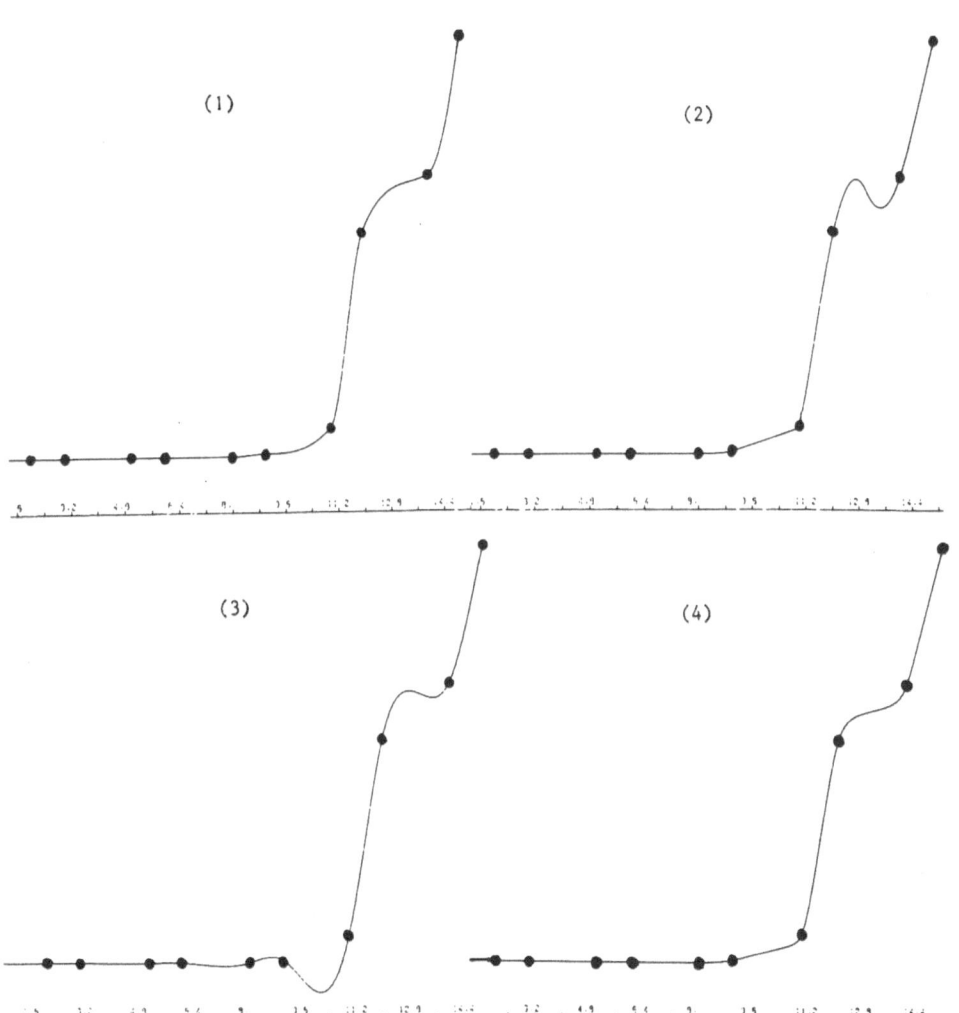

Figure 1: Example of Akima
       (1) Cubic Hermite polynomial, due to Fritsch, Carlson
       (2) Quadratic spline, due to Schumaker
       (3) Cubic Spline
       (4) Exponential Spline

Table 1: Example of Akima [1], see also [6], [13], [14]; n = 11

x_i	0	2	3	5	6	8	9	11	12	14	15
y_i	10	10	10	10	10	10	10.5	15	50	60	85
p_i*	15	30	15	30	15	3.5	9.3	0	5	10	

*tension parameters for the exponential spline due to (4.6), (4.7)

The Fritsch, Carlson interpolant looks quite visually pleasing. Only at 10.5 a small non-convex part is visible. The algorithm of Schumaker produces a sharp edge at x = 15 and suffers under the non-monotonicity of the interpo= lant in the interval [12,14]. In this interval the data switch from convex to concave shape. Because only co-convex parts are controlled, an interactive system must be used to adjust the slopes. The cubic spline does not work sa= tisfactory, because the scheme does not control monotonicity or convexity. The exponential spline is monotone and convex. The estimated values from (4.6) and (4.7) are a bit too strong. In the interval [9,11] the interpolant should be smoother.

The second example was introduced by McAllister, Roulier [9]. The data are given in Table 2, the interpolants are presented in Figure 2.

Table 2: Example of McAllister, Roulier; n = 12

x_i	0	1	2	3	4	5	6	7	8	9	10	11
y_i	0	0.3	0.5	0.2	0.6	1.2	1.3	1	1	1	0	-1
p_i*	5	5	4	3.5	0	0	0	30	30	30	30	

*tension parameters for the exponential spline due to (4.6), (4.7)

The Fritsch, Carlson interpolant seems to be too tough in the interval [5,7]. This is obviously due to the low class of smoothness. With the exception of a little bump in the interval [8,9] the scheme of Schumaker produces a quite nice interpolant. Though the cubic spline does not control convexity nor mo= notonicity, it is looking quite visually pleasing. The exponential spline

263

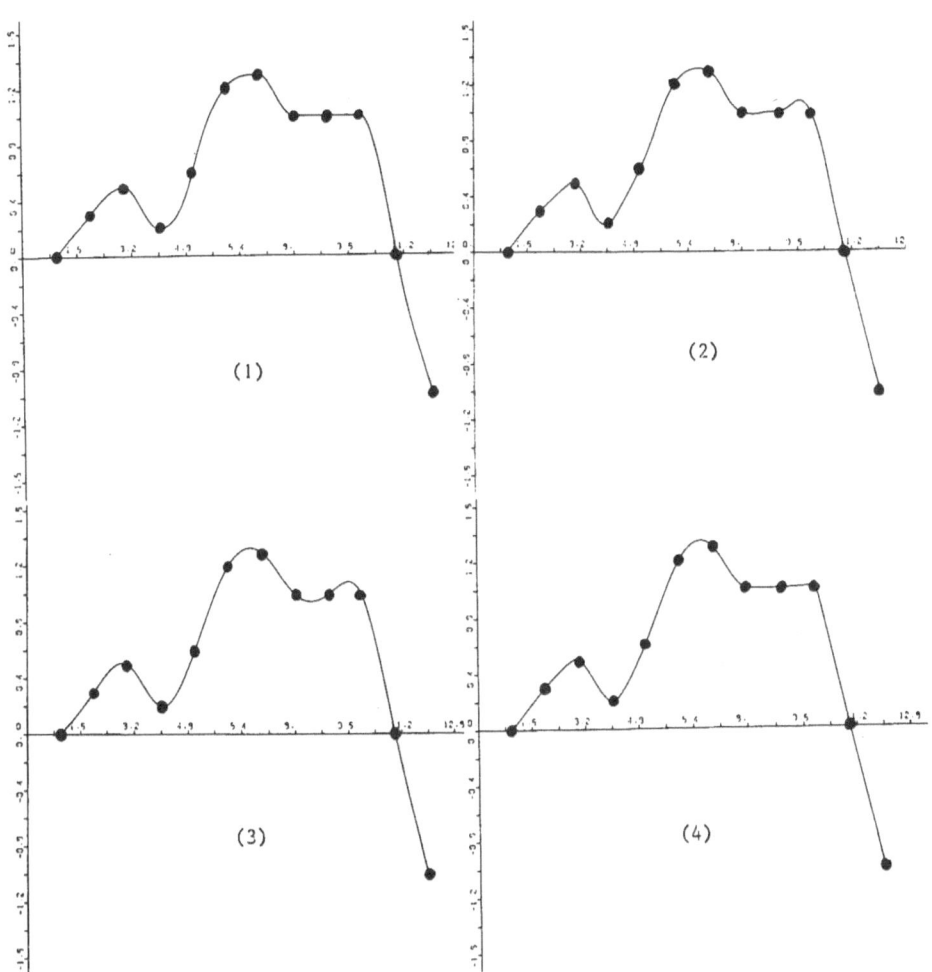

Figure 2: Example of McAllister, Roulier
      (1) Cubic Hermite polynomial, due to Fritsch, Carlson
      (2) Quadratic spline, due to Schumaker
      (3) Cubic Spline
      (4) Exponential Spline

produces a sharp edge at x = 9, because the only way for the interpolant to be concave is, to consist of two straight lines. This results in maximal va= lues for the tensile forces. This example demonstrates very nicely the dif= ficulties, to give a mathematical formulation for 'visually pleasing'.

In order to handle curves C = (x,y) in the plane, we have to introduce a pa= rametrization. We have chosen the standard non-weighted formula:

$$C(t) = (x(t),y(t)) \text{ with} \qquad (5.1)$$

$$t_o = 0$$

$$t_i = t_{i-1} + ((x_i - x_{i-1})^2 + (y_i - y_{i-1})^2)^{1/2}, \qquad i = 1, \ldots, n$$

Two typical examples of curve interpolation, whose data are listed in Table 3 and Table 4, have been tested.

Table 3: Closed Curve A, n = 27

x_i	0	1	2	2.8	4	5.5	8	10.5	13	12	10.5	8.5
y_i	3.5	2	-1	-3	-3.5	-3	-1.5	-0.5	0	-3	-5.5	-8

	4.5	0	-4.5	-8.5	-10.5	-12	-13	-10.5	-8	-5.5	-4	-2.8
	-10.5	-11	-10.5	-8	-5.5	-3	0	-0.5	-1.5	-3	-3.5	-3

	-2	-1	0
	-1	2	3.5

Table 4: Curve B, n = 33

x_i	2.5	3.5	5	7.5	9.5	11.8	13	11.5	9	6	2.5
y_i	-2.5	-0.5	2	4	4.5	3.5	0.5	-2	-3	-3.3	2.5

	0	-1.5	-3	-3.5	-2	0	2	3.5	3	1.5	0
	0	2	5	9	11	11.5	11	9	5	2	0

	-2.5	-6	-9	-11	-13	-11.8	-9.5	-7.5	-5	-3.5	-2.5
	-2.5	-3.5	-3	-2	0.5	3.5	4.5	4	2	-0.5	-2.5

The Figures 3 and 4 show that both examples are not critical against oscilla= tions. Each of the described algorithms produce satisfactory interpolating

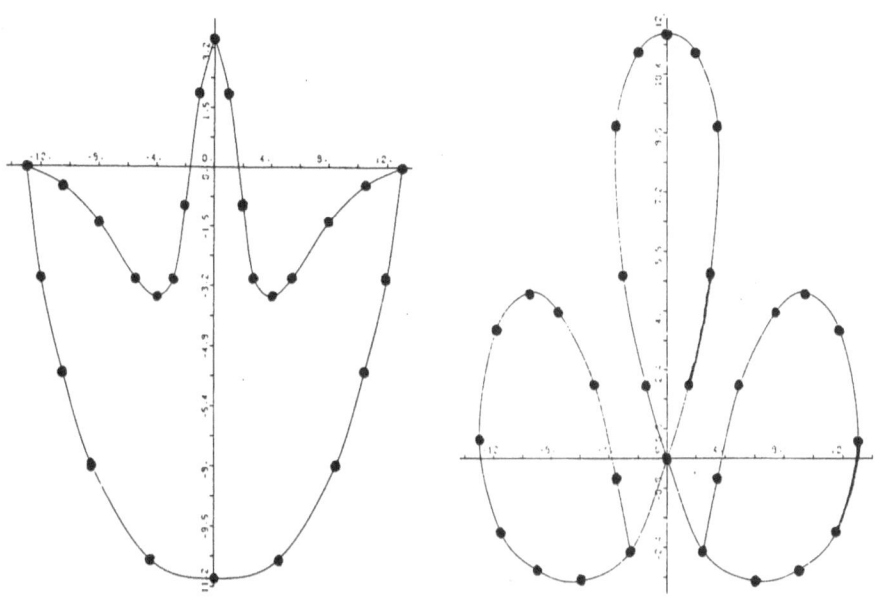

Figure 3: Curve A   Figure 4: Curve B

functions of the components (t_i, x_i) and (t_i, y_i), i=0,...,n. The curves of all
interpolation schemes are visually pleasing and cannot distinguished by eye.

6   Coordinates of the Needle Position

We look for an algorithm to determine the coordinates of the need=
le based on the computed curve $C(t) = (x(t), y(t))$ from part 5. The main hin=
drance is, that the curve representation is given in the parameter t. For the
desired constant stitch length we need a representation of the curve in the
arclength. But in general no explicit formulas exist to switch between diffe=
rent parametrizations. Because the algorithm should be cheap in terms of com=
putation time and complexity iteration techniques for nonlinear equations are
prohibited. Our method is based on a polygonal approximation of the curve and
works with an approximately constant stitch length NL. In sketch 4 the used

technique is demonstrated, the information in mathematical terms is given in the flow chart 3.

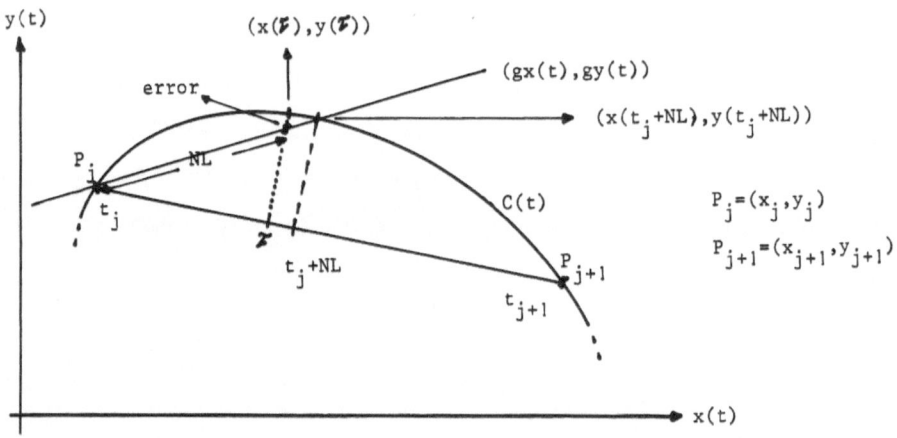

Sketch 4: Needle position

Flow Chart 3: Needle Position

Given: Data (x_j, y_j), $j=0, \ldots, n$

      Stitch length NL

      Curve representation $C(t) = (x(t), y(t))$

Evaluate: $(x(t_j+NL), y(t_j+NL))$

Compute the straight line between $P_j = (x_j, y_j)$ and $x(t_j+NL), y(t_j+NL))$

$$gx(t) = x_j + (t-t_j) \, Dx_j, \qquad gy(t) = y_j + (t-t_j) \, Dy_j$$

$$Dx_j = \frac{x(t_j+NL)-x_j}{NL}, \qquad Dy_j = \frac{y(t_j+NL)-y_j}{NL}$$

Take the stitch length NL on the straight line $(gx(t), gy(t))$ and project onto the interval $\left[t_j, t_{j+1}\right]$. This fixes the wanted value τ.

$$\text{with } NL^2 = (x_j - x(t_j+NL))^2 + (y_j - y(t_j+NL))^2 = (\tau - t_j)^2 (Dx_j^2 + Dy_j^2)$$

$$\implies \tau = t_j + \frac{NL}{\sqrt{Dx_j^2 + Dy_j^2}}$$

Replace (x_j, y_j) by $(x(\tau), y(\tau))$ and start again.

267

The last stitch lenght is always too short. This deviation is averaged to all intervals.

7   Conclusion

Because of limitations in time and page we have dropped many inte=
resting approaches for local C^1- and global C^2 - interpolation schemes. We
have neglected completly the global C^1-methods, because their advantages of
lower complexity are not paid off by their disadvantage of being global. Ne=
vertheless we hope that our survey allows a good choice of an interpolation
scheme for the control of a sewing machine. The cubic Hermite polynoms or an
interactive quadratic spline seem to be a fairly good choice if a local me=
thod is demanded. If the convexity/concavity of a sewing pattern should be as=
sured or periodic patterns are required, the global methods are necessary. In
this case spline methods like the exponential spline are useful.

Acknowledgement. We are indebted to H. Neunzert and the Technomathematic La=
boratory for the support. We thank C. Reinsch for the introduction to the ex=
ponential spline algorithm splex2 and helpful discussions. We express our ap=
preciation to R. Bulirsch and J. Stoer for their stimulating interest.

References

1   Akima, H.: A new method for interpolation and smooth curve fitting based
    on local procedures. J. ACM 17 (1970) 589-602

2   Bulirsch, R., Rutishauser, H.: Interpolation und genäherte Quadratur. In
    "Mathematische Hilfsmittel des Ingenieurs" Teil III, Hrsg.: R.Sauer, I.
    Szabo, Berlin-Heidelberg-New York: Springer Verlag 1968

3   Cline, A.: Scalar and planar-valued curve fitting in one and two-dimen=
    sional spaces using splines under tension. Comm. ACM 17 (1974) 218-223

4 DeBoor, C.: A practical guide to splines. Apl. Math. Sc. No.27
   Berlin-Heidelberg, New York: Springer 1978

5 Ellis, T.M.R., McLain, D.H.: Algorithm 514 - a new method of cubic curve
   fitting using local data. ACM TOMS 3 (1977) 175-178

6 Fritsch, F.N., Carlson, R.E.: Monotone piecewise cubic interpolation.
   SIAM J.Numer.Anal. 17 (1980) 283-246, see also SIAM J.Sci.Stat.Comp. 5
   (1984) 300-304

7 Heß, W., Schmidt, J.W.: Convexity preserving interpolation with exponen=
   tial splines. To appear in Computing

8 Knuth, D.E.: Mathematical Typography. Bullet.(New Series) of the AMS 1
   (1979) 337-372, see also $T_E X$ and Metafont - Part I. AMS - Digital Press
   1979

9 McAllister, D.F., Roulier, J.A.: An algorithm for computing a shape-pre=
   serving osculatory quadratic spline. ACM TOMS 7 (1981) 331-347

10 McAllister, D.F., Roulier, J.A.: Algorithm 574 - shape preserving oscula=
   tory quadratic splines. ACM TOMS 7 (1981) 384-386

11 Pruess, S.: Properties of splines in tension. J. Approx. Theory 17 (1976)
   86-96

12 Rentrop, P.: An algorithm for the computation of the exponential spline.
   Numer. Math. 35 (1980) 81-93

13 Rentrop, P., Wever, U.: Computational strategies for the tension parame=
   ters of the exponential spline. In "Lecture Notes in Control and Informa=
   tion Sciences" 95 (1987) 122-134, Ed. R.Bulirsch et al., Springer Verlag

14 Schumaker, L.L.: On shape preserving quadratic spline interpolation.
   SIAM J.Numer.Anal. 20 (1983) 854-864

15 Schweikert, D.G.: An interpolation curve using a spline in tension. J.
   Math. Phys. 45 (1965) 312-317

16 Späth, H.: Exponential spline interpolation. Computing 4 (1969) 225-233

17 Stoer, J., Bulirsch, R.: Introduction to numerical analysis. Berlin-Hei=
   delberg, New York: Springer 1980

Address: Prof. Dr. P. Rentrop, Dipl.-math. U. Wever
          Fachbereich Mathematik, Universität Kaiserslautern
          D-6750 Kaiserslautern

THE DEVELOPMENT OF A MECHANICAL MODEL FOR A TIRE:
A 15 YEAR STORY TO REPLACE TEST MACHINES

M. Bercovier, E. Jankovich, M. Durand

Introduction

In 1971 a research team in "tire mechanics" was setup at Kleber Colombes with the following goals:

1) Replace time consuming and costly trials (or reduce them) for the study of new tires.

2) Ultimately by a better knowledge of internal mechanical behavior get new technological directions.

At that time we thought that numerical simulations would

1) give a faster design cycle;

2) reduce tests to validity controls on conclusions drawn from computations;

3) get a fast parametric design tool (change of dimensions);

4) optimize ultimate product (material cost, life expectation, weight, energy consumption etc...);

"The final version of this paper will be submitted for publication elsewhere".

5) suppress some redundant or useless design rules;

6) bring new designs never thought of before.

Tire modelisation

Over the years many mechanical models of tire have been used (cf. Padovan [1] and Tanner [2] for some reviews); since our aim was the study of the "internal mechanics" (under working conditions) what was needed was a continuum mechanics approach and the 3D elasticity (Hookian) model was chosen. A first problem to overcome was the modelisation of composite materials (see belt and carcass, Fig. 1), this was to be the research subject of the second author for many years. Simple adhoc techniques were used at first and an initial axisymmetric continuum model was established, a simple axisymmetric pressure loading case was studied using a state of a art FEM code (Durand and Jankovich [3]).

The results were quite problematic (cf. Fig. 2):

1) To run a commercial code was expensive even for a 2D case.

2) Classical linear elasticity did not give coherent results.

The use of simple toroidal shapes under uniform pressure loads showed that the full non linear geometry due to large displacements had to be taken into account.

Let $x = (x_i)$ be the coordinate of an undeformed particle of the tire, $\Delta x = (u_i)$ the displacement, the deformation gradient is given by the matrix:

$$\nabla u = (\delta_{ij} + u_{i,j})$$

and the corresponding strain tensor is:

$$e_{ij} = 1/2 (\nabla u^T \nabla u - I)$$

in the classical (linear) theory this tensor is approximated by

$$e_{ij} = 1/2 (u_{i,j} + u_{j,i})$$

(i.e. all non linear terms in $u_{i,j}$ are dropped). Our own conclusion was that all terms had to be retained. But there were many more non linear problems in the modelisation:

1. A tire is in "contact" with the road/obstacle (with or without friction!), thus a one-sided non linear boundary condition;

2. The pressure in a tire is on the (final) deformed shape, not on the initial one;

3. Composite materials used have different moduli in the fiber direction wether in tension or compression (this

can be explained by some "micro" buckling of the fibers in compression);

4. Most of the filling material, the external envelop and so on is made of rubber, a material that is (nearly) incompressible and that obeys non linear constitutive laws or is defined by non linear energy equations;

5. Unstable geometries can appear (buckling of some layers, belt undergoing compression in the road contact zone...);

6. The final computational machine must take into account the dynamical behaviour of the tire;

7. Rubber is actually viscoelastic and energy losses are also due to heat generation from cyclic deformation.

Development of a computational test machine

A tool for the designer had to be easy to use, effective and give reliable results. Commercially available FEM software and hardware in 1972 did not lead to any of the above criterions: data preparation was time consuming, model (linear) was inadequate, costs of a 3D computation that would have between 10 to 20000 degrees of freedom was prohibitive (it was estimated even in 1974 at $2x10^5$ US dollars!).

The development of an in-house FEM software necessitated the research of new algorithms and new finite elements. Thus a "rubber" type element based on a penalty approach for the (quasi) incompressibility was devised [4, 5].

Newton-Raphson iterations had to be used since both the geometrical non linearity of the toroidal shape and pressure system and the rubber type elements non linearity proved to be unstable on all other quasi linearisation methods! This Newton-Raphson method had to be used together with a contact algorithm that was an original mixture of penalty (non linear springs) and Lagrangian approach (this contact algorithm has not yet been published). To get fast iterations and fast convergence (few iterations) the particular features of a tire/road contact were included in it (see Fig. 3 and Fig. 4).

The software design decisions followed extensive tests of available methods. The tire is a very stiff structure made of composite materials with extreme variations in directions. Thus the steel belt is made of steel cord and rubber; in the cord direction of the thread there is say a factor 10^5 for the Young modulus in extension, at 90° the Young modulus is coming from the rubber mainly and is of order 1. The same thing is true for the material cord in the side walls. Moreover the stiffest directions of the side wall and the steel belt are orthogonal! The steel belt main direction is perpendicular to any meridian plane while the side wall cord is entirely in that plane. Methods like the multigrid one or preconditionned conjugate gradient [6] were inadequate!

The first developments were done of course on an axisymmetric model. This was not appreciated by the tire designers who were closely following (and testing) the numerical model. One even told us that we were designing optimal spare tires! But this approach proved to be useful for the choices of the different algorithms and FEM approximations. Moreover the axisymmetric model was already 3D since the non symmetric behaviour of the belt induces a twist of the meridian planes under simple pressure loading. This simpler model could be used for material evaluation and measurement. Some industry tests such as bead resistance to high pressure (for truck and tractor tires) could be numerically done. Little by little development engineers came to the conclusion that many design problems could already be seen with these (simpler) axisymmetric loads (after all a bad spare tire will surely be a bad rolling tire!).

The first 3D numerical simulations that were done could simulate a tire at contact on the road in the static case, adding a constant rolling velocity was easy. By means of different boundary conditions some driving conditions could be simulated. The resulting numerical model could be used in design, the software developed proved to be up to 100 time faster than standard commercial codes. Moreover the advent of the super mini computer supressed the need for a large computer installation. By 1982 the computation case estimated earlier at $2x10^5$ US $ did not cost more than 500 US $!

It must also be pointed out that special input/output procedures were devised. Thus the data preparation is close to the way a development team defines a tire layout. The results are post processed to emphasize the crucial designer questions such as uniform (or not) cord tension under various loads, pressure at contact, intermaterial stresses etc...

Impact in the tire industry

Introduction of the FEM at the design stage has had a long lasting influence on the tire design. The industry had to devise new measurement tests for the basic materials it used. Some elasticity models such as the Mooney-Rivlin one of higher order had rarely been used and original measurements (not simple one dimension tractions!) had to be found. Modelling of (non linear) composite materials also proved to be a major research problem and great insight was gained there too. A data

base for the different materials had to be created. The
side effect of these required data was to give a better
knowledge of the micro and macro mechanical behaviour of
those materials.

Moreover each development team could use its own
(super) minicomputer installation as an immediate test
machine. In the old design cycle, numerous tires had to be
build (a long delay) and then tested (more time and very
costly!). The design engineer's tendency was to find some
variations on older design (thus same mould could be used)
and there would be less need for tests. Today a tire can
be designed mainly at the mechanical/computational level.
Parametric tests runs can be made on the computer and the
influence of the different parameters direcly analysed.
Thus the design team will build and test the few last
options after eliminating all bad solutions that were
highlighted by the numerical simulation. This gives more
freedom of choice/decision to the design team, and leads
to bolder decisions.

Limits and open problems

When the present work started the design engineers
put two main requirements on a numerical simulation: to
know the global dynamical behaviour (that is felt by the

user) and to have an estimate of the tire life in kilometers!

It is clear that the present numerical simulations do not answer to any of those questions! Global dynamical behaviour of such a non linear structure (imagine a shock on a stone and all the wave propagation problems!) is still a difficult problems, bigger computers and better algorithms for one sided hyperbolic problems [7] will certainly help in the future. Nevertheless a good qualitative indication is given by the numerical simulations: for instance a belt that will be in compression under the contact zone is certainly badly designed.

Tire modelisation offers still many challenging problems in numerical analysis, in mathematical theory and modelisation itself; to give a few in numerical analysis:
- Better approximations (based on mixed FEM) to take into account the quasi-shell structure.
- Dynamical (effective) numerical procedures valid for very stiff (anisotropic) materials.
- Fast solvers for very large unstable problems (buckling!).

On the theoretical sides much progress has been made in non linear analysis [8], but many questions remained unsolved as:

- A theory for bi-moduli non linear materials.

- Homogeneization of rubber type materials.

- Theory of one sided hyperbolic problems [7].

    In modelisation progress has to be made to get:

1. A "continuous" model taking into account the irregular grooves.

2. A non linear anisotropic incompressible materials.

3. A simple rheological model for energy loss.

Conclusion

    The implementation of the resulting software in the present work as a test machine for every design team could be seen as a first success. But the real success of this effort is in the publication of a press release on finite element [9]! Use of a standard commercial FEM code would have been limited due to high costs and absence of crucial modelisation options. Design of mechanical models and related software was successful because it was done at every stage in close cooperation with the design engineers (and sometimes under their sarcasm!) and thus answered real needs! New mathematical techniques had to be found (approximation of rubber materials), use of the physical nature of the problem had to be included in non linear algorithms and most of the existence and unicity problems had to be ignored!

References

[1] Padovan A., "Tire Mechanics", Lectures of the Dept. of Mech. Eng. University of Akton, 1977.

[2] Tanner John A., (ed.) "Tire Modelling", NASA Report CP2264, 1983.

[3] Durand M. and Jankovich E., "Non Applicability of Linear FEM Programs to the Stress Analysis of Tires", NASA, NASTRAN Conference, September 1982 (TM X 2637).

[4] Jankovich E., Durand M. and Bercovier M., "A Finite Element Method for the Analysis of Rubber Parts, Experimental and Analytical Assesment", Computers and Structures, Vol. 14, 1981, pp. 385-391.

[5] Bercovier M., "Pertubation of Mixed Variational Problems, Application to Mixed FEM", RAIRO, Vol. T2, 1978, pp. 211-236.

[6] Bercovier M. and Rosenthal A., "Using the Conjugate Gradient Method with Preconditioning for Solving FEM approximations of Elasticity Problems", Engineering Computations, Vol. 3, 1986, pp. 77-80.

[7] Schatzmann M. and Bercovier M., To appear.

[8] Ciarlet P.G., "Non Linear Elasticity", To appear, North Holland (1987).

[9] MICHELIN Co., Press Release, October 1985.

M. Bercovier
Department of Computer Science, The Hebrew University of Jerusalem, Jerusalem, Israel.

E. Jankowich
Hutchinson S.A., 2 Rue Balzac, Paris, France.

M. Durand
GAMD-BA, Saint-Cloud, France.

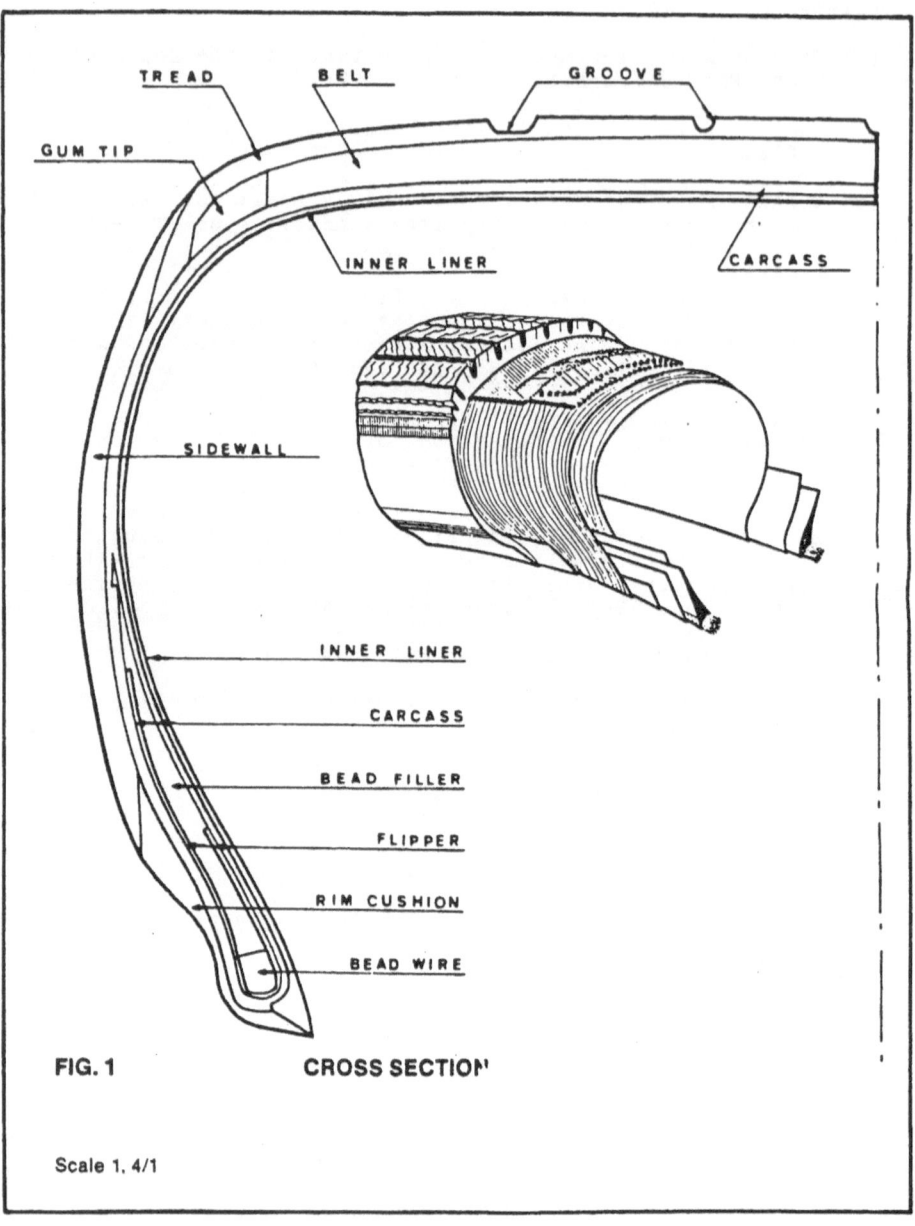

TREAD

BELT

GROOVE

GUM TIP

INNER LINER

CARCASS

SIDEWALL

INNER LINER

CARCASS

BEAD FILLER

FLIPPER

RIM CUSHION

BEAD WIRE

FIG. 1   **CROSS SECTION**

Scale 1, 4/1

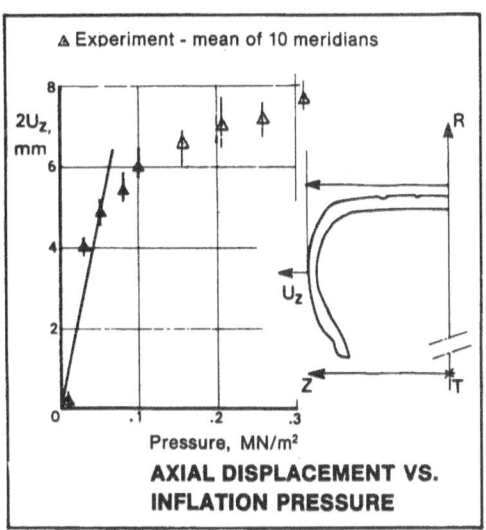

FIGURE 2.

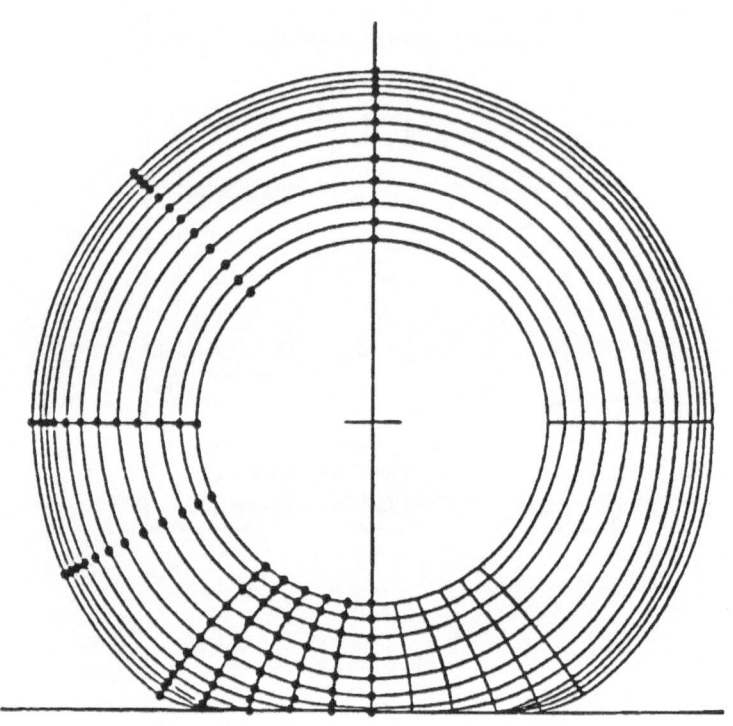

FIGURE 3.

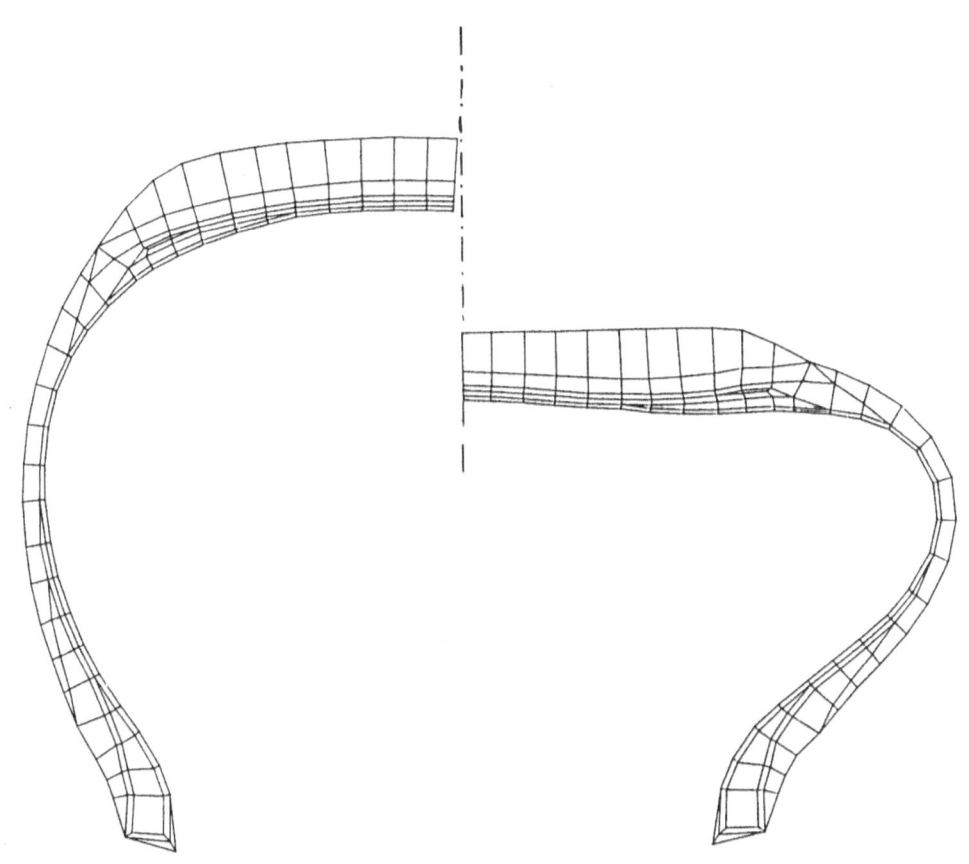

FIGURE 4.

Steady State Simulation of Chemical Plants - Flowsheeting *

D.Auzinger, L.Peer, Hj.Wacker

1 Introduction

This paper deals with flowsheeting problems, i.e. simulation/optimization of chemical plants. The practical impact of flowsheeting is, at least, twofold. The experienced engineer can check different configurations for a plant with respect both to the costs of constructing and running the plant. A second aspect is the possibility to analyse the system in dependence of the variation of some parameters. We confine ourselves to the steady state simulation of chemical plants with a main stress on plate separation columns. A plant consists of a number of units which we connected by mass and/or energy streams (see Fig. 1.1).

Figure 1.1: Acetic acid plant

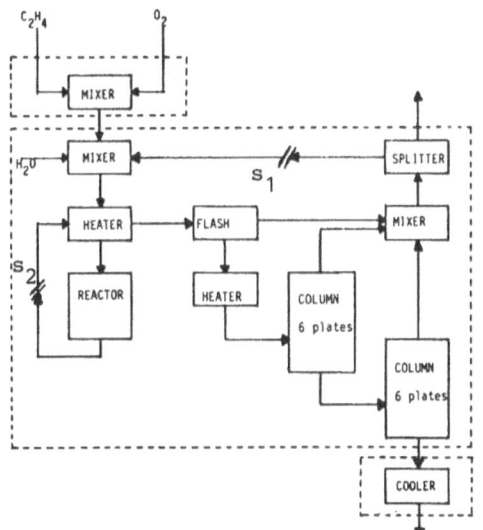

(the dotted line and the cuts s_1, s_2 will be discussed in Section 4)

At first we describe some special units: plate separation columns, flashes, heatexchangers. In all cases our numerical solution tries to make full use of the structure of each unit. This is explained in some detail for columns. A comparison with other techniques proposed in literature is presented.

---

* The final version of this paper will be submitted for publication elsewhere.

For flowsheeting problems there exist two different approaches: the equational approach and the sequential modular approach. We give some examples for both techniques. Finally, we conclude with some remarks concerning optimization with respect to some parameters of the system.

2 Plate Separation Columns

2.1 The Mathematical Model

The principle of a separation process is that some of the components of a two phase multicomponent mixture (e.g. liquid/vapor) change the phase and others do not because of different chemical or physical properties (volatility, solubility, ...). As this separation is mostly incomplete, plate separation columns (absorption and distillation columns) connect several of these processes in series (plates). The phases move through these plates countercurrently.

For our model we assume a thermodynamic equilibrium on each of the N plates (ideal plates) and we neglect the hydraulic by assuming a linear decreasing pressure. As Figure 2.2 shows, some of the liquid on plate n (molar composition $x_{n,j}, j = 1, \ldots, M$, M number of components) moves to plate $n-1$ (L_n total stream, $l_{n,j}$ component stream), and some of the vapor goes to plate $n + 1$ (composition $y_{n,j}$, stream V_n, $v_{n,j}$). There may be a feedstream on plate n (a given stream enters the column, F_n total feed, $f_{n,j}$ component feed) and a liquid or a vapor sidestream (a given quantity S_n^L resp. S_n^V leaves the column). The vapor stream leaving plate N is called top product, the liquid leaving plate 1 bottom product. An absorption column always has a liquid feed at plate N and a vapor feed at plate 1 (Fig. 2.1), while plate 1 of a distillation column is a reboiler and plate N a condensor.

Figure 2.1: absorption column   Figure 2.2: plate n

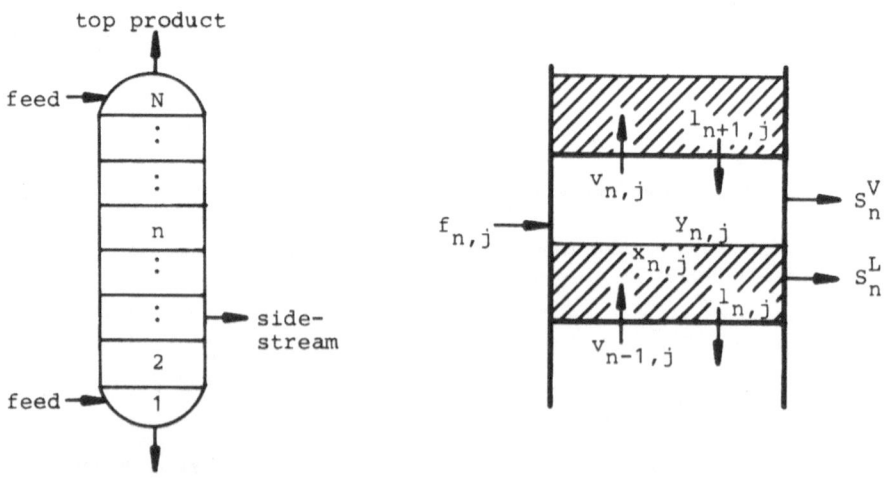

From our assumptions it follows that we have three sets of equations for our model:

1.) Component mass balances:

Balancing all the component streams leaving and entering a plate we obtain

$$b_{n,j} \equiv (1 + \frac{S_n^L}{L_n})l_{n,j} + (1 + \frac{S_n^V}{V_n})v_{n,j} - l_{n+1,j} - v_{n-1,j} - f_{n,j} = 0$$

$$\text{with} \quad L_n = \sum_j l_{n,j} \quad \text{and} \quad V_n = \sum_j v_{n,j}$$

2.) Enthalpy balances

Together with each stream heat is leaving or entering plate n, the sum of them must be zero. The quantity of heat (enthalpy) is proportional to the quantity of the stream and depends on its composition and on its temperature T_n. If H_n^L denotes the molar enthalpy of the liquid stream leaving plate n, H_n^V that of the vapor stream, H_n^F the molar enthalpy of the feed and $\triangle E$ a given heat transfer rate (heating or cooling), we obtain

$$E_n \equiv (L_n + S_n^L)H_n^L + (V_n + S_n^V)H_n^V - L_{n+1}H_{n+1}^L - V_{n-1}H_{n-1}^V - F_n H_n^F - \triangle E_n = 0$$

3.) Vapor-liquid-equilibrium relationship

The equilibrium coefficient $k_{n,j}$ gives the proportion of the vapor and liquid molar concentrations $y_{n,j}$ and $x_{n,j}$. Assuming that the molar composition on a plate and that of the stream leaving the plate are equal $\left(x_{n,j} = \frac{l_{n,j}}{L_n}, y_{n,j} = \frac{v_{n,j}}{V_n}\right)$, we obtain

$$g_{n,j} \equiv k_{n,j}\frac{l_{n,j}}{L_n} - \frac{v_{n,j}}{V_n} = 0$$

Like the molar enthalpies, also the equilibrium coefficient is a function of the plate temperature T_n and the molar composition. The specific form of the functions depends on the underlying thermodynamic model. For the UNIQUAC model (see [3]) we have:

$$H_n^L = \sum_j (a_j + b_j T_n)\frac{l_{n,i}}{L_n} + H^E(l_{n,1}, \ldots, l_{n,M}, T_n)$$

$$H_n^V = \sum_j (c_j + d_j T_n)\frac{v_{n,i}}{V_n}$$

$$k_{n,j} = \frac{par_j(T_n)}{P_n} \cdot \frac{\gamma_j(l_{n,1},\ldots,l_{n,M},T_n)}{\varphi_j(v_{n,1},\ldots,v_{n,M},T_n)}$$

The excess enthalpy H^E, the vapor pressure par_j, the activity coefficient γ_j and the fugacity coefficient φ_j are highly nonlinear functions, p_n is the given pressure of plate n.

The model developed up to now is a nonlinear set of $(2M + 1)N$ equations (M component mass balances, 1 enthalpy balance and M vapour-liquid-equilibrium-relationships for each of the N plates) for the $(2M + 1)N$ unknowns $l_{n,j}$, $v_{n,j}$, T_n. It describes an absorption column (with $V_0 := 0$ and $L_{N+1} := 0$), for a distillation column the enthalpy balances at plates 1 and N must be replaced. (see [1], [10]).

The basic model is now extended by considering pumparounds and side strippers: A pumparound removes a liquid stream of given quantity L^P from plate n_o and, possibly after cooling or heating, returns it at some plate n_i (Fig. 2.3). This changes the component mass balances of the involved plates:

$$b_{n,j} \to b_{n,j} + L^P \frac{l_{n_o,j}}{L_{n_o}} \qquad \text{for} \quad n = n_o$$

$$b_{n,j} \to b_{n,j} - L^P \frac{l_{n_o,j}}{L_{n_o}} \qquad \text{for} \quad n = n_i$$

The enthalpy balances are analogously modified (see [2]), the equilibrium relationships remain unchanged.

A side stripper is an absorption column linked to the main column in a certain way. Fig. 2.3 shows such a side stripper.

Figure 2.3: main column with pumparound and side stripper

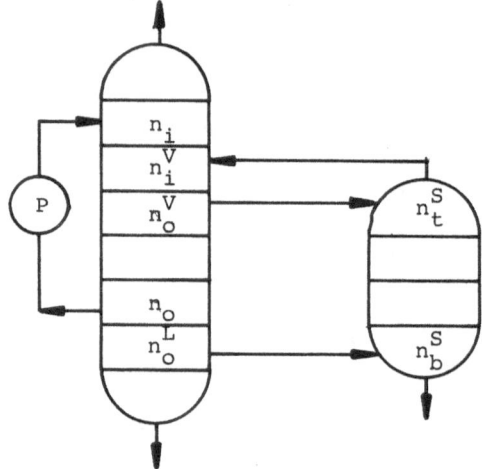

Liquid and vapor streams of given quantities L^S and V^S, removed from the plates n_o^L resp. n_o^V, enter a side stripper at the top plate n_t^S resp. bottom plate n_b^S. The top product of the side stripper is returned to the main column at plate n_i^V.

The model for the side stripper only is completely analogous to the basic model for the main column. The linkage between the two columns leads to the following changes: Component mass balances:

$$b_{n,j} \to b_{n,j} + L^S \cdot \frac{l_{n_o^L,j}}{L_{n_o^L}} \qquad \text{for} \quad n = n_o^L$$

$$b_{n,j} \to b_{n,j} + V^S \cdot \frac{v_{n_o^V,j}}{V_{n_o^V}} \qquad \text{for} \quad n = n_o^V$$

289

$$b_{n,j} \to b_{n,j} - v_{n_i^S,j} \qquad \text{for} \quad n = n_i^V$$

$$b_{n,j} \to b_{n,j} - L^S . \frac{l_{n_o^L,j}}{L_{n_o^L}} \qquad \text{for} \quad n = n_t^S$$

$$b_{n,j} \to b_{n,j} - V^S . \frac{v_{n_o^V,j}}{V_{n_o^V}} \qquad \text{for} \quad n = n_b^S$$

Enthalpy balances are analogously modified at the same plates.

The general system of nonlinear equations describing the full problem consists of $(2M + 1)(N + N^S)$ equations and unknowns, where N^S denotes the total number of plates occuring in sidestrippers. By ordering the unknowns resp. equations of plate n in the way $(l_{n,1}, \ldots, l_{n,M}, T_n, v_{n,1}, \ldots, v_{n,M})$ resp. $(b_{n,1}, \ldots, b_{n,M}, E_n, g_{n,1}, \ldots, g_{n,M})$ and by arranging these blocks according to the natural ordering of the plates (with the bottom plate of a side stripper immediately after the top plate of the main column resp. another side stripper), the Jacobian of the system is $(N + N^S)$-by-$(N + N^S)$ block tridiagonal with a few offtridiagonal blocks. All blocks are $(2M + 1)$-by-$(2M + 1)$ submatrices of a special partitioning in $M, 1$ and M rows resp. columns, see Fig. 2.4. Offtridiagonal blocks are either caused by pumparounds (D) or by side strippers (E,F,G).

Figure 2.4: Jacobian of the System of Figure 2.3

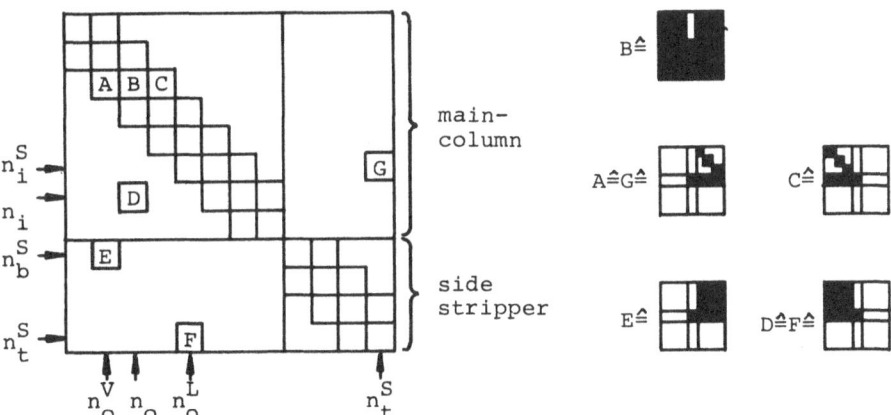

2.2 Numerical methods

The system of nonlinear equations of Section 2, shortly

$$f(x) = 0,$$

is usually solved by replacing it by a sequence of linear problems

$$J^k.(x^{k+1} - x^k) + f(x^k) = 0, \quad k = 0, 1, \dots$$

In particular, we discuss Newton's method and a quasi-Newton method. Direct as well as iterative procedures for solving the occuring linear systems are studied. Strategies for improving convergence complete the section.

Nonlinear solvers

For Newton's method, J^k coincides with the Jacobian $f'(x^k)$. The complexity of the underlying UNIQUAC model does not allow an efficient explicit evaluation of the Jacobian. Several variants of Newton's method have been suggested to overcome these difficulties. Westman et.al. [14] developed a hybrid method by splitting the Jacobian into two parts. The first part contains the easy-to-calculate derivatives and is computed analytically. The remaining part consists of all terms of the Jacobian, where derivatives of h_n^E, γ_j, φ_j occur. It is approximated either by the Schubert update (which is an adaptation of Broyden's rank-1 method to sparsity) or, as an alternative, by using finite differences for the derivatives of h_n^E, γ_j and φ_j.

Comparison of the two methods:

The numerical effort for one iterate of the hybrid quasi-Newton method (shortly: hQNM) is dominated by the amount of time t_l for solving the linear system. The hybrid finite difference Newton method (shortly: hFDNM) requires some additional time t_e for evaluating the approximate Jacobian. (Notice, that the numerical effort for solving the linear system is the same for hQNM and hFDNM, because hQNM does not allow an efficient direct update of the L-U factorization.) Numerical tests showed that the ratio $t_l{:}t_e$ is about 1:3 for the case of a main column only. Hence, about four times more iterates can be performed with hQNM than with hFDNM within the same time, which usually outweighs the lower rate of convergence for hQNM. Therefore, Westman et.al. [14] recommend hQNM in this case. However, the numerical effort for solving the linear systems increases considerably for main columns with pumparounds and side strippers. The ratio $t_l{:}t_e$ quickly changes from 1:3 to 1:1 for systems of moderate complexity, even to 3:1 for complex configurations. Therefore, we advocate for hFDNM in these cases.

Linear solvers

Either nonlinear solver generates a linear algebraic system at each iteration step.

A direct solver

In the absence of pumparounds and side strippers two approaches are usually taken to solve the linear system. One way is to stress the block tridiagonal structure of the matrix and use a block Thomas algorithm. For the other approach the matrix is considered as a band matrix and a band Gaussian L-U factorization is used. A general code for solving the linear system in the first way requires about $18\frac{2}{3}NM^3$ operations. The number of operations reduces to $8NM^3$ (resp. to $16NM^3$ with partial pivoting) for the second approach.

As an alternative, we propose a Gaussian L-U factorization with incomplete partial pivoting (shortly: ippLU) under exploitation of the special structure of the matrix. Roughly speaking, partial pivoting is used, whenever it does not essentially destroy the sparsity pattern. In particular, the diagonal elements in the right upper part of A_n, see Fig. 2.4, are chosen for pivoting. This method is reasonably stable and requires about $5\frac{5}{6}NM^3$ operations for problems with a main column only. It naturally extends to a method for the full problem of Section 2, when off-tridiagonal blocks lead to a fill-up of certain rows and columns.

Another solver for almost block tridiagonal systems was suggested by Kubiček et.al. in [9]: The original equation is transformed into a set of equations with a block tridiagonal structure. If we assume that these block tridiagonal systems are solved by ippLU then, for the example of a main column with one pumparound, the number of operations for Kubiček's method ranges from $10\frac{5}{6}NM^3$ to $13\frac{5}{6}NM^3$ depending on the position of the pumparound. The proposed extended ippLU requires between $5\frac{5}{6}NM^3$ and $10\frac{1}{3}NM^3$ operations for this example. The superiority of ippLU to Kubiček's method is even higher for more complex problems.

An iterative solver

It is not known to the authors of this paper that iterative methods have been studied for solving the linear systems in this context. A very natural splitting of the matrix J^k is suggested by the special structure of the problem:

$$J^k = T^k + R^k$$

with T^k is block tridiagonal and R^k is extremely sparse. This leads to the iterative method SPLIT

$$T^k.d^{k,l+1} + R^k.d^{k,l} + f(x^k) = 0, \quad l = 0, 1, \ldots$$

for determining $x^{k+1} - x^k$. Each step requires the solution of a block tridiagonal system, which is done by ippLU, with only one L-U factorization for all l.

Comparison between direct and iterative solver

The computational effort for the direct solver ippLU increases with the number of off-band blocks and their distances from the diagonal. Numerical tests show that the rate of convergence of the iterative solver SPLIT mainly depends on the magnitude of the off-band blocks but not on their position. The following example was chosen to demonstrate the influence of the magnitude of the off-band blocks on the rate of convergence.

Example: (for demonstrational purposes only)

Separation of methanol-water by a distillation column (main column: 30 plates, one pumparound from plate 2 to plate 28, see Fig. 7)

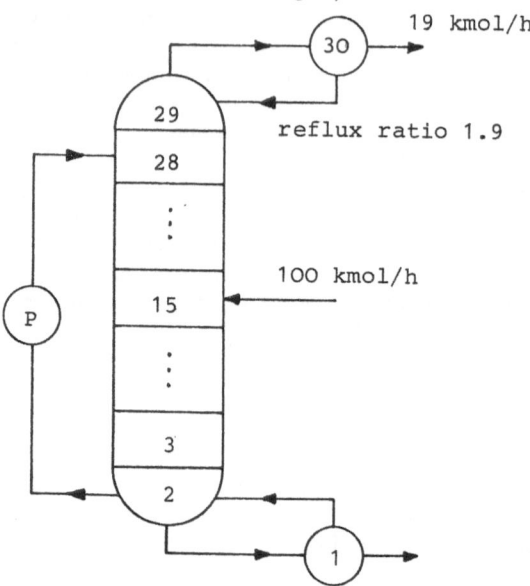

19 kmol/h

30

reflux ratio 1.9

100 kmol/h

P

Compared to ippLU, the iterative solver saves about 5 percent of total cpu-time if $L^P = 12$ kmol/h (L^P flow rate of pumped stream). For $L^P = 30$ kmol/h, the direct solver is faster by about 8 percent.

The described behaviour was typical for a series of further test problems, so we may say, that the iterative solver is recommended for systems with blocks far off a dominant block tridiagonal.

Improvements

Initial value calculations: Of great importance for the efficiency of either method (hQNM or hFDNM) is a good initial approximation of the solution. Using estimates of the temperature of top, bottom and feed stages, initial values for T_n are calculated by linear interpolation. A plate to plate overall mass balance calculation gives approximations to L_n and V_n. These total flow rates are partitioned into component flow rates $l_{n,j}$ and $v_{n,j}$ by means of a simplified model.

Damping: An inexact line search is incorporated in hQNM and hFDNM in order to increase the region of convergence.

Non-negativity conditions: There are natural lower bounds for the variables

$$l_{n,j} \geq 0, \quad v_{n,j} \geq 0, \quad T \geq 0 \quad \text{(in Kelvin)}$$

Whenever an iterate x^{k+1} violates these conditions, negative components are replaced by zero before continuing the method. This is particularly important to prevent the evaluation of h_n^E, γ_j and φ_j at meaningless points which could completely destroy convergence.

Homotopy methods: Even an improved initial value calculation does not necessarily ensure convergence. In this case homotopy methods may help to overcome the difficulties: The basic idea is to continuously deform the given problem into a simpler one. Then the required solution of the given problem is obtained by following a continuous path starting at a solution of the simpler problem. For details, see e.g. [13], [5]. The deformation can be done, for example, by changing some internal parameters of the problem. In particular the reflux ratio (for distillation columns), the heat transfer rate ΔE_n, flow rates of side streams, pumped streams or exchange streams between main column and side stripper are used.

3 Further Examples of Units

3.1 Flash

In principle a flash represents one single plate of a separation column:

Figure 3.1 Flash

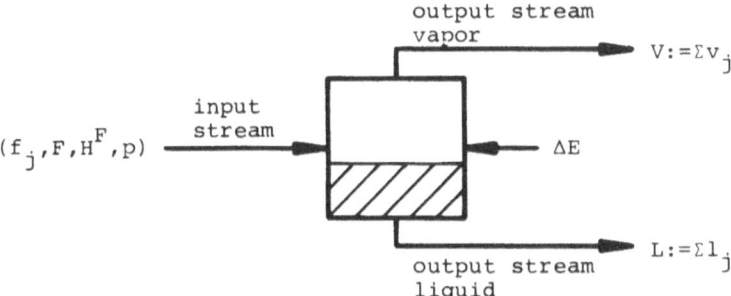

Depending on the input stream and on the heat transfer rate ΔE three cases may occur:

i) the liquid output stream vanishes: $L = 0 \Longrightarrow v_j = f_j$. This is the case if $FH^F + \Delta E \geq FH^V(f_j, T^d)$, which indicates that the temperature of the output stream is higher than the dew point T^d.

ii) the vapor output stream vanishes: $V = 0 \Longrightarrow l_j = f_j$. This case occurs if $FH^F + \Delta E \leq FH^L(f_j, T^b)$ (temperature of output stream lower than bubble point T^b)

iii) there is a liquid and a vapor output stream.

The general equations for the unknowns l_j, v_j and T are (compare section 2.1):

$$l_j + v_j - f_j = 0 \qquad\qquad j = 1, \ldots, M$$
$$LH^L + VH^V - FH^F - \Delta E = 0$$
$$k_j \frac{l_j}{L} - \frac{v_j}{V} = 0 \qquad\qquad j = 1, \ldots, M$$

For case i) the system reduces to one nonlinear equation (enthalpy balance) for the temperature T because of $l_j = 0$ and $v_j = f_j$. For case ii) the same holds, while for case iii) the whole system of $2M + 1$ equations must be solved. This is done by similar techniques as described in section 2.2. Again the structure of the Jacobian is exploited. The results of the dew and bubble point calculations can be used to get a good initial value T_0 for temperature T: $T_0 = T^b + \frac{H^F + \triangle E / F - H^b}{H^d - H^b} \cdot (T^d - T^b)$. Initial values for l_j and v_j are obtained by use of an idealized vapor-liquid relationship.

3.2   Counter-flow heat exchanger

A counter-flow heat exchanger consists of a system of N parallel tubes through which liquid and/or vapor mixtures of different temperatures flow countercurrently. The input temperature T_i^* of each stream is known, the output temperature must be calculated. Figure 3.2 shows the principle of this type of heat exchanger.

Figure 3.2

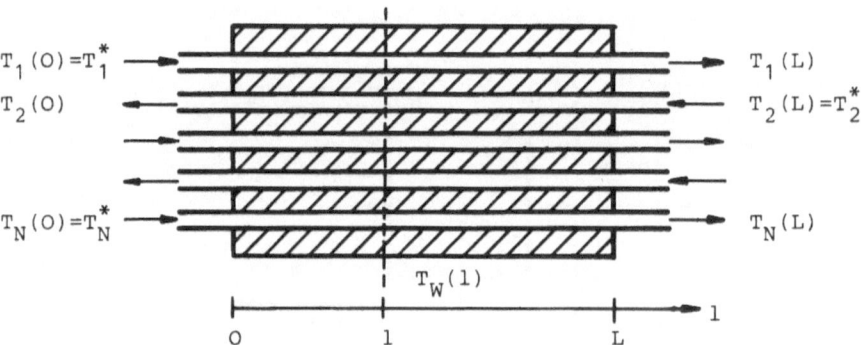

The main assumptions for our model are:
(i) There is no heat transport in length direction inside the tubes except the heat transport connected to the mass flow.
(ii) For l given we have one joint temperature T_W of the apparatus.

Again considering the steady state we get a nonlinear boundary value problem by balancing the heat fluxes for each mixture:

$$\frac{d}{dl}[m_i H_i(z_{i,j}, T_i)] = \triangle Q_i = k_i \frac{F_i}{L}[T_W(l) - T_i(l)]$$

$z_{i,j}$ is the molar composition of stream i and $k_i \frac{F_i}{L}$ the heat exchange coefficient of stream i and the heat exchanger material.
The temperature of the heat exchanger $T_W(l)$ is given by

$$\sum_i \triangle Q_i + k_W \frac{F_W}{L}[T_W(l) - T_0] = 0$$

with T_0 the outside temperature. Using this equation for eliminating the temperature $T_W(l)$ we get the following nonlinear BVP with separated linear boundary values

$$\frac{dT_i(l)}{dl} = \frac{1}{H_i(z_{i,j}, T_i)} \frac{k_i F_i}{m_i L} \left[\frac{\sum_{i=1}^{N} T_i(l) k_i F_i + T_0 k_W F_W}{\sum_{i=1}^{N} k_i F_i + k_W F_W} \right]$$

$$T_i(0) = T_i^* \qquad i \in J_A$$
$$T_i(L) = T_i^* \qquad i \in J_E = \{1, \dots, N\} \setminus J_A$$

The system can be solved by multiple shooting techniques (MST). But as there might be phase transitions within the heat exchanger - at most two for each stream - the right hand side of the ODE's may jump at any a priori unknown point l. Even the number of jumps may change during the MST-iteration. We cope with this problem by the following procedure:

a) Starting process: for each stream we calculate the dew and bubble point curves $T^d(l)$ and $T^b(l)$. A suitable approximation for them is stored.

b) MST-iteration: let us assume that we have a phase transition during an iteration, say: $T_i(l_k) < T^d(l_k)$ and $T_i(l_{k+1}) > T^d(l_{k+1})$. To avoid problems with the convergence of the ODE-integration we must use $l_k + \delta$ as an integration point. The unknown stepsize δ is determined by solving the equation $T_i(l_k + \delta) = T_i^d(l_k + \delta)$. As an alternative one might use switching functions as proposed by Bulirsch (private communication).

3.3 Other Units

Under certain assumptions - vapor only, equal velocity of all components, a special law for the intensity of reactions - we get an initial value problem for a reactor (steady state!). Depending on the velocity of the reactions some stiffness might be involved, in some cases it might even lead to algebro differential equations. Other units which are easily to be solved are: compressor, mixer, splitter,.. . In his book King [8] gives 6 pages of separation units, one unit a line. From this it follows that either one specializes and/or one gives the user a possibility to write a program for his special unit(s) and insert it into a general flowsheeting program.

4 Flowsheeting

Mathematically a flowsheet can be described as a directed graph. To solve for the steady state we proceed as follows:

Step 1: It is obvious that a flowsheet should be determined sequentially as far as possible. Therefore, one first determines all irreducible subsystems (Figure 1.1: dotted system) - see Tiernan [12]. An irreducible subsystem is characterized by its connected graph. The corresponding model is a simultaneous equational system.

Step 2: To solve the irreducible subsystems we have two alternatives:

    (i) Equational approach: The whole large scale nonlinear set of equations describing a subsystem is solved simultanously.

        Example: QUASILIN [6,7]. Of course, the computer storage demand is very high. Also problems might arise for units which, though essentially stationary, are described by initial or boundary value problems. On the other hand the solution is performed quite efficiently.

    (ii) Sequential modular approach: one determines "optimal" cuts such that the resulting subsystems can be computed sequentially, provided the necessary input streams (= cut streams) are defined. "Optimal" here means that we may put weights on the streams, f.e. depending on the number of components. This can be done by the procedure proposed by Pho/Lapidus [11]. Then we have to solve a two stage iteration process

        (Int) With the (estimated) input streams given the subsystem can be computed sequentially, the calculation of the involved units may be done iterative.

        (Ext) At the cuts we have to demand continuity. This condition leads to a fixed point equation

        Example:

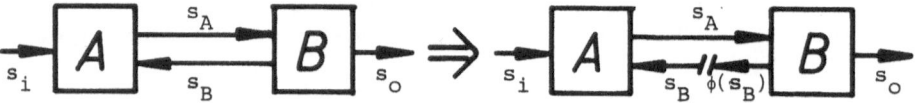

We get: $s_B = B[s_A] = B[A[s_i, s_B]] =: \phi(s_B)$ where A resp. B also stands for the mathematical description of units A and B and s_B is a stream with, in general, both vapor and liquid components. This fixed point equation can be solved by direct substitution, by Quasi-Newton or by Newton's Method.

4.1 Numerical Examples

The moduls described in the two stage iteration process above are by no means uniquely determined. It depends on the software facilities of the user what can be defined as a unit. We present two examples which at the same time give hints for a comparison between the equational approach and the sequential modular approach. To our knowledge results for large scaled problems ($n > 500$) by the equational approach are not yet available.

Example 1:

Crude Oil Cracking: 4 oil fractions $R_1, \ldots, R_n$ and water in a plate distillation column

Figure 4.1 (a)   (b)

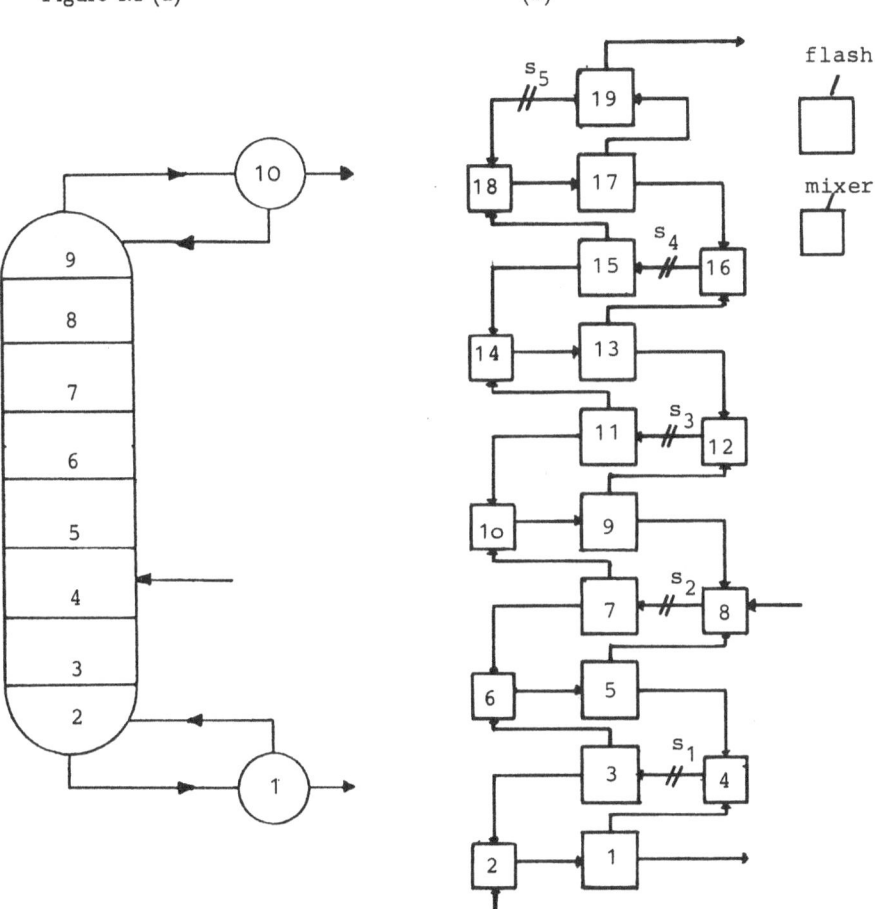

By exchanging each plate of the column (a) against a combination of a mixer and a flash one obtains a model (b), which is completely equivalent to that of chapter 2. While (a) is a plant consisting of one unit, plant (b) consists of 19 units and is one irreducible subsystem. Our algorithm defines five cut streams: $\{s_1, s_2, s_3, s_4, s_5\}$. We give a comparison between (a), (b): CPU [sec] BASF/7/78:

(a) Newton's Method NM (see section2)   2''
(b) Sucessive Approximation   1'29''
     QNM   1'21''
     NM   14''

Example 2: Argon-Nitrogen-Oxygen Separations

Figure 4.2

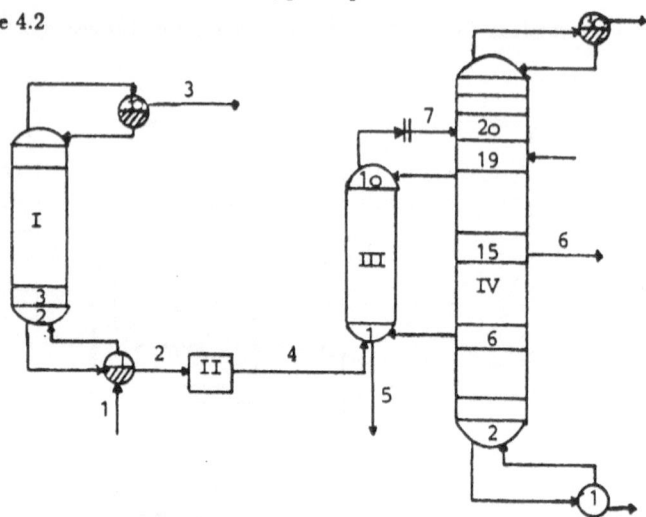

The irreducible subsystems are: $I, II, \{III, IV\}$
To reduce the complexity of $\{III, IV\}$ we may use stream 7 as a cut stream.

(a) $I, II, \{III, IV\}$
NM (see section 2)   18″

(b) I, II, III, IV
sucessive approximation 1′22″
QNM   41″
NV   31″

In these simple cases the equational approach proves superior. In literature one finds a tendency to give the user a possibility to define his units in a modular way. Then the program constructs the full system (see, f.e., QUASILIN [6, 7]). However, we point out that there might be some difficulties with the equational approach when more complicated units are involved.

4.3   Optimization and other Aspects

In almost all applications something can be made "optimal". In our case we want to have a high degree of purity for the, say, top product of a plate separation column. This can be performed by, say again, inserting heat into a certain plate. In the general case we have:

$$F\big(x(p), p\big) = 0, x \in \mathbb{R}^n, x \geq 0, p \in \mathbb{R}^p$$
$$\phi\big(x(p), p\big) = Extr!$$

For the optimization process we have in mind a Quasi Newton method. The gradients of the objective can be determined in a way where we directly use the results of the function evaluations. For the gradients which must be determined up to a certain accuracy we have:

$$\nabla_p \phi = \phi_x x_p + \phi_p$$

The vector x_p can be determined by help of the solution of F:

$$\frac{d}{dp} F(x(p), p) = 0 \implies F_x x_p + F_p = 0$$

F_x is already known when we use Newton's Method. We may even use our L-U decomposition. Therefore, one Newton step gives at the same time the new solution $x^+ := x + \triangle x$ and x_p at only little additional costs. In practical cases p will have a small number of components compared to n.

Another import aspect is the transition between two steady states of a system depending on a change of certain parameters. This problem is solved by using dynamic models which mathematically lead to algebro differential equations [4].

References

[1] Auzinger, D.: Berechnung stationärer Zustände von Destillations- und Absorptionskolonnen, Diplomarbeit, Linz 1985, pp.10-16

[2] Auzinger, D.; Kokert, F.; Peer, L.; Wacker, Hj.; Zulehner, W.: Adapted numerical methods for the steady state simulation of plate separation columns, Institutsbericht Nr. 325, Institut für Mathematik, Universität Linz, März 1987, p.6

[3] Fredenslund, A.; Gmehling, J.; Resmussen, P.: Vapor-liquid equilibria using UNIFAC, ELSEVIER, Amsterdam 1977

[4] Gani, R.; Ruiz, C.A.; Cameron, I.T.: A generalized model for distillation columns - I. Model description and applications, Comp.Chem. Engng., Vol. 10, Nr. 3 (1986), pp. 181–198

[5] Hackl, J.; Wacker, Hj.; Zulehner, W.: An efficient step size control for continuation methods, BIT 20, 1980, pp. 475–485

[6] Hutchinson, H.P.; Jackson, D.J.; Morton, W.: The development of an equation-oriented flowsheet simulation and optimization package - I. The Quasilin program, Comp.Chem.Engng., Vol 10, Nr.1 (1986), pp.19–33

[7] Hutchinson, H.P.; Jackson, D.J.; Morton, W.: The development of an equation-oriented flowsheet simulation and optimization package - II. Examples and results, Comp.Chem.Engng. Vol 10, Nr. 1(1986) pp.31–48

[8] King, J.C.: Separation Process, McGraw Hill Book Comp., New York 1980, 2nd edition, pp.??

[9] Kubicek, M.; Hlavacek, V., Prochaska, F.: Global modular Newton-??-technique for simulation of an interconnected plant applied to complex rectification columns, Chem.Engng.Sci, 31, pp.227–284

[10] Peer, L.: Berechnung stationärer Zustände von Destillations- und Absorptionskolonnen, Diplomarbeit, Linz 1985, pp.13–15

[11] Pho, T.K.; Lapidus, L.: An Optimum Tearing Algorithm for Recycle Systems, AICHE Journal, Vol.19, Nr.6, 1973, pp.1170–1180

[12] Tiernan, J.C.: An efficient search algorithm to find the elementary Circuits of a Graph, Commof the ACM, Vol.13, Nr.12, 1970, pp.722–726

[13] Wacker, Hj.: Continuation Methods, Academic Press, New York, 1978

[14] Westman K.R.; Lucia, A,; Miller, D.C.: Flash and distillation calculation by a Newton-like method, Comp.Chem.Engng. 8, 1984, pp.219–228

Acknowlegements:

We want to give our thanks both to Dipl.-Ing. F. Kokert (VOEST) and Dr. W. Zulehner for their valuable contributions.
The work of one of the authors was supported by the VOEST-ALPINE AG.

Address:

Institut für Mathematik
Johannes-Kepler-Universität Linz
Altenbergerstraße 69, A-4040 Linz, Austria

A Scheduling Problem in the Production

Line "Steel Making - Continuous Casting -

Hot Rolling" *

Heinz W. Engl and Gerhard Landl

Summary: If a given product mix of rolled steel of different widths is to be produced via direct hot rolling, it is important to schedule the relevant processes in the continuous casting machines and the rolling mill in such a way that the life span of the rolls is maximal. In this paper, we describe a mathematical model for this scheduling problem. The model leads to a hierarchical optimization problem; we describe an algorithm for its numerical solution and report about its performance.

1.) Introduction

The problem treated in this paper was posed to us by VOEST-Alpine AG and is only one of several projects in connection with steel making our department worked on. See [4], especially the articles by Auzinger/Wacker and Engl/Langthaler there, and [3] for other projects we were involved in jointly with VOEST-Alpine AG.

The project described here is concerned with scheduling the different processes in steel making, continuous casting and direct hot rolling, the combination of which constitutes the most modern and most efficient process route in steel making. As opposed to older technologies, in this process route, the three main steps of steel production (namely, steel making, casting, and rolling) are closely interconnected. Especially, since usu-

* Supported in part by the Austrian Fonds zur Förderung der wissenschaftlichen Forschung (project S32/03). This paper is in final form.

ally steel of different widths is to be produced, the widths
in the rolling mill, in the continuous casting machines and the
amount of steel coming from the steel mill have to be coordina-
ted. Due to the wear of the rolls, the order in which the different widths have to be rolled has to obey severe restrictions.
Worn out rolls have to be replaced. The aim of our project is
the following: Given a "product-mix" of different widths of steel
to be produced, the times when each pot of steel arriving from
the steel mill is cast into the continuous casting machine, when
the rolls are replaced, when the continuous casting machines are
readjusted, and the order of production and lenght of each required width of steel to be cast and rolled have to be determined.
At first glance, it would be attractive to determine these quantities in some optimal way. However, we will see that it is already hard enough to determine a feasible solution to our problem.

2.) The Mathematical Model

We assume that a given order by one or more customers
contains j_0 different widths (given in mm)

(2.1) $0 < b_1 < b_2 < \ldots < b_{j_0}$;

for $j \in \{1, \ldots, j_0\}$, L_j meters of width b_j have to be produced.
All lengths (here and below) refer to steel after the rolling
process.

The quantities b_j and L_j are also referred to as "product-mix".

Due to the wear, a set of rolls in the rolling mill has
to be replaced after about 80 km of steel have been rolled.
Thus, the number i_0 of necessary rolls can be predetermined as

(2.2) $i_0 = -[-(\sum_{j=1}^{j_0} L_j)/80000]$.

The mentioned life span of $a_0 := 80$ km of a set of rolls can only
be reached if the rolls wear in a uniform way. This can be achieved by ordering the widths to be rolled with the same rolls according to the following "grouping":

If b_O denotes the maximal width to be rolled by a given set of rolls (b_O need not be b_{j_O} !), then with this set of rolls, one has to start with a width b_A, which is between 60 % and 80 % of b_O. Then, the widths have to be increased monotonically until they reach b_O, which has to happen between a_1 and a_2 meters of rolling. By a, we denote the length rolled (in m) until the width b_O is reached. From then on, the widths have to decrease monotonically. In both parts of this grouping, the difference between two consecutive widths should not exceed a given quantity Δb_{max}. In our model, a_1 and a_2 were given as 8 and 16 km, respectively.

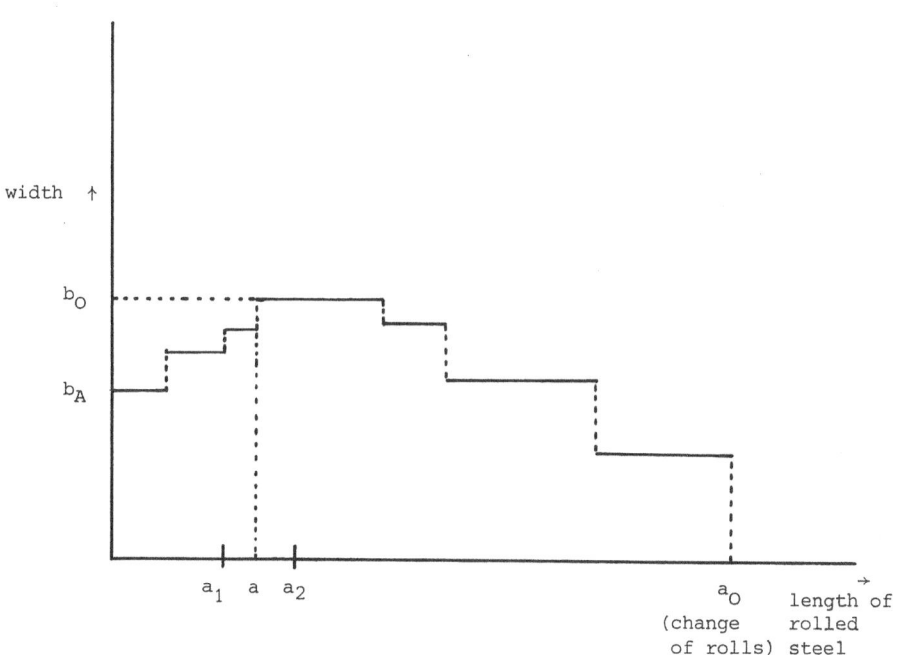

Grouping

As we will see, these requirements about the grouping will pose major problems in the model.

Now, we start introducing the variables of the model. In choosing the variables, our aim was to make the model as easily understandable as possible.

One set of variables refers to the lenghts of each given width rolled with each of the i_0 sets of rolls, where we distinguish between the first (increasing) and second (decreasing) parts of the grouping as indicated above. Also, since there may be more than one, say n_0, continuous casting machines in parallel, a fourth index has to refer to the casting machine where the respective slabs come from.

Thus, by $l_{ijk}^{(n)}$ we denote the lenght (in m) of steel with width b_j coming from the n-th casting machine and rolled in the k-th part of the grouping by the i-th roll. Here, $i \in \{1, \ldots, i_0\}$, $j \in \{1, \ldots, j_0\}$, $n \in \{1, \ldots, n_0\}$; k=1 or 2 for the first or second part of the grouping. This gives us already $2 \cdot i_0 j_0 n_0$ variables.

Now, not all widths have to be present for each roll. For $i \in \{1, \ldots, i_0\}$, $n \in \{1, \ldots, n_0\}$, we use the notations

(2.3) $m_i^{(n)} := \min \{p/l_{ip1}^{(n)} > 0\}$,

(2.4) $M_i^{(n)} := \max \{p/l_{ip2}^{(n)} > 0\}$.

Thus, $m_i^{(n)}$ and $M_i^{(n)}$ are integer variables referring to the smallest or largest width coming from the n-th casting machine to be rolled with the i-th roll in the first or second phase of the grouping, respectively. Thus, for the i-th roll, $\min \{m_i^{(n)}/n \in \{1, \ldots, n_0\}\}$ and $\max \{M_i(n)/n \in \{1, \ldots, n_0\}\}$ play the role of b_A and b_O as described above.

The first restriction is of course that the given product mix should be produced, i. e.,

(2.5) $\sum_{n=1}^{n_0} \sum_{j=1}^{j_0} \sum_{k=1}^{2} l_{ijk}^{(n)} = L_j$ $(j \in \{1, \ldots, j_0\})$.

Because of the restricted life span of each roll, we have

(2.6) $\quad \sum_{n=1}^{n_0} \sum_{j=1}^{j_0} \sum_{k=1}^{2} l_{ijk}^{(n)} \le a_0 \quad (i \in \{1, \ldots, i_0\})$.

The requirements for the grouping of the widths as described
above lead to the restrictions

(2.7) $\quad a_1 \le \sum_{n=1}^{n_0} \sum_{j=1}^{j_0} l_{ij1}^{(n)} \le a_2 \quad (i \in \{1, \ldots, i_0\})$

and

$$0.6 \max \{b_{M_i}(n)/n \in \{1, \ldots, n_0\}\} \le \min \{b_{m_i}(n)/n \in \{1, \ldots, n_0\}\} \le$$

(2.8)

$$\le 0.8 \max \{b_{M_i}(n)/n \in \{1, \ldots, n_0\}\} \quad (i \in \{1, \ldots, i_0\}).$$

By definition, the integer variables $m_i^{(n)}$ and $M_i^{(n)}$ have to fulfill that $m_i^{(n)} < M_i^{(n)}$, which is contained in (2.8) because of (2.1).
By definition of the $m_i^{(n)}$ and $M_i^{(n)}$, the following has to hold
for $i \in \{1, \ldots, i_0\}$ and $n \in \{1, \ldots, n_0\}$:

(2.9) $\quad l_{ij1}^{(n)} = 0 \quad (j \in \{1, \ldots, m_i^{(n)} - 1\})$

(2.10) $\quad l_{ij1}^{(n)} = 0 \quad (j \in \{M_i^{(n)}, \ldots, j_0\})$

(2.11) $\quad l_{ij2}^{(n)} = 0 \quad (j \in \{M_i^{(n)} + 1, \ldots, j_0\})$.

If we looked only at the rolling mill separately, there would be
no reason for a relation between the last width rolled before
and the first width rolled after a change of rolls. However, the
slabs come immediately from the continuous casting machines,
where a change of widths can be done without stopping the casting, but only at a rate of about $\alpha := 18$ mm/min. Thus, we have
to assume a relation between the last width rolled before and
the first width rolled after a change of rolls, provided the
slabs come from the same casting machine, which we do by requiring that

(2.12) $\quad l_{ij2}^{(n)} = 0 \quad (i \in \{1, \ldots, m_{i+1}^{(n)} - 1\})$.

To formulate (2.12), we need n_0 additional variables

$m_{i_0+1}^{(n)} \quad (n \in \{1, \ldots, n_0\})$.

Since we have for all indices that

(2.13) $\quad l_{ijk}^{(n)} \geq 0$,

we can replace (2.9) - (2.12) equivalently by the single restriction

$\quad \Sigma l_{ijk}^{(n)} = 0$, summed over all indices as they appear in

(2.14) $\quad$ (2.9) - (2.12).

Since, as mentioned above, a change of widths can be done in the casting machine only at the relatively slow rate α, we have to make sure that for each width, at least the length necessary to change the width to or from the next larger one (whose actual length is non-zero) is produced. This would certainly be guaranteed by requiring that for all indices,

(2.15) $\quad l_{ijk}^{(n)} > 0$ implies $l_{ijk}^{(n)} \geq \dfrac{\Delta b_{max}}{\alpha} v$,

where v is the casting speed (scaled according to the fact that lengths refer to slabs after rolling). In our model, we assume a fixed casting speed. In reality, the casting speed during a change of widths is slightly smaller, which could be incorporated easily into our model (and program). We remark that the slabs (of varying widths) produced during a change of widths can be cut down to the smaller width (say, b_j) and are hence incorporated into the corresponding length $l_{ijk}^{(n)}$. It turned out that (2.15) posed enormous algorithmic difficulties (to us). Hence, we did not use (2.15), but the similar (and weaker) set of restrictions, which have to hold for $i \in \{1, \ldots, i_0\}$, $n \in \{1, \ldots, n_0\}$

$$\sum_{p=J(j)}^{j-1} l_{ipk}^{(n)} \geq \frac{b_{J(j)} - b_j}{\alpha} v \qquad \qquad \text{for}$$

(2.16)
$\qquad$ ($k=1$ and $j \in \{m_i^{(n)}+1, \ldots, M_i^{(n)}\}$) and for
$\qquad$ ($k=2$ and $j \in \{m_{i+1}^{(n)}+1, \ldots, M_i^{(n)}\}$),

where

(2.17) $\quad J(j) := \min \{p / b_p \geq b_j - \Delta b_{max}\}$.

Note that if Δb_{max} is chosen such that no intermediate width can be left out, then (2.16) essentially reduces to (2.15).

From the steel mill, pots containing around 250 metric tons of liquid steel arrive in fixed intervals of about $T_f := 20$ min. We assume that the steel contained in each pot is poured wholly into one of the continuous casting machines. From the given product mix, the total number H of pots to be considered is given, if, as has to be checked in the beginning, the amount of steel necessary for producing this product mix constitutes an integer number of pots. Then, the number of pots used for each casting machine is set a priori: Let for $n \in \{1, \ldots, n_0\}$, $H_n \in \mathbb{N}$ be the number of pots used for the n-th casting machine (with $\sum_{n=1}^{n_0} H_n = H$; usually, $H_n \approx \frac{H}{n_0}$). Then, with ρ as density of steel, G denoting the mass contained in one pot, and d as the thickness of the rolled steel, we have the restrictions that for $n \in \{1, \ldots, n_0\}$,

$$(2.18) \quad \sum_{i=1}^{i_0} \sum_{j=1}^{j_0} \sum_{k=1}^{2} l_{ijk}^{(n)} \cdot b_j \cdot d \cdot \rho = G \cdot H_n.$$

Usually, several pots of steel are poured into the casting machines directly following each other to form a continuous strand, which is cut into slabs after leaving the casting machines. These consecutive pots are called a "sequence". Let $s_q^{(n)}$ be the number of pots forming the q-th sequence in the n-th casting machine. Since this number is bounded from above (because of the necessity to adjust the casting machines from time to time) we have the restrictions

$$(2.19) \quad s_q^{(n)} \le s_{max} \qquad (n \in \{1, \ldots, n_0\}, q \in \{1, \ldots, q_0^{(n)}\}),$$

where $q_0^{(n)}$ is at most H_n.

After each of these sequences, the continuous casting machines have to be adjusted. As opposed to the change of rolls, the duration of which is negligible, this adjustment takes considerable time, during which this casting machine cannot be used. This is one reason why usually $n_0 > 1$.

In order to model this, we need a new set of variables, namely times when certain relevant processes start. We emphasize that most of these variables are introduced here only for clarity, since they are (as described below) heavily connected with the other variables used so far.

For this purpose, it makes sense to compress the indices (i, j, k) into one new index ν in such a way that the (i, j, k) are ordered according to their chronological sequence, i. e.: $\nu = 1, 2, 3, \ldots, 2i_0 j_0$ corresponds to the order

$$(i,j,k) = (1,1,1), (1,2,1), \ldots, (1,j_0,1), (1,j_0,2), (1,j_0-1,2), \ldots$$
$$\ldots, (1,1,2), (2,1,1), \ldots, (i_0,1,2). \text{ This transformation of indices is bijective.}$$

For $n \in \{1, \ldots, n_0\}$, $\nu \in \{1, \ldots, 2i_0 j_0\}$, let $t_\nu^{(n)}$ be the time when production of the collection of slabs corresponding to the index ν starts in the n-th casting machine (cf. the remark following (2.25)). By $\tau_\nu^{(n)}$, we denote the time needed for casting this collection of slabs (including intermediate adjustment times of the casting machine). As we will describe below, $\tau_\nu^{(n)}$ depends on $l_\nu^{(n)}$, but also on other quantities, especially the waiting times $w_q^{(n)}$ of the n-th casting machine before the q-th sequence in the n-th casting machine starts. These new variables have to obey the restrictions

(2.20) $w_q^{(n)} \geq w_0$ $\qquad (q \in \{2, \ldots, q_0\}, n \in \{1, \ldots, n_0\})$

and

(2.21) $w_1^{(n)} \geq 0$ $\qquad (n \in \{1, \ldots, n_0\})$,

where w_0 denotes the minimal adjustment time for a casting machine between the sequences (which is usually around 25 min).

For $h \in \{1, \ldots, H\}$, let $T(h)$ be the time when the casting of the h-th pot commences.

We number the casting machines in such a way that

(2.22) $t_1^{(n)} := (n-1)T_f + w_1^{(n)}$ $\qquad (n \in \{1, \ldots, n_0\})$.

Given the actual values for the variables $l_\nu^{(n)}$, $s_q^{(n)}$ and $w_q^{(n)}$, one can (by simulating the actual production process corresponding to these values for the variables $1_\nu^{(n)}$, $s_q^{(n)}$ and $w_q^{(n)}$)

compute step by step the corresponding values for the $\tau_\nu^{(n)}$

and $T(h)$. Thus,

(2.23) $\tau_\nu^{(n)} = \tau_\nu^{(n)}(1,s,w)$ $(n \in \{1, \ldots, n_0\}, \nu \in \{1, \ldots, 2i_0j_0\})$,

(2.24) $T(h) = T(h)(1,s,w)$ $(h \in \{1, \ldots, H\})$,

where the functions appearing in (2.23) and (2.24) are quite
complicated and are implemented as a subroutine in our model.
This subroutine of course also needs the values of the constants v, T_f, b_j, ρ, G mentioned above. The time scale is
changed in such a way that the waiting times of the casting
machines do not give rise to discontinuities in these functions.
Because of this, $\tau_\nu^{(n)}$ depends also on values of $1_\nu^{(\mu)}$, $s_q^{(\mu)}$,
$w_q^{(\mu)}$ for $\mu \neq n$. For the exact form of these functions see [6].

Now, for $\nu \in \{2, \ldots, 2i_0j_0\}$, $n \in \{1, \ldots n_0\}$,

(2.25) $t_\nu^{(n)} = t_{\nu-1}^{(n)} + \tau_{\nu-1}^{(n)}$.

Note that usually, many $\tau_\nu^{(n)}$ will vanish, namely those,
for which the slabs described by the index ν are not actually
produced, i. e., $1_\nu^{(n)} = 0$. Thus, for such indices, $t_\nu^{(n)} = t_{\nu-1}^{(n)}$.

These new variables have to obey the following restrictions:
Since there is only one rolling mill, which cannot handle slabs
of different widths simultaneously, each casting machine has to
complete the casting of any given width before at any other
casting machine, slabs of a different width become available
for rolling. This leads to the set of constraints

(2.26) $t_\nu^{(n)} + \tau_\nu^{(n)} \leq \min \{t_\mu^{(p)} + T_s, t_\mu^{(p)} + \tau_\mu^{(p)}\}$,

which have to hold for all (ν,n) such that $1_\nu^{(n)} > 0$ and all p

such that $1_\mu^{(p)} > 0$, where $\mu = \mu(\nu,p) := \min \{r > \nu / 1_r^{(p)} > 0\}$.

T_s denotes the time necessary to cast one single slab of maximal length. Note that (2.26) can be thought of as constraints involving the original variables $l_v^{(n)}$, $s_q^{(n)}$, $w_q^{(n)}$ via (2.23) (in a quite complicated way). The same is true (via 2.24)) for the last set of constraints

(2.27) $T_f \cdot (h-1) \leq T(h) \leq T_f \cdot (h-1) + T_w$ ($h \in \{1, \ldots, H\}$),

where T_w denotes the maximal time a pot of liquid steel can be allowed to wait before casting of its contents begins; in practice, T_w can be up to 45 min, which introduces some flexibility into the model.

The restrictions described so far constitute our model for a feasible operating policy of the system "steel making - continuous casting - direct hot rolling" in order to produce the given product mix. In addition, one could want to impose various optimization criteria like minimal production time or minimal fluctuations in the throughput (per unit time) in the rolling mill. However, these optimization criteria are practically much less relevant than the aim of finding a feasible solution. Thus, we decided to aim just at finding a feasible solution, which turned out to be complicated enough (see Section 3). Thus, we have the following

Problem: Find values for the (indexed) variables l, w, s, m, M (as described above) such that the constraints (2.5) - (2.8), (2.13), (2.14), (2.16), (2.18) - (2.21) and (using the definitions (2.22) - (2.25) for the dependent variables t, τ, T(h)) (2.26), (2.27) are fulfilled.

Note that it follows from (2.22) that we assume that before the production period we consider, all casting machines are empty. Also, we do not take into account what happens after the production period we consider. One could impose conditions on the state of the system at the beginning and the end of the production period (or a periodicity condition linking these times). Since this would considerably complicate the model, we did not take conditions of this type into account so far.

3.) Numerical Solution of the Problem

The main difficulties of our Problem as described in
Section 2 are the following:
There are continuous variables denoted by l and w (with suitable
indices) and discrete (integer) variables denoted by m, M, s.
Moreover, the number of variables $s_q^{(n)}$ and $w_q^{(n)}$ for each n is
also a variable (denoted by q_0 in Section 2, but also depending
on n). The formulation of the constraints (2.14) and (2.16) depends on the actual values of the variables $m_i^{(n)}$ and $M_i^{(n)}$. Thus,
the constraints of the model cannot be formulated a priori,
they change with the variables. The same is true for the constraints (2.26), since the indices for which these constraints
have to hold (and hence their number) can be determined only
relative to the actual values of the variables l. Moreover,
(2.23) and (2.24) are quite complicated.

We tried to overcome these difficulties by the following
"hierarchical" approach:
The variables and the restrictions were divided into two "layers":

Layer 1 contains the (indexed) variables m, M, s∈ℕ (together
with q_0) and the restrictions (2.8) and (2.19).

Layer 2 contains (for any fixed setting of m, M, s coming from
layer 1) the (indexed) variables l and w and all the constraints
of our Problem except (2.8) and (2.19).

Note that layer 2 depends on the actual setting of the variables
in layer 1, which is the reason for the use of the term "hierarchical approach" above. Since the values for m and M are inputs
for layer 2, (2.3) and (2.4) cannot be part of the model. However, (2.3) is enforced by (2.14) and (2.16); (2.14) also im-
plies that $M_i^{(n)} \geq \max\{p/l_{ip2}^{(n)} > 0\}$, which is of course weaker than
(2.4). One could enforce (2.4) by imposing a restriction as it
appears in the second inequality of (2.15) for $j=M_i^{(n)}$. We plan
to incorporate these additional restrictions into our model,
although this omission caused no problems so far; (2.4) was always fulfilled in the final result. However, also the results

of the test example below indicate that such a constraint
should be added.

Now, the aim is to set the variables in layer 1 in such
a way that a feasible solution can be found in layer 2. Then,
all the variables form a solution of our Problem, a feasible
operation policy for the whole system can easily be deduced.

Most likely, for an initial choice of the variables in
layer 1, no feasible solution in layer 2 will exist. In this
case, we will have to change the variables in layer 1 in such
a way that in the next step, we come "closer" to a feasible solution in layer 2. Thus, we need a "measure of closeness" to a
feasible solution in layer 2. Since in layer 2, we have linear
and nonlinear constraints, it is reasonable to reformulate the
problem of finding a feasible solution in layer 2 in such a
way that the sum of the squares of the defects in the nonlinear
constraints is minimized subject to the linear constraints. The
final value of this (suitably weighted) sum of squares is the
desired "measure of closeness". More precisely:
In layer 2, we minimize (for given values of the variables
from layer 1) the sum of the squares of the defects in (2.26)
and (2.27) (referred to as "auxiliary objective function") with
respect to the linear constraints (2.5) - (2.7), (2.13), (2.14),
(2.16), (2.18), (2.20), (2.21). This is a nonlinear, but linearly
constrained optimization problem; its constraints and objective
function depend on the values of the variables in layer 1. We
sketch how this problem is treated; details can be found in
[6]:

First, a feasible solution for the set of linear constraints is computed by the simplex method. Since the simplex
matrix is quite sparse and has a specific block structure, we
used the version of the simplex method due to C. Winkler described in [1,pp.77-92]. Note that by using this method, we can
also obtain the projections onto sets of active constraints to
be used later. As starting values, the values obtained in the
last (accepted, see below) call of layer 2 were used.

The first attempt to minimize the auxiliary objective
function subject to the linear constraints was a reduced gradient
method (combined with an active set strategy) as described in

[5,ch.11], where the matrix Z describing the reduced variables can be calculated from the version of the simplex method described above. Of course, at this point the question of differentiability of the auxiliary objective function has to be discussed. This is intimately connected with the differentiability properties of the functions defined in (2.23) and (2.24). It turns out (see [6]) that these functions are continuous and piecewise differentiable; e. g., if the variables (l, s, w) are such that a change of widths takes place at the same time as the beginning or end of an adjustment of a casting machine, then these functions need not be differentiable at this point. At such points, gradients have to be replaced by subgradients (cf. [10, Theorem 1.11]), if the auxiliary objective function is convex there; otherwise, one leaves this point by a slight perturbation, which usually leads back to a differentiability region. In this sense, "gradient" has to be understood. Using the specific form of the functions (2.23), (2.24) given by the subroutine mentioned above, it is possible to compute these "gradients" of the auxiliary objective function explicitly, also in the form of a subroutine (see [6]).

In numerical tests, this (modified) reduced gradient method did not work for our Problem; it was much too slow. Thus, we replaced it by the conjugate gradient method due to Beale as described in [7, (1.6), (3.4), (3.5)] (with restart as described in [7,p.252]). Of course, this CG-method has to be combined with an active set strategy as described above (cf. [9]); if the set of active constraints changes, a restart is necessary.

This CG-method proved quite effective in numerical tests for layer 2 of our Problem.

The result given back from layer 2 to layer 1 is the information if the linear constraints are compatible and (if yes) the achieved value of the auxiliary objective function. If this value is (close to) 0, then we have found a feasible solution for our Problem. Otherwise, the variables in layer 1 are changed, and layer 2 is entered with these new values again.

As a first attempt, we used a random search method, more precisely, the "evolutionary strategy" described in [8, ch.5].

There, the configuration of variables in layer 1 is changed
randomly (according to a multi-dimensional Gaussian distribution) in such a way that the constraints (2.8) and (2.19) remain fulfilled (and that the relative values of the variables
make sense, i. e., $m_i^{(n)} < M_i^{(n)}$ holds). If with these values,
layer 2 gives a feasible solution for the linear constraints
and a reduced value for the auxiliary objective function, this
new configuration of variables is accepted. The covariance matrix
of the Gaussian distribution (which we always assumed to be diagonal) was used as a "step-size control" as described in [8,pp.
129 - 131]. For our Problem, this method did not yield values
for the variables in layer 1 giving rise to a feasible solution
in layer 2.

The key to success was to allow only certain variables
or sets of variables to be changed by the evolutionary strategy
in each step. These variables were chosen as it seemed plausible from their meaning for our Problem; e. g., it seems not to
make sense to use vastly different values for $m_i^{(n)}$ for the same
i and different n, since all casting machines feed into the same
rolling mill. If one restricts the changes the evolutionary
strategy is allowed to make, one should not insist that in each
step, the auxiliary objective function in layer 2 is reduced,
but should also allow an increase to be accepted with a probability depending on the amount of increase. We used this idea as
described in [2].

With this modification of layer 1, we managed to compute
a configuration of variables in layer 1 leading to a feasible
solution for layer 2 and thus for the whole Problem in the
following

Test Example: We took a production period of 10 pots of liquid
steel, i. e., totally 2500 metric tons of steel, to be cast on
$n_0 = 2$ parallel continuous casting machines. The $j_0 = 4$ widths to
be produced were $b_1 = 750$, $b_2 = 850$, $b_3 = 950$, $b_4 = 1000$ mm. Of the
total weigth, 30 % should be rolled with widths b_1 and b_4 each,
20 % with widths b_2 and b_3 each. From this, the lengths
$L_1, \ldots, L_4$ can be easily calculated; their sum turns out to
be 146 km. Thus, we needed $i_0 = 2$ rolls. We decided a priori to

cast 5 pots on each of the casting machines.

Most constants were chosen as mentioned above. However, since the widths were relatively small, so that the casting takes longer, we increased T_f to 32.5 min. In order to really test our program, we decreased T_w to 16 min, which makes it much harder to find a feasible solution.

For this configuration, the number of variables $l_{ijk}^{(n)}$ is 32, q_0 is ≤ 5, so that there are at most 10 variables $w_q^{(n)}$ and $s_q^{(n)}$ each. Finally, there are 6 variables $m_i^{(n)}$ and 4 variables $M_i^{(n)}$. Thus, layer 1 contains at most 20, usually (since q_0 fluctuates between 2 and 3) around 15 variables. Layer 2 contains up to 45 linear constraints.

We took the following starting values for layer 1: $q_0=2$ for both casting machines, $s_1^{(1)}=s_1^{(2)}=3$, $s_2^{(1)}=s_2^{(2)}=2$; all $m_i^{(n)}$ were taken as 1, all $M_i^{(n)}$ as 4. The algorithm performed as follows:

In layer 1, 114 random changes of the variables as described above were made. Among them, 63 were not feasible for layer 1, so that the program entered layer 2 51 times. In layer 2, twice no feasible solution for the linear constraints existed. In 27 calls of layer 2, the new solution was accepted. In 10 of these cases, the auxiliary objective function actually increased; this shows the effect of using the algorithm described in [2] as opposed to the standard evolution strategy!

The number of CG-steps (as described above) varied greatly around an average of 50 per call of layer 2; for the stopping criterion, we used the norm of the gradient and the relative change between consecutive values of the auxiliary objective function.

After a total CPU-time of 16.4 min on a BASF 7/78, a feasible solution was found. We describe this feasible solution:

For both casting machines, $q_0=3$, $s_1^{(1)}=s_1^{(2)}=s_3^{(1)}=s_3^{(2)}=1$, $s_2^{(1)}=s_2^{(2)}=3$. All $M_i^{(n)}$ were 4, while $m_1^{(1)}=m_2^{(1)}=m_2^{(2)}=1$, $m_3^{(1)}==m_1^{(2)}=m_3^{(2)}=3$.

The following lengths (in m after rolling) $l_{ijk}^{(n)}$ are to be produced in the order as indicated:

width	1st casting machine, 1st roll	2nd casting machine, 1st roll	1st casting machine, 2nd roll	2nd casting machine, 2nd roll
b_1	9152	0	3352	2919
b_2	1533	0	1612	1722
b_3	1089	1436	688	873
b_4	21	342	18940	18920
b_3	3526	2446	4700	12060
b_2	11810	13290	0	0
b_1	17520	18010	0	0

The waiting times $w_q^{(n)}$ were only up to $\frac{1}{2}$ min larger than w_0 except $w_1^{(2)}$, the waiting time for the second casting machine between the arrival of its first pot and the actual commencement of the casting (cf.(2.22)), which was 4.5 min. Because of this and the necessary synchronisation, the second casting machine starts the casting with width b_3. From these results, all times when relevant processes start are easily computable.

By adding up the total lengths of each width and weighting the results with the widths, one can check that the distribution of the widths is as required in the product mix.

Although the computed solution is mathematically feasible, the result still is not practically usable in this form, since it does not make sense to change the width from b_3 to b_4, then to produce just 21 and 342 m of rolled steel of this width and then to change the width back to b_3. However, this problem can easily be overcome by adding linear constraints in the form of lower bounds for some $l_{ijk}^{(n)}$ as indicated above.

For practical problems, the number of variables will be larger: In a real example from VOEST-Alpine AG, $i_0=3$, $j_0=9$, $n_0=2$, so that the number of variables $l_{ijk}^{(n)}$ is 108. While the $s_q^{(n)}$ and $w_q^{(n)}$ will stay roughly as above, we will have 8 $m_i^{(n)}$ and 6 $M_i^{(n)}$; since these numbers also have a greater range of

317

variation, the number of possible configurations in layer 1 increases significantly. In layer 2, we will have around 130 linear constraints. Thus, because of the special structure of these constraints mentioned above, we can expect that layer 2 of the algorithm works without major problems. However, layer 2 will be called much more often. Thus, we think that for layer 1, an interactive approach which also allows the user to specify plausible values for the variables in layer 1 should be used if one wants to avoid excessively long computer runs.

References:

[1] Bastian, M.: Lineare Optimierung großer Systeme. Königstein/Ts.: Verlagsgruppe Athenäum, Hain, Scriptor, Hanstein 1980.

[2] Burkard, R. E.; Rendl, F.: A thermodynamically motivated simulation procedure for combinatorial optimization problems. Europ. Jour. of Operational Research 17 (1984) 169 - 174.

[3] Engl, H. W.; Langthaler, Th.: Numerical solution of an inverse problem connected with continuous casting of steel. Zeitsch.f.Oper.Res.29 (1985) B 185 - B 199.

[4] Engl, H. W.; Wacker, Hj.; Zulehner, W. (eds.): Case Studies in Applied Mathematics. To appear with Teubner - Reidel (this series).

[5] Fletcher, R.: Practical Methods of Optimization, vol. 2. Chichester: Wiley 1981.

[6] Landl, G.: Mathematische Modellierung und numerische Behandlung eines Produktionssteuerungsproblems im System "Stahlwerk-Stranggußanlage-Warmwalzwerk". Diplomarbeit Universität Linz 1987.

[7] Powell, M. J. D.: Restart procedures for the conjugate gradient method. Math. Programming 12 (1977) 241 - 254.

[8] Schwefel, H. P.: Numerische Optimierung von Computer-Modellen mittels der Evolutionsstrategie. Basel: Birkhäuser 1977

[9] Shanno, D. F.; Marsten, R. E.: Conjugate gradient methods for linearly constrained nonlinear programming. In:Buckley, A. G.; Goffin, J.-L. (eds.): Algorithms for Constrained Minimization of Smooth Nonlinear Functions. Amsterdam: North Holland= Math. Programming Study 16 (1982) 149 - 161.

[10] Shor, N . Z.: Minimization Methods for Non-Differentiable Functions. Berlin: Springer 1985.

Authors' address: Heinz W. Engl and Gerhard Landl, Institut für Mathematik, Johannes-Kepler-Universität, A-4040 Linz, Austria.

A REFERENCE MODEL FOR LASER OPTO-ELECTRONICS, DIGITAL SIGNAL PROCESSING, AND HOLOGRAPHY

Walter Schempp

ABSTRACT

For studying the application of mathematics in engineering and industry, a key strategy is to exploit the mathematical structure. Following this strategy, Part I of the paper develops a group theoretic model of axisymmetric optical fiber waveguides and their semiconductor laser sources. The reference model which exploits the symmetries of these opto-electronic devices via two fundamental postulates will provide the reference refractive index profile to compute explicitly the quantized transverse eigenmode spectra of circularly symmetric opto-couplers and rectangular semiconductor injection laser diodes (ILDs). The procedure is based on harmonic analysis on the diamond solvable Lie group which is a toroidal extension of the real Heisenberg nilpotent Lie group. As a result, the coupling coefficients of these opto-electronic devices are expressed in terms of Krawtchouk polynomials. Moreover, the paper presents for the first time a group theoretic approach to the cleaved-coupled-cavity semiconductor ILDs. Furthermore, a two-stage scheme for adapting the group theoretic model to actual graded-index multimode optical fiber waveguides will be discussed. - In Part II of the paper, the group theoretic model will be applied to digital signal processing. The representation of digital signals in the complex plane leads via the important holographic identities to a series of identities for theta-null values. Their groups of invariants are classified by the cyclic groups $\mathbb{Z}/m\mathbb{Z}$ of order m satisfying the crystallographic restriction $m \in \{1,2,3,4,6\}$. In this way a classification of the radial information cells of digital signals in loss-free transmission channels arises. The spatial coded case $m = 6$ yields a particularly important example of radial information cells since the retinal receptors of most animals are arranged in a hexagonal pattern.

This paper is in final form and no version of it will submitted for publication.

320

0.   Contents

1.   Introduction
2.   The Economic Power of Laser Opto-Electronic Systems

1.   Introduction

      Ever since ancient times, one of the principal interests
of human beeings has been to devise communication systems for
sending messages from one distant place to another [1]. Many
forms of communication systems have appeared over the years. The
principal motivations behind each new form of communication
systems are either to improve the transmission fidelity, to increase the data rate (measured in bits/s) so that more information (telephone conversations, television programming, computer
and peripheral equipment hookup) could be sent, or to increase
the transmission distance between repeater units. In the ensuing
years after the beginning of the era of electrical communications an increasingly larger portion of the electromagnetic
spectrum was utilized for transmitting information. An important
portion of the electromagnetic spectrum encompasses the optical
region. The optical spectrum ranges from about 50 nm (ultra-
violet) to approximately 100 µm (far infrared). In recent years
a great interest in communication at the optical frequencies was
created with the dramatic increase in the physics and industrial
applications of what might be termed opto-electronic devices.
This has been brought about both by the development of the laser,
with its rapidly growing list of applications, and the stag-
gering growth in semiconductor device electronics and photonic
material technology. The narrow beams of highly coherent, nearly

monochromatic light with almost perfectly plane wavefronts
emitted by the semi-conductor ILD eliminated a major obstacle to
light-wave communication: ordinary sources of light cannot be
modulated rapidly enough to carry large amounts of information.
Since optical frequencies are on the order of 5×10^{14} Hz, the
semiconductor ILD has a theoretical information capacity ex-
ceeding that of microwave systems by a factor of 10^5, which is
approximately equal to 10 million television channels. Finally,
a central point about the semiconductor ILDs is that these are
the only lasers pumped directly by an electric current. Consequently they are easily incorporated into the electronic cir-
cuitry used in the communications industry. As a result, the
semiconductor ILDs form, by economic standards and the degree
of their industrial applications, the most important of all
lasers.

What was lacking in the early years of opto-electronics
was a practical set of photonic materials that could meet its
technical demands. Because of these demands the optical losses
in silica glass caused by absorption have been reduced in the
past 25 years by a factor of 10^4. A window made by the best
optical waveguide glass 3 kilometers thick would be more transparent to ligth than an ordinary window only 3 millimeters
thick. The advent of optical fiber waveguides with losses considerably less than 20 dB/km and with high information carry-
ing capacity (e.g., bandwidths > 100 MHz) has meant that they
have become viable and cost-effective alternatives to coaxial
cable as well as satellite communications systems.

In any practical application of optical waveguide technology, the hair-thin silica fibers (of diameter 10-100 μm) need
to be incorporated in some type of cable structure. The immense
information capacity previously undreamed of, which opto-electronic signal processing made available due to the highly
coherent, nearly monochromatic optical frequency carriers of
laser sources like cleaved-coupled-cavity semiconductor ILDs
(cf. Section 6) and the wideband transmission media of monomode
optical fiber waveguides will be illustrated by the following
facts: In 1956 the first transatlantic wire cable TAT-1 connecting the United States and Europe had the ability to transmit

36 telephone conversations simultaneously whereas the latest coaxial cable TAT-7 which was laid in the mid-1970's has a capacity of 10.000 conversations. The first intercontinental fiber optical cable TAT-8 which has been installed in 1986 and which is scheduled to begin operating in 1988 provides a communication capacity of nearly 40.000 individual telephone conversations. The estimated costs of TAT-8 will be about 35 million US-$ shared by 30 telephone companies located in U.S.A., Canada, and West-Europe. Essentially all new long-distance telephone cables in the U.S. are optical ones. Opto-couplers are even being employed to connect adjacent electronic machines or equipment frames within a machine. In such applications they are desirable for their immunity from electromagnetic interference as well as for their extremly high pulse rates exceeding 10^{10} bits/s and their small size. However, engineers suggest that the biggest potential market of wideband fiber optical communication systems presently lies in the field of civil telecommunication.

The basic elements of an optical transmission system encompass the emitter, the optical fiber waveguide, and the receiver. The emitter is usual a semiconductor ILD or a light emitting diode (LED), whilst the receiver may be a PIN photodetector or avalanche photodiode (APD). The transmission channels are formed by the optical fiber waveguide. Electro-optical repeater units may be necessary over relative long links to counter the effects of fiber transmission losses and intermodal dispersion. In these opto-couplers the signal is detected, amplified and then re-emitted. A separate power supply line must be provided for the repeater units.

If an eigenmode emerging from one of the components, say a semiconductor ILD is injected into another component of the transmission system, a graded-index optical fiber waveguide, for example, a set of the eigenmodes of the opto-coupler is excited. The coupling to the various transverse eigenmodes as well as the power transfer can be described by a mathematical model based on the harmonic analysis on the diamond solvable Lie group D(R) and evaluated explicitly in terms of the Krawtchouk polynomials (Section 7). In a second step, however, the group theoretic model has to be refined in order to get agreement with the measured data of actual optical fiber waveguides (cf. Section 8).

The signals transmitted in optical fiber waveguides are usually formed by digital pulses. The group theoretical model discussed in Section 3 supra actually encompasses the treatment of digital signals in loss-free transmission channels. The basic idea of holography to consider the time domain and the frequency domain of digital signals simultaneously leads to the holographic identities (cf. Section 9). In this way, a series of new identities for theta-null values arises which are of their own mathematical interest. Finally it provides a classification of the information cells of digital signals in loss-free transmission channels which is included in Section 10 infra.

2. The Economic Power of Laser Opto-Electronic Systems

The technological advances in the fabrication of laser sources, optical fiber waveguides, and photodetectors described in the preceding section have propelled the field of fiber optic communication that was in its infancy in the early 1970's into a major industry of the 1980's. In all the fields of laser opto-electronics, new research results are translated into technological developments and industrial applications at a breathtaking pace. The list displayed below shows the sales of laser opto-electronic systems of the Western industrial countries in 1986 and makes the economic power of this field of High Tech evident.

Laser systems (100%)	10.6 billions US-$
Laser systems for print processing (52%)	5.5 billions US-$
Laser systems for signal processing (20%)	2.1 billions US-$
Laser systems for medicine (3.7%)	0.4 billions US-$

*****

Leading nations in laser technology	U.S.A., Japan
West Europe	10% of the total sale
Federal Republic of Germany	<3% of the total sale

(Sources: Institut der Deutschen Wirtschaft, Cologne. Laser 87, Munich)

There is no doubt that laser opto-electronics forms presently one of the most promising fields of High Tech from the scientific as well as from the industrial point of view. In particular, holography is at the beginning of a rapid development. The latest American market studies (Newport Corporation, Spectron Development Laboratories,...) refer to holography as one of the most important key technologies of the future and attest it an enormous growth.

Part I - Laser Opto-Electronics

3.   The Group Theoretic Model

Since the central technologies of optical communication - the semiconductor ILD, or signal producer, and the optical fiber waveguide, or signal transmission medium - are intimately related, an explanation of the mathematical model begins not with the semiconductor ILD but with the fiber. This model of a circularly symmetric (graded-index) optical fiber waveguide discussed in [15] is based on the following two postulates:

(i)   The standard basis of each plane transverse to the fiber axis satisfies the canonical commutation relations of quantum mechanics.

(ii) The intensity distribution of the laser light in each plane transverse to the fiber axis is radial with respect to a scalar product.

In order to combine the postulates (i) and (ii) supras, let us identify the peripheral circle T of the circularly symmetric optical fiber waveguide with the maximal compact subgroup $SO(2,R)$ of $Sp(1,R) = SL(2,R)$ the elements of which act as automorphisms of the three-dimensional real Heisenberg two-step nilpotent Lie group $\tilde{A}(R)$ by rotating the first two coordinates of the elements $(x,y,z) \in \tilde{A}(R)$ and leaving the third "central" coordinate fixed. Then the external semi-direct product

$$D(R) = T \ltimes \widetilde{A}(R)$$

is called the diamond group. It forms a four-dimensional, real, connected, non-exponential, solvable Lie group. Let

$$U_\nu : \widetilde{A}(R) \longrightarrow U(L^2(R)), \quad \nu \neq 0$$

denote the Stone-von Neumann representation and H_ν the Hamiltonian of the harmonic oscillator with eigenvalues

$$-2\pi(\text{sign } \nu).(2n+1), \quad n \in \mathbf{N}.$$

Then U_ν acts on the wave functions $\psi \in L^2(R)$ according to prescription

$$U_\nu \psi(x,y,z) = e^{2\pi i \nu(z+yt)} \psi(x+t) \quad (t \in R).$$

Notice that the topologically irreducible, continuous, unitary, linear representation U_ν of $\widetilde{A}(R)$ acting in the complex Hilbert space $L^2(R)$ is uniquely determined up to isomorphy by its restriction to the center

$$Z = \{(0,0,z) \in \widetilde{A}(R) \mid z \in R\}$$

of $\widetilde{A}(R)$. Let H_1 form the infinitesimal generator of $SO(2,R)$ in $Sp(1,R)$ and let

$$\widetilde{T} = \{e^{\pi i \theta} \mid \theta \in R\}$$

denote the double covering of the torus group $T = \{e^{2\pi i \theta} \mid \theta \in R\}$. Then we may state the following

Theorem 1. The subgroup $\widetilde{T} \times Z$ of the diamond solvable Lie group $D(R)$ forms a group theoretic model of a circularly symmetric optical fiber waveguide.

Define a unitary linear representation S_ν of $\widetilde{T} \times Z$ by

$$S_\nu(e^{\pi i \theta}, z) = e^{i\theta H_\nu}$$

and the characters of $\widetilde{T} \times Z$ by the prescription

$$\chi_\nu^n(e^{\pi i\theta}, z) = e^{2\pi i((2n+1)\theta + \nu z)} id_{L^2(R)} \qquad (n \in Z).$$

Then $S_\nu U_\nu \otimes \chi_\nu^n$ $(n \in Z)$ forms a family of topologically irreducible, continuous, unitary, linear representations of $D(R)$
having U_ν as their restrictions to $\widetilde{A}(R)$. Conversely, each representation of $D(R)$ having these properties is unitarily isomor-
phic to exactly one of the representations $S_\nu U_\nu \otimes \chi_\nu^n$ $(n \in Z)$. See
Lion [8].

4.   Quantized Transverse Eigenmode Spectra

A wave function $\psi \in L^2(R)$ is called a quantized transverse eigenmode of a circularly symmetric optical fiber waveguide if the analog auto-correlation function

$$H(\psi; x, y) = \langle U_1(x, y, 0)\psi | \psi \rangle = \int_R \psi(t+x)\overline{\psi}(t) e^{2\pi i y t} \, dt$$

is radial on the plane $R \oplus R$ with respect to a scalar product.
The following theorems characterize the quantized transverse
eigenmodes and give the explicit forms of the associated analog
cross-correlation function

$$H(\psi, \varphi; x, y) = \langle U_1(x, y, 0)\psi | \varphi \rangle = \int_R \psi(t+x)\overline{\varphi}(t) e^{2\pi i y t} dt,$$

respectively (cf. [12] and [13]).

Theorem 2.   Let $(H_n)_{n \geq 0}$ denote the sequence of Hermite functions. The waveform $\psi \in L^2(R)$ is a quantized transverse eigenmode of a circularly symmetric optical fiber waveguide if and
only if

$$\psi = \xi_n H_n \qquad (\xi_n \in C)$$

for an integer $n \geq 0$.

Proof. Apply Schur's lemma to the representations $S_1 \cdot U_1 \otimes X_1^n$
$(n \in \mathbf{N})$ of $D(R)$ in $L^2(R)$. -

Theorem 3. The wave function $\psi \in L^2(R \oplus R)$ is a resonant
quantized eigenmode of a rectangular semiconductor ILD if and
only if

$$\psi = \xi_{n,m} H_n \otimes H_m \qquad (\xi_{n,m} \in \mathbf{C}),$$

where $n \geq 0$ denotes the lateral mode number, $m \geq 0$ denotes the
transversal mode number, and the resonant frequency is stan-
dardized by $\nu = 1$.

Proof. Since the topologically irreducible, continuous, unitary,
linear representation U_1 of $\tilde{A}(R)$ in $L^2(R)$ is square integrable
mod Z, the result follows from the orthogonality relation

$$\langle H(\psi',\varphi';.,.)|H(\psi,\varphi;.,.)\rangle = \langle\psi'|\psi\rangle\langle\varphi|\varphi'\rangle$$

where ψ',φ' and ψ,φ are wave functions in $L^2(R)$. -

　　　　For a treatment of the resonant transverse eigenmodes of
semiconductor ILDs, see, for instance, [20]. In Section 6 infra
we will see that the mode partition within the gain profile of
the semiconductor ILD requires a third mode number, the axial
mode number p, to characterize completely the resonant quantized
eigenmodes of an ILD resonator.

Theorem 4. Let $(L_n^{(\alpha)})_{n\geq 0}$ denote the sequence of Laguerre functions of order $\alpha > -1$. Then Schwinger's formula

$$H(H_m,H_n;x,y) = \sqrt{\frac{n!}{m!}} \ (\sqrt{\pi}(x+iy))^{m-n} L_n^{(m-n)}(\pi(x^2+y^2)) \qquad (m \geq n \geq 0)$$

holds for $(x,y) \in R \oplus R$.

Proof. Switch from the Schrödinger realization of the Stone-von
Neumann representation U_1 to the Bargmann-Fock realization. -

It is noteworthy that the interference term on the right hand side of Schwinger's formula can be expressed in terms of the value at the point $m \in \mathbb{N}$ of the Charlier-Poisson polynomial of degree n and parameter value $\pi(x^2+y^2)$.

5.   The Orbit Alternative

The coadjoint action of the diamond solvable Lie group D(R) as introduced in Section 3 supra on the dual of its four-dimensional Lie algebra takes the form:

$$
D(R) \ni (e^{2\pi i\theta}, x, y, z) \longmapsto
\begin{bmatrix}
1 & x\sin 2\pi\theta - y\cos 2\pi\theta & x\cos 2\pi\theta + y\sin 2\pi\theta & -\frac{1}{2}r^2 \\
0 & \cos 2\pi\theta & -\sin 2\pi\theta & y \\
0 & \sin 2\pi\theta & \cos 2\pi\theta & -x \\
0 & 0 & 0 & 1
\end{bmatrix}
$$

where $r^2 = r^2(x,y) = x^2+y^2$ (cf. [15], [18]). Therefore D(R) admits two kinds of non-trivial coadjoint orbits:

(I)   The paraboloids of revolution

$$
\mathcal{O}_{\mu,\nu} = \{ (f_0, f_1, f_2, f_3) \mid f_3 = \nu, \ f_1^2 + f_2^2 + 2\nu f_0 = \mu \}
$$

where $(\mu,\nu) \in \mathbb{R} \times \mathbb{R}^{\times}$.

(II) The circular cylinders

$$
\mathcal{O}_{\mu,0} = \{ (f_0, f_1, f_2, f_3) \mid f_3 = 0, \ f_1^2 + f_2^2 = \mu \}
$$

where $\mu > 0$.

The Kirillov correspondence associates to the sequence of paraboloids of revolution

$$
\mathcal{O}_{(2n+1),1} \qquad\qquad (n \in \mathbb{N})
$$

(notice the standardization $\nu = 1$) the sequence of quantized transverse eigenmodes $(\xi_n H_n)_{n \geq 0}$ of a circularly symmetric optical fiber waveguide as determined in Theorem 2 supra. The double covering $\tilde{T}$ describes the phase retardation suffered from the skew rays when their paths meet the inner and outer caustic as they travel along the optical fiber waveguide. Each smooth wave function Ψ on R^3 gives a section, $d\Psi$, of the cotangent bundle of R^3. Since the refractive index N of a transparent material is the ratio of the speed c_o of light in vacuum to the speed c of light in material we get in the case of an optical fiber waveguide $\tilde{T} \times Z$

$$N(e^{\pi i \theta}, z) = c_o / c (e^{\pi i \theta}, z)$$

for all $(e^{\pi i \theta}, z) \in \tilde{T} \times Z$. Consequently we have

Theorem 5. The group theoretic model of Theorems 1 and 2 based on D(R) via the postulates (i) and (ii) of Section 3 corresponds to a circularly symmetric graded-index multimode optical fiber waveguide with parabolic refractive index profile in the fiber core. The index N of refraction is assumed not to vary along the fiber axis.

If the core diameter allows only the choice n = 0, the group theoretic model corresponds to a circularly symmetric graded-index monomode optical fiber waveguide with parabolic refractive index profile. The degenerate case of circular cylinders $\mathcal{O}_{\mu, 0}$ $(\mu > 0)$ corresponds to the step-index optical fiber waveguides (cf. Section 7 infra).

In the case of buried heterostructure injection laser diodes (BH-ILDs for short) with weak lateral and transversal confinement the active region is surrounded on all sides by the wide-band-gap material of lower refractive index, so that the rectangular tubelike active layer acts like the core of a rectangular optical fiber waveguide.

Theorem 6. The group theoretic model of Theorem 3 based on D(R) corresponds to an oxyde-isolated stripe-geometry configuration having the maximum value of the refractive index at a point located below the stripe contact in the active region of the semiconductor ILD. It is assumed that the index of refraction is spatially symmetric about the peak and has quadratic decay rates in the lateral and transversal directions. Furthermore it is assumed that the refractive index does not vary along the direction of the optical resonator axis.

In actual BH-ILDs the lateral and transversal confinements of the active region are usually chosen in such a way that only the fundamental quantized transverse eigenmode (n=m=0) and no of the hybride eigenmodes can propagate (monomode BH-ILD).

6.   The Cleaved-Coupled-Cavity Semiconductor Laser Diode

Throughout Section 4 the standardization ν = 1 of the laser oscillation frequency has been assumed to be valid. In the case of optical frequency modulation, of course, different oscillation frequencies have to be taken into account. Then the analog cross-correlation function is defined for wave functions ψ and φ in $L^2(\mathbf{R})$ according to the prescription

$$_\nu H(\psi,\varphi;x,y) = \langle U_\nu(x,y,0)\psi|\varphi\rangle = \int_{\mathbf{R}} \psi(t+x)\bar{\varphi}(t)e^{2\pi i\nu yt}dt$$

where $(x,y)\in\mathbf{R}\oplus\mathbf{R}$ and $\nu\in\mathbf{R}^x$. Similarly, the analog auto-correlation function takes the form

$$_\nu H(\psi;x,y) = {_\nu H}(\psi,\psi;x,y) \quad ((x,y)\in\mathbf{R}\oplus\mathbf{R}).$$

Consider a cleaved-coupled-cavity semiconductor laser diode (C^3-ILD for short). It consists of a pair of electrically isolated standard BH-ILDs that differ slightly in length, say by 20 %, which were self-aligned and very closely coupled to form a two-cavity resonator. Each of the two BH-ILDs is controlled by its own injection current. The active strip from each BH-ILD is precisely aligned with respect to the other on a straight line and they are separated from each other by a distance < 5 μm. All the reflecting facets are formed by cleaving along crystallographic planes and hence are perfectly mirror-flat and parallel to each other. Typically, the total length of a C^3-ILD is 200-400 μm.

Let one of the coaxial BH-ILDs be biased with an injection current level above its lasing threshold. Then this ILD
is operated as a laser (L) and the positions of its resonant
quantized eigenmodes $\psi'_{n'm'p'} \in L^2(R \oplus R)$ are fixed. These can be
computed explicitly by rescaling the resonant quantized eigenmodes of the rectangular semiconductor ILDs (cf. Theorem 3) by
those resonant frequencies $\nu'_{p'}$ that are positioned within the
gain profile of (L). If L'_{eff} denotes the effective length of the
BH-ILD then the resonant oscillation frequencies are given by

$$\nu'_{p'} = p'c_0/2L'_{eff} \qquad (p' \in N)$$

and we have

Theorem 7. The resonant quantized eigenmodes in $L^2(R \oplus R)$ of a
rectangular BH-ILD take the form

$$\psi'_{n'm'p'}(x,y) = \xi'_{n'm'p'} H_{n'}(\sqrt{\nu'_p},x) H_{m'}(\sqrt{\nu'_p},y) \qquad ((x,y) \in R \oplus R)$$

where $n' \geq 0$ denotes the lateral, $m' \geq 0$ the transversal, and
$p' \geq 0$ the axial mode numbers, and $\xi'_{n'm'p'} \in C$ is a constant.

Let the second BH-ILD be biased with some injection
current below lasing threshold. Thus it acts as a frequency
modulator (M). If $\psi_{nmp} \in L^2(R \oplus R)$ denotes the resonant quantized
eigenmodes of (M), the coaxial coupling coefficient of the
C^3-ILD is defined by the overlap integral

$$_{\nu'_{p'},\nu_p}C(\psi'_{n'm'p'},\psi_{nmp}) = \int\int_{R \oplus R} {}_{\nu'_p}H(\psi'_{n'm'p'};x,y) \cdot {}_{\nu_p}\bar{H}(\psi_{nmp};x,y)dxdy.$$

The orthogonality relations of the coefficient functions
on $\tilde{A}(R)/Z$ of the non-isomorphic, topologically irreducible,
continuous, unitary linear representations of $\tilde{A}(R)$ (cf. [13])
establish the destructive interference of the different resonant
quantized eigenmodes within the C^3-ILDs.

Theorem 8. Keep to the preceding notations and let $p' \geq 0$ and $p \geq 0$ denote axial mode numbers of the pair of BH-ILDs forming a cleaved-coupled-cavity semiconductor laser diode. If $\nu'_{p'} \neq \nu_p$ are resonant oscillation frequencies within the gain profiles, then there is no coaxial coupling between the resonant quantized eigenmodes $\psi'_{n'm'p'} \in L^2(R \oplus R)$ and $\psi_{nmp} \in L^2(R \oplus R)$, i.e.,

$$\mathop{C}\limits_{\nu'_{p'}, \nu_p}(\psi'_{n'm'p'}, \psi_{nmp}) = 0$$

for all lateral and transversal mode numbers $n' \geq 0$, $m' \geq 0$, and $n \geq 0$, $m \geq 0$, respectively.

Proof. The coadjoint orbits associated with the isomorphy classes of the topologically irreducible, continuous, unitary, linear representations $U_{\nu'_{p'}}$ and U_{ν_p} of $\widetilde{A}(R)$ in $L^2(R)$ under the Kirillov correspondence are different affine planes in the dual of the real Heisenberg Lie algebra. Therefore the isomorphy classes of $U_{\nu'_{p'}}$ and U_{ν_p} are disjoint.-

Theorem 8 supra shows that the C^3-ILD emits a radiation of high spectral purity. This is in complete agreement with the experimental results. Since various spectra can be obtained by applying different injection current levels to the frequency modulator (M), and no eigenmode partition was observed under turning off and on with ultrahigh speed, cleaved-coupled-cavity semiconductor laser diodes are particularly suitable to serve as optical sources for ultrahigh-capacity, long-distance optical communication systems having monomode optical fiber waveguides as transmission lines of virtually absent chromatic dispersion and vanishing intermodal dispersion (cf. Section 8 infra).

7.   Opto-Coupling

Let us return to the standardization $\nu = 1$ of the wave length. If the wave functions ψ', φ' and ψ, φ belonging to $L^2(R)$ represent two quantized transverse eigenmodes of two axi-symmetric optical devices like laser resonators or optical fiber

waveguides, the reasoning of the preceding section suggests to
define their coaxial coupling coefficient according to the prescription

$$_pC(\psi',\varphi',\psi,\varphi) = \int_P H(\psi',\varphi';x,y).\bar{H}(\psi,\varphi;x,y)dxdy.$$

The overlap integral defining the coupling functional $_pC$ has to
be evaluated at the coupling plane P transverse to the common
beam axis of the optical devices. If P coincides with the reference plane, we set $C = {}_pC$. From the proof of Theorem 3 we
then infer

Theorem 9. Let the wave functions $\psi',\varphi',\psi,\varphi$ belong to $L^2(R)$ -
then the identity

$$C(\psi',\varphi',\psi,\varphi) = <\psi'|\psi><\varphi|\varphi'>$$

holds.

      Our aim is to calculate the coaxial coupling coeffi-
cients explicitly in the circularly symmetric as well as in the
rectangular case in terms of the Gaussian beam parameters α and
q (cf. Kogelnik [7]). If w denotes the beam width at the coupling plane P, define

$$\alpha = \frac{2}{w^2}$$

and the associated transfer matrix

$$\begin{bmatrix} \frac{1}{\sqrt{\alpha}} & 0 \\ 0 & \sqrt{\alpha} \end{bmatrix}$$

Furthermore, denote by R the radius of curvature of the phase
front at the coupling plane P and by

$$\begin{bmatrix} 1 & 0 \\ \dfrac{2\pi\nu}{Rc_0} & 1 \end{bmatrix}$$

the associated transfer matrix. Then the complex beam radius

$$q = (\frac{1}{w'^2} + \frac{1}{w^2} + \frac{\pi}{c_0}i\nu(\frac{1}{R'}-\frac{1}{R}))$$

combines the beam widths and the radii of the phase fronts at P
(notice the signs!) of the incoming and the outgoing Gaussian
beams.

Recall the definition of the Krawtchouk polynomials
$K_n(x;p,N)$. For all integers $N \geq 0$ these hypergeometric polynomials are given by

$$K_n(x;p,N) = {}_2F_1(-n,-x;-N;p^{-1})$$

where $0 \leq n \leq N$ and $p \in C^x$. In terms of shifted raising facto-
rials the polynomial $K_n(x;p,N)$ in x of degree n admits the
following expression

$$K_n(x;p,N) = \sum_{0 \leq k \leq n} \frac{(-n)_k(-x)_k}{(-N)_k k!} (\frac{1}{p})^k \qquad (0 \leq n \leq N).$$

For $p \in [0,1[$ the Krawtchouk polynomials are orthogonal on the
set $\{0,1,\ldots,N\}$ with respect to the binomial distribution. Thus
their orthogonality relations take the discrete form

$$\sum_{0 \leq x \leq N} K_n(x;p,N)K_m(x;p,N)\binom{N}{x}p^x(1-p)^{N-x} = \binom{N}{m}^{-1}(\frac{p}{1-p})^{-m}\delta_{n,m}$$

for $n \leq N$, $m \leq N$ and $p \in [0,1[$.

Theorem 10. Let $m \geq n \geq 0$, $m' \geq n' \geq 0$ be integers. Keep to the
preceding notations.

a) In the circular case let m = p, m' = p' be the radial eigenmode numbers and m - n = l, m' - n' = l' the azimuthal eigenmode
numbers. Then

$$_p C_{p,l,p',l'} = 0 \quad \text{for} \quad l \neq l',$$

i.e., there is no coaxial coupling at P between quantized
transverse eigenmodes labeled by different azimuthal numbers.
Moreover,

$$_p C_{p,l,p',l} = \left(\frac{2}{ww'q}\right)^{l+1} \frac{(p+p'+1)!}{\sqrt{p!p'!(p+l)!(p'+l)!}} \; (1-\frac{\alpha}{q})^p (1-\frac{\alpha'}{q})^{p'} \cdot$$

$$K_p(p'; \frac{(q-\alpha)(q-\alpha')}{q(q-\alpha-\alpha')}, p+p'+1)$$

b) In the rectangular case we have

$$_p C_{m,n,m',n'} = 0 \quad \text{for} \quad \begin{cases} m+m' \equiv 1 \bmod 2, \\ \\ n+n' \equiv 1 \bmod 2, \end{cases}$$

i.e., there is no coaxial coupling at P between even and odd
quantized transverse eigenmodes. In other words, the parity is
preserved under the coaxial coupling of quantized eigenmodes.
In the case

$$m' = 2\mu', \quad m = 2\mu, \quad n' = 2\nu', \quad n = 2\nu$$

we get in terms of the Krawtchouk polynomials

$$_p C_{m,n,m',n'} = (-\frac{1}{2})^{\mu+\nu+\mu'+\nu'} \left(\frac{2}{ww'q}\right) \frac{(2\mu+2\mu')!(2\nu+2\nu')!}{(\mu+\mu')!(\nu+\nu')!\sqrt{2\mu!2\mu'!2\nu!2\nu'!}}$$

$$(1-\frac{\alpha}{q})^{\mu+\nu} (1-\frac{\alpha'}{q})^{\mu'+\nu'} \cdot K_\mu(\mu'; \frac{(q-\alpha)(q-\alpha')}{q(q-\alpha-\alpha')}, \mu+\mu'-\frac{1}{2})$$

$$K_\nu(\nu'; \frac{(q-\alpha)(q-\alpha')}{q(q-\alpha-\alpha')}; \nu+\nu'-\frac{1}{2})$$

A similar result holds in the case

m' = 2μ'+1,   m = 2μ+1,   n' = 2ν'+1,   n = 2ν+1.

The fact that the weight function associated with the Krawtchouk polynomials is discrete reflects the fact that the quantized transverse eigenmode spectra of the circular and rectangular fiber optical waveguides form a discrete series. For details see [15] and [18].

Opto-couplers are used as repeaters in opto-electronic telecommunication systems. Moreover they are used to separate electrically analog and digital circuits, for instance, in high quality CD (= Compact Disc) players.

8.   Adaption of the Model to Actual Optical Fibers

Graded-index optical fiber waveguides with parabolic refractive index profiles have the advantage that collimated laser beams launched into the fibers remain well collimated. It is important to observe that this stability property can exist only in dielectric transmission media whose refractive index obeys the strict parabolic law. Any higher order terms in an expansion of the index of refraction causes a beam break-up in the perturbed medium of the optical fiber waveguide.

A disadvantage of multimode optical fiber waveguide is that they suffer from intermodal dispersion. This means that the eigenmodes in a given optical pulse arrive at the fiber end at slightly different times, thus causing the pulse to spread out in time as it travels along the optical fiber waveguide. This effect can be reduced by using a graded-index profile in fiber core. This allows graded-index optical fiber waveguides to have much larger bandwidths, i.e., data rate transmission capabilities, than step-index fiber waveguides, which are not particularly suitable for wideband communications. On the other hand, the graded-index optical fiber waveguides with parabolic refractive index profile are not optimal in the sense of smallest intermodal dispersion. Therefore practicing engineers dealing

with optical fiber telecommunication system designs consider
circularly symmetric optical fiber waveguides having a power law
refractive index profile with characteristic exponent $\alpha \geq 1$.
Step-index optical fiber waveguides can be dealt with by letting
$\alpha \rightarrow \infty$. Starting with the parabolic refractive index profile
($\alpha = 2$) of the group theoretic model (Theorem 1) as the reference
profile, the Rayleigh-Ritz approximation procedure allows us to
calculate numerically the intensity distributions and propagation constants for weakly guiding optical fiber waveguides
having more general circularly symmetric refractive index profiles than the reference profile.

The Rayleigh-Ritz minimization method proceeds in the
following way. In view of Theorem 9 supra the assignment

$$\langle \psi' \otimes \varphi' | \psi \otimes \varphi \rangle = C(\psi', \psi, \varphi, \varphi')$$

defines by continuous linear extension a scalar product on the
complex Hilbert space $L^2(\mathbf{R}) \, \hat{\otimes}_2 \, L^2(\mathbf{R})$ of Hilbert-Schmidt operators on $L^2(\mathbf{R})$. Include a strictly positive linear operator into
the scalar product and let Q be the associated quadratic functional. Then solve the problem

$$\min_{a_j \in \mathbf{C}} Q\left(\sum_j a_j \Phi_j\right)$$

by choosing the functions $H_n \otimes H_m$ ($n \geq 0$, $m \geq 0$) of the reference model as Ansatz functions Φ_j. In this way, we may calculate, for instance, α profiles and the influence on the eigenmodes of a central refractive index dip, which accompanies the
collapse of the fiber preform during the CVD (Chemical Vapour
Deposition) process [3]. The computer plot of Figure 1 displays
the refractive index profile of a graded index optical fiber
waveguide. Details of these techniques will be included in a
forthcoming paper.

9.   The Holographic Identities of Digital Signals

Most applications of optical fiber waveguides in opto-
electronic communication systems use some form of digital en-

velope modulation of an optical signal. The most commonly modulation format is the pulse code modulation (PCM). If the transmitted signal is formed by digital pulses, the digital cross-correlation function (cf. [11]) takes the form

$$(x,y) \longrightarrow \sum_{\mu \in Z} \psi(\mu+x)\bar{\varphi}(\mu)e^{2\pi iy\mu}.$$

On the other hand, a band-limited analog signal (voice or video, for example) can be reconstructed from its "samples" by an ideal low-pass filter. This is the contents of the Whittaker-Nyquist-Shannon-Kotel'nikov sampling theorem. If the analog cross-correlation function H is restricted to the quadratic lattice Z ⊕ Z in the time-frequency plane R ⊕ R we get the following result:

Theorem 11. Let the wave functions ψ and φ belong to $L^2(R)$ - then the holographic identity

$$\sum_{(\mu,\nu) \in Z \oplus Z} H(\psi;\mu,\nu) \cdot \bar{H}(\varphi;\mu,\nu) = \sum_{(\mu,\nu) \in Z \oplus Z} |H(\psi,\varphi;\mu,\nu)|^2$$

holds.

It should be observed that on the left hand side of the holographic identity the signal terms and on the right hand side the interference terms occur. This explains the name and suggests to call the analog cross-correlation function $H(\psi,\varphi;.,.)$ the holographic transform of the pair (ψ,φ). In view of Theorem 4 we get the following form of the holographic identity by choosing for ψ and φ the Hermite functions:

Theorem 12. Let m,n be integers such that $m \geq n \geq 0$ - then the identity

$$\sum_{(\mu,\nu) \in Z \oplus Z} L_m^{(0)}(\pi(\mu^2+\nu^2)) \cdot L_n^{(0)}(\pi(\mu^2+\nu^2)) =$$

$$\frac{n!}{m!} \pi^{m-n} \sum_{(\mu,\nu) \in Z \oplus Z} (\mu^2+\nu^2)^{m-n} (L_n^{(m-n)}(\pi(\mu^2+\nu^2)))^2$$

holds.

The preceding theorem implies a series of remarkable identities for theta-null values:

m = 1,   n = 0

$$\pi = \frac{\sum\limits_{\mu \in Z} e^{-\pi\mu^2}}{4 \sum\limits_{\mu \in Z} \mu^2 e^{-\pi\mu^2}}$$

m = 2,   n = 1

$$\pi^3 = \frac{15 \sum\limits_{\mu \in Z} (8\pi^2\mu^4 - 1) e^{-\pi\mu^2}}{32 \sum\limits_{\mu \in Z} \mu^6 e^{-\pi\mu^2}}$$

m = 3,   n = 2

$$\pi^5 = \frac{45 \sum\limits_{\mu \in Z} (16\pi^4\mu^8 - 140\pi^2\mu^4 + 21) e^{-\pi\mu^2}}{64 \sum\limits_{\mu \in Z} \mu^{10} e^{-\pi\mu^2}}$$

m = 4,   n = 3

$$\pi^7 = \frac{91 \sum\limits_{\mu \in Z} (256\pi^6\mu^{12} - 15840\pi^4\mu^8 + 166320\pi^2\mu^4 - 25245) e^{-\pi\mu^2}}{1024 \sum\limits_{\mu \in Z} \mu^{14} e^{-\pi\mu^2}}$$

.
.
.

For additional material in the field of theta inversion, the reader is referred to the papers [2], [4], and [14].

10.   Information Cells of Digital Signals in Loss-Free

        Transmission Channels and Holographic Coding

    A mapping

$$\sigma : R \oplus R \longrightarrow R \oplus R$$

is said to be an invariant of the analog auto-correlation function H if the identity

$$H(\psi ; \sigma(x,y)) = H(\psi_\sigma ; x,y)$$

holds for all wave functions $\psi \in L^2(R)$ and all pairs $(x,y) \in R \oplus R$
such that $\psi \longrightarrow \psi_\sigma$ is a unitary linear operator of $L^2(R)$. Thus the
energy is preserved under this mapping. The metaplectic repre-
sentation of the double covering of $SL(2,R)$ which has certain
formal resemblances to the spin representation of $\tilde{T}$ yields

Theorem 13. A mapping $\sigma : R \oplus R \longrightarrow R \oplus R$ is an invariant of the
analog auto-correlation function H if and only if $\sigma \in SL(2,R)$.
For all wave functions $\psi \in L^2(R)$ the associated wave function
$\psi_\sigma \in L^2(R)$ is uniquely defined up to a phase factor $\xi_\sigma \in T$.

    For example the rescaling transformations with matrices

$$\begin{bmatrix} \dfrac{1}{\sqrt{\nu_p}} & 0 \\ 0 & \sqrt{\nu_p} \end{bmatrix} \in SL(2,R)$$

yielding the transition of Theorem 3 to Theorem 7 by means of
the resonant oscillation frequencies ν_p within the gain profile
of the ILD are invariants of the function H. More precisely, the
rescaling transformations belong to the neutral component of the
maximal torus of $SL(2,R)$. Similarly the transfer transformations
used in Section 7 supra are invariants of the auto-correlation

function H(;.,.).

The preceding theorem is of importance for a treatment of optical resonators (Fabry-Pérot étalons). Details will be presented in a forthcoming paper.

Theorem 14. Let $\sigma \in$ SL(2,Z) be an invariant of the holographic identity displayed in Theorem 12 supra. Then σ must have (at least) one of σ^2, σ^3, σ^4, σ^6 equal to $\mathrm{id}_{R \oplus R}$, independently of the transversal mode numbers m,n.

Proof. If $\sigma \in$ SO(2,R) $\cap$ SL(2,Z) and $\sigma \notin \{-\mathrm{id}_{R \oplus R}, +\mathrm{id}_{R \oplus R}\}$ then $|\operatorname{tr} \sigma| < 2$. Therefore tr $\sigma \in$ Z implies tr $\sigma \in \{-1,0,+1\}$. Hence with respect to some scalar product of R $\oplus$ R (unique up to a scalar) the mapping σ is a turn through $\pi/3$, $\pi/4$, or $2\pi/3$, respectively.-

It follows that the holographic identity has the cyclic groups Z/mZ (m $\in \{1,2,3,4,6\}$) as their groups of radial invariants. See Weyl [19] for the crystallographic restriction. Nothing like a turn through $\pi/5$ is possible. For radial information cells of digital signals only the classical planar crystal symmetries and no Penrose pattern, well known from the theory of quasi-crystals, are allowed (cf. [17]). Indeed, the planar crystal structures displayed in Figure 2 can be discovered in the fine structure of holograms, i.e., in the interference pattern produced by the reference laser beam. The hologram itself records the superposition of the reference pattern with the scattered field pattern generated by the signal laser beam. In a more general way, we obtain the following result on the holographic coding of informations.

Theorem 15. The radial information cells of sequences of digital signals in a loss-free phase-plane coded optical transmission channel admit the symmetries of two-dimensional crystal structures.

Figure 3 shows the information density extracted by the extra-foveal retina (see Koenderink-van Doorn [6], Porat-

Zeevi [10], and [17]). Each of the hexagonal information cells
represents the same number of sample points and contains about
200 ganglion cells in macaque monkey. Logarithmic scaling along
the radial coordinate axis results in the regular honeycomb
lattice (m = 6) corresponding to the fact well known from neuro-
anatomy that the transformation from visual field to the cortex
is logarithmic.

It is the purpose of a forthcoming paper to study the
implications of Theorem 15 supra to the exciting holographic
brain model developed by the neurophysiologist Karl H. Pribram
at Stanford.

Acknowledgments. Part of the work for this paper was
done while the author enjoyed the hospitality of the Centre de
Physique Théorique, Marseille-Luminy (France) and the Tel Aviv
University (Israel). The author wishes to thank Professors
Richard Askey (Madison, Wisconsin), George Gasper (Evanston,
Illinois), Alexandre Grossmann (Marseille), S.K. Suslov (Moscow),
Yehoshua Y. Zeevi (Haifa), and the Electrical Engineer Miklós
Nyari (Budapest) for helpful discussions and valuable hints.
Finally, he is grateful to Deutscher Akademischer Austausch-
dienst and Deutsche Forschungsgemeinschaft for financial support.

References

1. Aschoff, V., Geschichte der Nachrichtentechnik. Berlin-
    Heidelberg-New York-Tokyo: Springer 1984

2. Borwein, J., Advanced problem # 6491. Amer. Math. Monthly
    92 (1985), 217. Solutions of advanced problems: Theta
    inversion, Heisenberg uncertainty, and radar. Amer. Math.
    Monthly 93 (1986), 822-823

3. Geckeler, S., Lichtwellenleiter für die optische Nach-
    richtenübertragung. Nachrichtentechnik 16. Berlin-Heidel-
    berg-New York-Tokyo: Springer 1986

4. Grosjean, C.C., Note on two identities mentioned by
    Professor Dr. W. Schempp near the end of the presentation
    of his paper. Polynômes Orthogonaux et Applications.
    Proceedings, Bar-le-Duc 1984, pp. 553-554. Lecture Notes
    in Math., Vol. 1171. Berlin-Heidelberg-New York-Tokyo:
    Springer 1985

5. Higgins, J.R., Five short stories about the cardinal
    series. Bull. (New Series) Amer. Math. Soc. 12 (1985),
    45-89

343

6. Koenderink, J.J., van Doorn, A.J., Visual detection of spatial contrast; influence of location in the visual field, target extent and illuminance level. Biol. Cybernetics **30** (1978), 157-167

7. Kogelnik, H., Coupling and conversion coefficients for optical modes. Proc. of the Symposium on Quasi-Optics, New York 1964, pp. 333-347. Microwave Research Institute Symposia, Vol. **14**. Brooklyn, N.Y.: Polytechnic Press 1964

8. Lion, G., Extensions de représentations de groupes de Lie nilpotents et indices de Maslov. C.R. Acad. Sc. Paris **288** (1979), Série A, pp. 615 - 618

9. Mehta, M.L., Eigenvalues and eigenvectors of the finite Fourier transform. J. Math. Phys. **28** (1987), 781-785

10. Porat, M., Zeevi, Y.Y., The generalized Gabor scheme of image representation in biological and machine vision. IEEE Trans. on PAMI, 1987 (in press)

11. Rabiner, L.R., Gold, B., Theory and Application of Digital Signal Processing. Englewood Cliffs, N.J.: Prentice-Hall 1975

12. Schempp, W., Radar ambiguity functions, the Heisenberg group, and holomorphic theta series. Proc. Amer. Math. Soc. **92** (1984), 103-110

13. Schempp, W., Harmonic analysis on the Heisenberg nilpotent Lie group, with applications to signal theory. Pitman Research Notes in Mathematics, Vol. **147**. Harlow, Essex: Longman Scientific & Technical. New York: John Wiley & Sons 1986

14. Schempp, W., Group theoretical methods in approximation theory, elementary number theory, and computational signal geometry. Approximation Theory V. Proc. of the Fifth International Symposium on Approximation Theory, held at Texas A&M University 1986, pp. 129-171. Boston-Orlando-San Diego-New York-Austin-London-Sydney-Tokyo-Toronto: Academic Press 1986

15. Schempp, W., The oscillator representation of the meta-plectic group applied to quantum electronics and compu-terized tomography. Dordrecht-Boston-Lancaster-Tokyo: D.Reidel (in print)

16. Schempp W., Nonlinear laser optics I: Duality of semisimple rings and phase matching. C.R. Math. Rep. Acad. Sci. Canada **9** (1987)

17. Schempp W., Information cells: Classification and reali-zations. C.R. Math. Rep. Acad. Sci. Canada **9** (1987)

18. Schempp, W., Signal geometry (to appear)

19. Weyl, H., Symmetry. J. Washington Acad. Sci. **28** (1938), 253-271. Collected papers, Vol. III, pp. 592-610. Berlin-Heidelberg-New York: Springer 1968

20. Winstel, G., Weyrich, C., Optoelektronik I. Berlin-Heidelberg-New York: Springer 1981

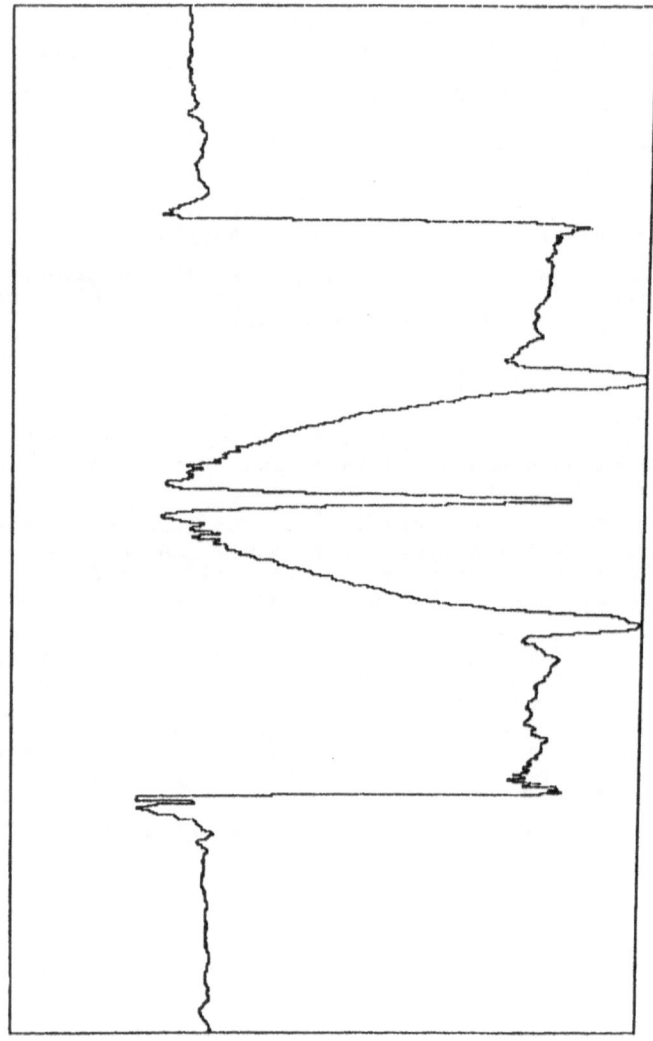

Figure 1

Figure 2

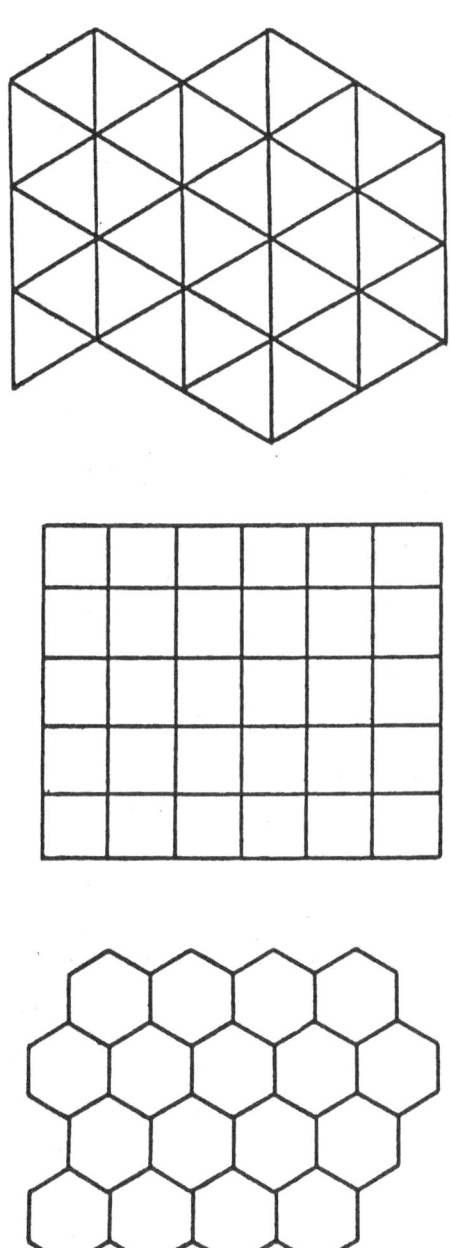

346

Figure 3

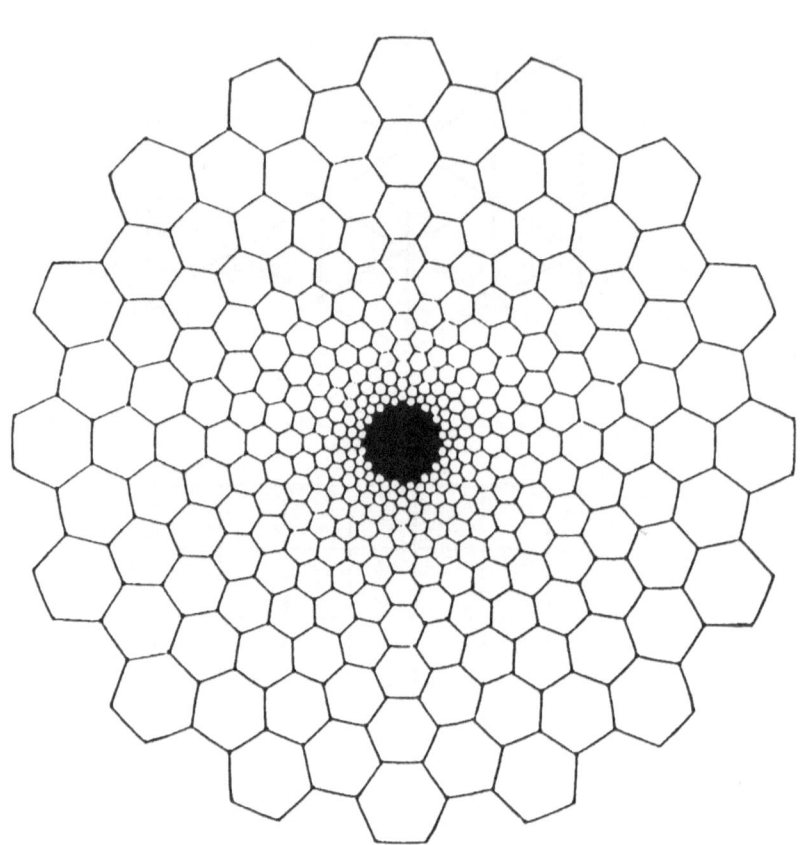

Lehrstuhl fuer Mathematik I
University of Siegen
D-5900 Siegen
Federal Republic of Germany

Application of the Projected Radon Transform
in Picture Processing

Ulrich Eckhardt

Institut für Angewandte Mathematik

der Universität Hamburg

Bundesstraße 55

D-2000 Hamburg 13

Gerd Maderlechner

ZT ZTI INF 12

SIEMENS AG

Otto-Hahn-Ring 6

8000 München 83

Abstract

P.V.C. Hough (1962) observed that in binary pictures lines correspond to maxima of the Radon transform. For this reason the Radon
(or Hough) transform is frequently used in picture processing applications. Since the amount of data to be processed in performing
this transformation can become very large, numerous approaches were
proposed in the literature for data reduction by means of projection. In the present paper conditions are given for a generalized
version of the Radon transform to yield useful results.

1. Introduction

Due to the wide availability of cheap and powerful computers, picture processing becomes more and more available for different practical applications. A specially important role is played by automatic analysis of printed or typewritten documents. In this and numerous other applications one is concerned with binary pictures (i.e. black-and-white pictures).

It was an observation made by P.V.C. Hough (1962) that in binary pictures lines or more generally linear structures are mapped into maxima of the Radon transform (Radon 1917). This latter transform associates to each function of two variables the set of all line integrals

$$H_o(\beta)(p,\alpha) = \int\limits_{-\infty}^{\infty} \beta\left(\begin{matrix} t\cos\alpha + p\sin\alpha \\ t\sin\alpha - p\cos\alpha \end{matrix}\right) dt \qquad (1)$$

$\beta(x,y)$ is the grey value function of the picture and consequently it takes only values 0 or 1 for binary pictures (for the Radon transform see Deans 1983, Helgason 1980 or Natterer 1986). In the picture processing literature the Radon transform is referred to as Hough transform. There exist numerous generalizations of this transform, specifically for picture processing applications, e.g. for detection of circles (Kimme, Ballard, Sklansky 1975), ellipses (Tsuji, Matsumoto 1978) or more general structures (Ballard 1981).

In order to carry out the Radon transform numerically, the integrals (1) must be evaluated on a suitably dense finite set of parameter values. On non-parallel computers the computational effort can become prohibitive, specifically if generalizations of the transform with a higher-dimensional parameter space (the so-called accumulator) are considered. This is the main reason for many approaches reported in the literature for reducing the computational effort which is necessary to transform the picture and to search for maxima in the accumulator.

Since the contribution of each point in the discretized picture to the integrals (1) is independent of the function values at other points, the method can be implemented as an "evidence accumulation" process and it fits very well to parallel computers (Merlin, Farber 1975, Yalamanchili, Aggarwal 1985, Sanz, Hinkle 1987).

There exist numerous approaches in the literature for handling the large amount of data produced by the Radon transform by applying clever storage techniques to the accumulator (Brown, Curtiss, Sher 1983, Li, Lavin, Le Master 1985, Neveu, Dyer, Chin 1986, O'Rourke, Sloan 1984). The computational amount for performing the transform can be reduced efficiently by reducing the number of relevant points in the picture. The latter can be achieved by skeletonization (Kushnir, Abe, Matsumoto 1985) or else by concentration on boundary points of black regions only (Eckhardt, Scherl, Yu 1987).

The aim of this paper is to investigate a specific approach for data reduction which proved useful in different practical applications. This approach is specifically useful if one is not interested in individual lines in the picture but in collective properties of them. Such a collective property can be the common direction of written lines in a document or the property of straight lines in a diagram of being parallel. For finding such properties it is useful to project the accumulator onto a hyperplane contained in it. This results in an efficient data reduction and at least the effort for finding the maxima in the accumulator is reduced. The cost one has to pay is a loss of information. This approach was proposed by different authors, specifically for higher dimensional accumulators (Ballard, Sabbah 1983, Biland, Wahl 1986, Gerig, Klein 1986, Silberberg, Davis, Harwood 1984).

In this paper, a more theoretical investigation is carried out in order to isolate useful properties of these projections and to explain for some of the effects one observes in the literature. We concentrate completely on the classical Radon transform for lines although the results of the paper can be easily extended to more general variants of the Radon transform.

2. Basic Definitions

Given a compact set R in the plane. An *image* is a function $\beta : \mathbb{R}^2 \to \mathbb{R}^k$ which is an element of some function space X and whose support is contained in R. Some examples for images are: *Binary images* (k = 1 and β is the characteristic function of some set), *gray tone images* (k = 1, $0 \leq \beta \leq 1$), *colour images* (k = 3), *gradient images* (k = 2, or k = 1 if β is the length of the gradient vector), *multi spectral images* (k > 1) etc. In the context of this paper we are only interested in binary images.

For analyzing and understanding the content of an image, the identification of certain *features* is important. In this paper we con-

centrate ourselves on *linear features*. For investigating whether
a linear feature is contained in the image, we parametrize all
lines in the plane by number pairs (a,b). The set of all parameter pairs belonging to lines, the so-called *accumulator* is denoted by A. We assume that there is a one-to-one correspandence between parameter pairs in A and lines in the plane.

Let ℓ(a,b) be the line corresponding to (a,b) ε A. The line integral (whenever it exists)

$$H_o(\beta)(a,b) = \int_{\ell(a,b)} \beta(x,y) \, d\sigma \qquad (2)$$

(σ denoting the arc length on the line ℓ(a,b); see also (1))
is the *Radon transform* (or *Hough transform*) of the image β.

For the purpose of this paper and also for practical applications
we generalize the concept of the classical transform by introducing a *filter* F : X × A → $L^1(R_1)$ which is defined as a (generally
nonlinear) operator depending on the parameters in A and mapping
X into $L^1(R_1)$ where R_1 is a compact set containing R.

The *generalized Radon transform* (or simply **Radon** transform in the
context of this paper is defined as follows:

$$H(\beta)(a,b) = H_o(F(\beta;a,b))(a,b) =$$
$$= \int_{\ell(a,b)} F(\beta;a,b)(x,y) \, d\sigma \qquad (3)$$

3. Traces

Without loss of generality we assume that the Radon transform (3)
is to be projected with respect to parameter b onto the hyperplane
b = 0 in the accumulator. Other parametrizations can be easily
treated by reparametrizing A.

There exists a one-to-one correspondence between line pairs and
motions of the plane. For fixed a we denote by L_a the set of all
lines ℓ(a,b) and by M_a the set of all motions of the plane defined
by line pairs in L_a. If we assume that for any two motions in M_a
the product is also in M_a, the set of motions in M_a is a subgroup
of all motions of the plane.

Given P ε $\mathbb{R}^2$ fixed. The *trace* t(P) generated by P is the set of
all images of P under the motions in M_a. We assume that the traces
are continuous curves in the plane which is certainly true if the

motions in M_a depend continuously on the defining line pairs.

In the following Theorem some obvious properties of traces are collected:

<u>Theorem 3.1</u>: 1. $P \in t(P)$ for all $P \in \mathbb{R}^2$.

      2. $Q \in t(P)$ if and only if $P \in t(Q)$.

      3. $Q \in t(P)$ and $R \in t(Q)$ implies $R \in t(P)$.

      4. If $Q \in t(P)$ then $t(P)$ and $t(Q)$ are disjoint.

For future use we add two definitions. Given a motion M of the plane, the *moved picture* is defined by $\beta'(P) = \beta(MP)$. The filter F is *motion invariant* if and only if

$$F(\beta';a',b')(P) = F(\beta;a,b)(MP) \quad \text{for all } P \in \mathbb{R}^2 \quad (4)$$

for each motion M, where a' and b' are defined by the requirement $M\ell(a',b') = \ell(a,b)$. Most filters in practical applications are motion invariant.

Given an operator G which is defined on X and a fixed line ℓ. Then G is *line translation invariant* with respect to ℓ if and only if for all vectors Q parallel to ℓ and all translated images $\beta'(P) = \beta(P + Q)$ $(\beta \in X)$ one has $G(\beta') = G(\beta)$.

4. Projections

For fixed parameter a denote by

$$A_a = \{b \mid (a,b) \in A\} \quad (5)$$

and

$$\Omega_a = \{Q \in \mathbb{R}^2 \mid Q \in M\ell(a,b) \text{ for any } M \in M_a\}. \quad (6)$$

As a consequence of assertion 3 of Theorem 3.1, Ω_a does not depend on the choice of b in (6).

The *projected Radon transform* is defined by

$$H_b(\beta)(a) = (\Pi H(\beta))(a), \quad (7)$$

where Π is a (linear or nonlinear) *projector* assigning to each function of two variables $f(a,b)$ a function of one variable $(\Pi f)(a)$.

The projected Radon transform should exhibit some obvious invariance properties in order to lead to meaningful results. In addition to the line translation invariance with respect to certain lines also invariance with respect to motions in M_a should be required.

Given $M \in M_a$. Then for each $b \in A_a$ there exists a uniquely determined $b_M \in A_a$ such that

$$M\ell(a,b) = \ell(a,b_M(b)). \tag{8}$$

The projector Π is *trace invariant* if for all $f \in Y$ the function $f'(a,b) := f(a,b_M(b))$ is also in Y and $(\Pi f')(a) = (\Pi f)(a)$.
If a projector is not trace invariant, then a motion of the plane which maps lines in L_a into lines in L_a can lead to a different projected Radon transform. This is certainly not desirable.

Now we formulate the main Theorem:

Theorem 4.1: Assume that
1. F does not depend on b and $\Pi H_o(\phi)(a)$ is a continuous and linear functional of $\phi \in L^1(R_1)$.
2. The projector Π is trace invariant.
3. There exists a line in L_a such that the operator ΠH is line translation invariant with respect to this line.
4. F is separable with respect to a, i.e.

$$F(u;a) = \sum_{i=1}^{n} F_i(u) \cdot \gamma_i(a) \tag{9}$$

with translation invariant operators F_i.

Then the projected Radon transform

$$\Pi H(\beta)(a) = C(a) \cdot \sum_{i=1}^{n} \gamma_i(a) \cdot \int_{\mathbb{R}^2} F_i(\beta)(P)\ dP \tag{10}$$

is motion invariant with respect to all motions of the plane.

The proof of the Theorem rests on a straightforward application of Riesz' representation theorem for continuous linear functionals of L^1 (Steinhaus 1919) applied to the continuous linear functional in assumption 1.

When the operator ΠH is invariant with respect to all motions of the plane, it cannot exhibit any useful information about the contents of the image. Hence, in order to get useful results, we are required to drop either the linearity condition of the projection or its translation invariance or else to use a filter which depends severely nonlinearly on a.

5. Linear Projections

Assume that β represents a binary picture containing prominent
directions, for example a document, a flow diagram etc. We want
to adjust this picture by identifying the prominent directions.
In contrast to the situation with the ordinary Radon transform
for lines we are not interested in the individual lines in the
picture.

In the examples of this paragraph we use the filter function
$F(\beta;a,b) \equiv \beta$ which clearly meets conditions 1 and 4 of Theorem
4.1. Any projection of the form

$$(\Pi f)(a) = \int_{A_a} g(b) \cdot f(a,b)\ db \tag{11}$$

with suitable g is of course continuous and linear. It is trace
invariant as required by condition 2 of Theorem 4.1 if and only
if g(b) is constant. Without loss of generality we assume that
$g \equiv 1$. There remains only conditon 3 which needs closer investigation.

Example 1. $(a,b) = (m,b)$ and $\ell(m,b)$ is given by the *slope-intercept*
representation of a line

$$y = m \cdot x + b. \tag{12}$$

This parametrization was originally used by Hough (1962). It has
the disadvantage that parallels to the y-axis cannot be represented but it has the advantage of exhibiting the point-line duality
of projective geometry. Therefore it is recommended by different
authors (see e. g. Biland, Wahl 1986).
a) for fixed m, the group of motions defining the traces is given
by all translations parallel to the y-axis. Consequently, the traces are parallels to the y-axis and L_m is the set of all lines
with direction arctan m with respect to the x-axis. We have
$\Omega_m = \mathbb{R}^2$ and

$$H_b(\beta)(m) := \int_{-\infty}^{\infty} H(\beta)(m,b)\ db = \sqrt{1+m^2} \cdot \int_R \beta(P)\ dP. \tag{13}$$

Obviously, condition 3 of the Theorem is fulfilled for any translation vector with direction arctan m.
b) If b is fixed, the traces are circles with center $\binom{0}{b}$. L_b is
the set of all lines through the point $\binom{0}{b}$ and $\Omega_b = \mathbb{R}^2$. We have

354

$$H_m(\beta)(b) := \int_{-\infty}^{\infty} H(\beta)(m,b)\ dm = \int_R \frac{\sqrt{x^2 + (b-y)^2}}{x^2} \cdot \beta(P)\ dP.$$
(14)

This expression can become singular when there are features on the y-axis. Since this projection does not fulfill the line invariance condition, it becomes hard to interpret the results of its application.

Example 2. $(a,b) = (\alpha,p)$ and $\ell(\alpha,p)$ is given by *Hesse's normal form*

$$x \cdot \sin \alpha - y \cdot \cos \alpha = p.$$
(15)

This parametrization is usually applied in the Radon transform.
a) For fixed α, the traces are lines perpendicular to direction α. Obviously all conditions of Theorem 4.1 are met, hence

$$H_p(\beta)(\alpha) := \int_{-\infty}^{\infty} H(\beta)(\alpha,p)\ dp = \int_R \beta(P)\ dP.$$
(16)

This result is trivial and well known (see e.g. Helgason 1980, proof of Lemma 5.1).
b) For fixed p, the traces are circles around the origin. One has

$$\Omega_p = \{ \binom{x}{y} \in \mathbb{R}^2 \mid x^2 + y^2 \geq p^2 \},$$
(17)

L_p consists of all lines tangential to Ω_p and

$$H_\alpha(\beta)(p) := \int_0^{2\pi} H(\beta)(\alpha,p)\ d\alpha = \int_{\Omega_p} \frac{\beta(P)}{x^2 + y^2 - p^2}\ dP.$$
(18)

This projection is not line translation invariant. If the picture contains only a line segment with fixed center, then ΠH will be maximal if the line segment points towards the origin. ΠH will become larger as the line segment is moved from outside towards the periphery of a circle around the origin with radius p; it becomes singular, if the line segment touches the periphery. The interior of the disc with radius p around the origin is invisible for ΠH$(\beta)(p)$.

Example 3. For a fixed number S with $x^2 + y^2 < S^2$ for all $\binom{x}{y} \in R$ a line can be parametrized by (α,ϕ):

$$x \cdot \sin \alpha - y \cdot \cos \alpha = S \cdot \sin(\phi - \alpha).$$
(19)

This parametrization is related to Wallace's "Muff"-transformation (Wallace 1985).

a) For fixed α the traces are lines with direction perpendicular to direction α as in Example 2.a. L_α is the set of all lines with direction α meeting the circle $x^2 + y^2 = S^2$.

$$H_\phi(\beta)(\alpha) := \int_0^{2\pi} H(\beta)(\alpha,\phi)\; d\phi =$$

$$= \int_R \frac{\beta(P)}{\sqrt{S^2 - (y \cdot \cos \alpha - x \cdot \sin \alpha)^2}}\; dP. \qquad (20)$$

Condition 2 of Theorem 4.1 is fulfilled for all lines with direction α.

The projection with respect to α is not interesting here.

At a first glance, the projection of Example 3 looks very promising. Numerical experiments were performed with pictures containing only single line segments having different lenghts, orientations and positions in the picture. In the projection (20) these lines were clearly indicated by a peak. Experiments with more realistic pictures, however, were not encouraging. The more details were present in the picture, the more constant its projection becomes.

We can conclude from the results of the examples that the severely non translation invariant projections (Examples 1.b and 2.b) are difficult to interpret. If a projection is non translation invariant, the contribution of a specific feature in the image to the projected Radon transform depends on its location within the picture. It is therefore not possible to distinguish faint details at a good position from strong details at a position of poor visibility. In this situation it seems more preferable to apply the ordinary Radon transform to a "window" within the image which gives controlled visibility and also reduces the sampling effort considerably.

If on the other hand the projection is only moderately non translation invariant (Example 3), it approaches the conditions of Theorem 4.1 and therefore it ought to be nearly constant.

We can therefore conclude in the light of Theorem 4.1 that the condition of translation invariance should not be abandoned. There remain two possibilities: Applying a nonlinear projection or using a nonlinear filter.

6. Nonlinear Approaches

There exists a multitude of approaches in the literature using nonlinear projectors. Most authors prefer the *maximum projector* which assigns to each function $f(a,b)$ of two variables the projection

$$(\Pi f)(a) = \max_b f(a,b). \tag{21}$$

This projector was successfully applied by Ballard and Sabbah (1983) for reducing the dimensionality of a high-dimensional accumulator. Gerig and Klein (1986) used it in the context of a Radon transform for detecting circles. Silberberg, Davis and Harwood (1984) projected a 5-dimensional accumulator by this projector.

A different possibility for introducing nonlinearity into the projection is given by

$$(\Pi f)(a) = \int_{A_a} |f(a,b)|^m \, db. \tag{22}$$

Biland and Wahl (1986) reported good results with m between 1.5 and 4.0.

Rubart (1986) used a projection involving the derivative of f with respect to b

$$(\Pi f)(a) = \int_{A_a} \left| \frac{\partial f}{\partial b}(a,b) \right| \, db. \tag{23}$$

This approach proved extremely useful for finding prominent directions within a picture.

All these approaches involving nonlinear projectors have in common that it is necessary to calculate the full accumulator array in advance before projecting it because the projector and the integral operator of the Radon transform do not commute. In the preceding paragraph we used a linear projector and therefore the projection of the Radon transform had an integral representation. This fact is very important when the dimension of a high-dimensional accumulator is reduced by iterated projection.

There exists in the literature a very remarkable example for a linear projector applied to a nonlinear filter which was proposed by Bernhardt (1984) and which was reinvented quite recently by Sinden (1985). This projection also starts with the gradient image (k = 2). The lines are parametrized in the usual way as in Example 2 above. A nonlinear filter operating on the gradient image is defined by

$$F(\beta;\alpha) = \left| \frac{\partial\beta}{\partial x} \cdot \cos\,\alpha + \frac{\partial\beta}{\partial y} \cdot \sin\,\alpha \right| \qquad (24)$$

and

$$H_p(\beta)(\alpha) := \int_{-\infty}^{\infty} H(\beta)(\alpha,p)\ dp =$$

$$= \int_{-\infty}^{\infty}\int_{-\infty}^{\infty} \left| \beta_x \begin{pmatrix} t\,\cos\,\alpha + p\,\sin\,\alpha \\ t\,\sin\,\alpha - p\,\cos\,\alpha \end{pmatrix} \cdot \cos\,\alpha\ + \right. \qquad (25)$$

$$\left. +\ \beta_y \begin{pmatrix} t\,\cos\,\alpha + p\,\sin\,\alpha \\ t\,\sin\,\alpha - p\,\cos\,\alpha \end{pmatrix} \cdot \sin\,\alpha \right|\ dt\ dp.$$

In binary images the integration along a line over $F(\beta;\alpha)$ means counting the black-white and white-black transitions along this line. Therefore this method can be implemented quite easily.

Since in Bernhardt's method the projector is linear, the operations of projection and integration can be interchanged which will be very attractive for applications to higher-dimensional accumulators. It seems necessary to investigate nonlinear filters in connection with the Radon transform. Bernhardt reports in his paper about descriptors extracted from the projected Radon transform. These descriptors were used together with descriptors from the original Radon transform for the recognition of hand-written block letters which, however, were already segmented. The method was implemented into the SIEMENS Computer Multifont Reader CSL 2610. This system works in industrial applications since 1984.

7. Conclusions

In the present paper, a theoretical framework was given for understanding different known approaches for reducing the amount of data in the Radon (or Hough) transform by means of projection. The main Theorem states that linear and translation invariant projections of the Radon transformation necessarily yield useless results. There remain three possibilities for reducing the dimension of the accumulator of the Radon transformation by projection:

- Using a non translation invariant projection leads to results which are hard to interpret.
- Using a nonlinear projection operator has the disadvantage

that the projector and the integral operator of the Radon transformation do not necessarily commute, which might be difficult
when high-dimensional generalizations of the Radon transform
are considered. There are, nevertheless, many nonlinear projection operators recommended in the literature.
- A completely different approach was proposed by Bernhardt (1984).
It consists in using a linear projection together with a nonlinear filter. Also this approach can be understood in the light
of the theory developed here.

A large number of numerical experiments was performed by the authors on realistic documents. Some of the results of these experiments together with further details were documented elsewhere
(Eckhardt, Maderlechner 1987).

References

Ballard DH (1981) Generalizing the Hough transform to detect arbitrary shapes.
Pattern Recognition 13:111-122

Ballard DH and Sabbah D (1983) Viewer independent shape recognition.
IEEE Trans. PAMI-5:653-660

Bernhardt L (1984) Three classical character recognition problems, three new
solutions. Siemens Forsch.- u. Entwickl.-Ber. 13:114-117

Biland HP, Wahl FM (1986) Understanding Hough space for polyhedral scene
decomposition. IBM Zürich Research Laboratory, RZ 1458 (# 52978) 3/35/86

Brown CM, Curtiss MB, Sher DB (1983) Advanced Hough transform implementations.
In: Proceedings of the Eighth International Joint Conference on Artificial
Intelligence, 8-12 August 1983, Karlsruhe, Germany, pp. 1081-1085

Deans SR (1983) The Radon Transform and Some of Its Applications.
New York etc.: John Wiley and Sons

Eckhardt U, Scherl W, Yu Z (1987) Representation of plane curves by means of
descriptors in Hough space. I. Continuous theory. Berichte des Rechenzentrums
der Universität Hamburg

Eckhardt U, Maderlechner G (1987) Projections of the Hough transform.
Institut für Angewandte Mathematik der Universität Hamburg, Preprint

Gerig G, Klein F (1986) Fast contour identification through efficient Hough
transform and simplified interpretation strategy.
IAPR - afcet: Eighth International Conference on Pattern Recognition. Paris,
France, October 27-31, 1986

Helgason S (1980) The Radon Transform. Progress in Mathematics 5.
Boston, Basel, Stuttgart: Birkhäuser Verlag

Hough PVC (1962) Method and means for recognizing complex patterns.
U.S. Patent 3,069,654. Washington: United States Patent Office, December 18, 1962

Kimme C, Ballard D, Sklansky J (1975) Finding circles by an array of accumulators.
Comm. Assoc. Comput. Machinery 18:120-122

Kushnir M, Abe K, Matsumoto K (1985) Recognition of handprinted Hebrew characters
using features selected in the Hough transform space.
Pattern Recognition 18:103-114

Li H, Lavin MA, LeMaster RJ (1985) Fast Hough transform.
IBM Research Report RC 11080 (# 49754) 3/29/85

Merlin PM, Farber DJ (1975) A parallel mechanism for detecting curves in pictures. IEEE Trans C-24:96-98

Natterer F (1986) The Mathematics of Computerized Tomography.
Stuttgart: B.G. Teubner

Neveu CF, Dyer CR, Chin RT (1986) Two-dimensional object recognition using multiresolution models. Computer Vision, Graphics, and Image Processing 34:52-65

O'Rourke J, Sloan KR (1984) Dynamic quantization: Two adaptive data structures for multidimensional spaces. IEEE Trans. PAMI-6:266-280

Radon J (1917) Über die Bestimmung von Funktionen durch ihre Integralwerte längs gewisser Mannigfaltigkeiten.
Ber. Verh. Sächs. Akad. Wiss. Leipzig, Math.-Nat. Kl. 69:262-277

Rubart L (1986) Ermittlung von Geraden in Binärbildern durch die Hough-Transformation. Diplomarbeit Universität Hamburg, Januar 1986

Sanz JL, Hinkle EB (1987) Computing projections of digital images in image processing pipeline architectures. IEEE Trans. ASSP-35:198-207

Silberberg TM, Davis L, Harwood D (1984) An iterative Hough procedure for three-dimensional object recognition. Pattern Recognition 17:621-629

Sinden FW (1985) Shape information from rotated scans.
IEEE Trans. PAMI-17:726-730

Steinhaus H (1918) Additive und stetige Funktionaloperationen.
Math. Zeitschr. 5:186-221

Tsuji S, Matsumoto F (1978) Detection of ellipses by a modified Hough transform.
IEEE Trans. C-27:777-781

Wallace RS (1985) A modified Hough transform for lines.
IEEE Computer Society Conference on Computer Vision and Pattern Recognition,
June 19-23, 1985, San Francisco, California, pp. 665-667.
Silver Spring: IEEE Computer Society Press
Amsterdam: North-Holland Publishing Company

Yalamanchili S, Aggarwal JK (1985) A system organization for parallel image processing. Pattern Recognition 18:17-29

Proceedings of the Conference

Mathematics in Industry

October 24—28, 1983 Oberwolfach

Edited by Prof. Dr. H. NEUNZERT, University of Kaiserslautern, W.-Germany

1984. 287 pages. 16,2 × 23,5 cm. ISBN 3-519-02610-4. Paper DM 52,—

Contents

ORGANIZED COOPERATION BETWEEN UNIVERSITY AND INDUSTRY

R. S. Anderssen and F. R. de Hoog: A Framework for Studying the Application of Mathematics in Industry / A. B. Tayler: Oxford Study Groups with Industry; 1967—1983 / H. Wacker: Hydro Energy Optimization / J. Spanier: Applied Mathematics Education at the Claremont Colleges / H. Neunzert: Mathematic in the University and Mathematics in Industry — Complement or Contrast? / K. Hoffmann: On Establishing Contacts with Industry / M. Schulz-Reese: A Report of the „Kaiserslauterer Modellversuch": Continuing Mathematical Education / H.-E. Gross and U. Knauer: University Education as Preparation for Professional Praxis / A. M. Kempf: Mathematical Modelling in the French Grandes Ecoles. The Particular Case of the E.S.I.E.A.

INDIVIDUAL PROJECTS AT THE UNIVERSITIES

C. Cercignani: Mathematics and Fluiddynamics / M. Primicerio: Sorption of Swelling Solvents by Glassy Polymers / M. Shinbrot: Icebreaking by Hovercraft / B. Rihtarsic, F. Krmelj and I. Kuscer: Oscillations in Pipelines of Hydroelectric Power Plants / A. K. Louis: The Limited Angle Problem in Computerized Tomography / H. Frank: Computer Aided Design in Piping of Chemical Plants / W. Krüger: The Trippstadt-Problem / B. Aulbach: Trouble with Linearization

PROBLEMS POSED BY INDUSTRY

J. Bukovics: Oscillations of a Gasbody with Absorbant Walls (A Problem Occuring in Structural Acoustics of Passenger Cars) / P. Causemann: Repuirements for a Calculating Program Regarding a Two-Mass Vibration System to Optimize Damping Force Characteristics for Vehicle Shock Absorbers / A. Gamst: Geometric Design of Mobile Radio Telephone Systems / U. Pallaske: Large Systems of Stiff Ordinary Differential Equations. Numerical Treatment by System Reduction / R. Zobel: Validation of a Vehicle Crash Model

B. G. Teubner Stuttgart